An Introduction to Formal Logic

Formal logic provides us with a powerful set of techniques for criticizing some arguments and showing others to be valid. These techniques are relevant to all of us with an interest in being skilful and accurate reasoners. In this highly accessible book, Peter Smith presents a guide to the fundamental aims and basic elements of formal logic. He introduces the reader to the languages of propositional and predicate logic, and then develops formal systems for evaluating arguments translated into these languages, concentrating on the easily comprehensible 'tree' method. His discussion is richly illustrated with worked examples and exercises. A distinctive feature is that, alongside the formal work, there is illuminating philosophical commentary. This book will make an ideal text for a first logic course, and will provide a firm basis for further work in formal and philosophical logic.

PETER SMITH was formerly Senior Lecturer in Philosophy at the University of Cambridge. His other books include *Explaining Chaos* (1998) and *An Introduction to Gödel's Theorems* (2007), and he is a former editor of the journal *Analysis*.

An Introduction to
Formal Logic

Peter Smith

University of Cambridge

CAMBRIDGE
UNIVERSITY PRESS

CAMBRIDGE
UNIVERSITY PRESS

University Printing House, Cambridge CB2 8BS, United Kingdom

Cambridge University Press is part of the University of Cambridge.

It furthers the University's mission by disseminating knowledge in the pursuit of education, learning and research at the highest international levels of excellence.

www.cambridge.org
Information on this title: www.cambridge.org/9780521008044

First published 2003
7th printing with corrections 2013
9th printing 2016

10 0766240X

Printed in the United Kingdom by Clays, St Ives plc.

A catalogue record for this publication is available from the British Library

ISBN 978-0-521-80133-3 Hardback
ISNB 978-0-521-00804-4 Paperback

Contents

Preface

The world is not short of good introductions to logic. They differ widely in pace, style, the coverage of topics, and the ratio of formal work to philosophical commentary. My only excuse for writing another text is that I didn't find one that offered quite the mix that I wanted for my own students (first-year philosophy undergraduates doing a compulsory logic course). I hope that some other logic teachers and their students will find my particular combination of topics and approach useful.

This book starts from scratch, and initially goes quite slowly. There is little point in teaching students to be proficient at playing with formal systems if they still go badly astray when faced with ground-level questions about the whole aim of the exercise. So I make no apology for working hard at the outset to nail down some basic ideas.

The pace picks up as the book proceeds and readers get used to the idea of a formal logic. But even the more symbol-phobic students should be able to cope with most of the book, at least with a bit of judicious skipping. For enthusiasts, I give soundness and completeness proofs (for propositional trees in Chapter 19, and for quantifier trees in Chapter 30). The proofs can certainly be skipped: but I like to think that, if explained in a reasonably relaxed and accessible way, even these more 'advanced' results can in fact be grasped by bright beginners.

I have kept the text uncluttered by avoiding footnotes. You can follow up some of the occasional allusions to the work of various logicians and philosophers (such as Frege or Russell) by looking at the concluding notes on further reading.

The book has a web-site at www.logicbook.net. You will find there some supplementary teaching materials, and answers to the modest crop of exercises at the end of chapters. (And I'd like to hear about errors in the book, again via the web-site, where corrections will be posted.)

I am very grateful to colleagues for feed-back, and to the generations of students who have more or less willingly road-tested versions of most of the following chapters. Special thanks are due to Hilary Gaskin of Cambridge University Press, who first encouraged my plan to write this book, and then insisted that I didn't keep revising it for ever; to Dominic Gregory and Alexander Paseau, who read late drafts of parts of the book, and provided many corrections; and to Laurence Goldstein, who did much more than it was reasonable to expect of a publisher's reader.

Not least, I must thank Patsy and Zoë Wilson-Smith, without whose love and support this book would never have been finished.

Additional warm thanks are due to all those who kindly told me about mistakes in the first printed version of the book. I took the opportunity of an initial reprint to make the needed corrections and to make many other minor changes to improve clarity. Joseph Jedwab then gave me a long list of further errors, which have now also been corrected.

What is logic?

The business of logic is the *systematic evaluation of arguments for internal cogency*. And the kind of internal cogency that will especially concern us is *deductive validity*. But these brief headlines leave everything to be explained.

- What do we mean here by 'argument'?
- What do we mean by evaluation for 'internal cogency'?
- What do we mean, more particularly, by 'deductive validity'?
- What sorts of 'systematic' evaluation of arguments are possible?

This introductory chapter makes a start on answering these questions.

1.1 What is an argument?

By 'argument' we mean, roughly, a chain of reasoning in support of a certain conclusion. So we must distinguish arguments from mere disagreements and disputes. The children who shout at each other 'You did', 'I didn't', 'Oh yes, you did', 'Oh no, I didn't', are certainly disagreeing: but they are not *arguing* in our sense, i.e. they are not yet giving any reasons in support of one claim or the other.

Reason-giving arguments are the very stuff of serious enquiry, whether it is philosophy or physics, literary criticism or experimental psychology. But of course, episodes of reasoning equally feature in everyday, street-level, enquiry into what explains our team's losing streak, the likely winner of next month's election, or the best place to train as a lawyer. For we quite generally want our opinions to be true and to be defensible in the market-place of ideas, and that means that we should aim to have good reasons backing up our opinions. That in turn means that we have an interest in being skilful and accurate reasoners, using arguments which really do support their conclusions.

1.2 What sort of evaluation?

The business of logic, then, is the evaluation of stretches of reasoning. Let's take a very simple case (call it argument **A**).

Suppose you hold

(1) All philosophers are eccentric.

I then introduce you to Jack, who I tell you is a philosopher. So you come to believe

(2) Jack is a philosopher.

Putting these two thoughts together, you can obviously draw the conclusion

(3) Jack is eccentric.

This little bit of reasoning can now be evaluated along two quite independent dimensions.

- First, we can ask whether the *premisses* (1) and (2) are true: are the 'inputs' to your reasoning correct? (1) in fact looks very disputable. And maybe Jack's reflective skills are so limited that we'd want to dispute the truth of (2) as well.

- Second, we can ask about the quality of the *inference* from the premisses (1) and (2) to the *conclusion* (3). In this case, the movement of thought from the premisses (1) and (2) to the conclusion (3) is surely absolutely compelling. We have agreed that it may be open to question whether (1) and (2) are both true. However, *if* (1) and (2) are granted to be true (granted 'for the sake of argument', as we say), then (3) has got to be true too. There's just no way that (1) and (2) could be true and yet (3) false. Someone who asserted that Jack is a philosopher, and that all philosophers are eccentric, yet went on to *deny* that Jack is eccentric, would be implicitly contradicting himself.

In brief, it is one thing to consider whether an argument starts from true premisses; it is another thing entirely to consider whether it moves on by reliable inferential steps. To be sure, we normally want our arguments to pass muster on both counts. We normally want to start from true premisses *and* to reason by steps which will take us on to further truths. But it is important to emphasize that these are distinct aims.

The premisses (and conclusions) of arguments can be about all sorts of topics: their truth is usually no business of the logician. If we are arguing about historical matters, then it is the historian who is the expert about the truth of our premisses; if we are arguing about some matter of physics, then the physicist is the one who can help us about the truth of our premisses; and so on. The specific concern of logic, by contrast, is not the truth of initial premisses but *the way we argue from a given starting point*. Logic is not about whether our premisses are true, i.e. match up to the world, but about whether our inferences really do support our conclusions once the premisses are granted. It is in this sense that logic is concerned with the 'internal cogency' of our reasoning.

1.3 Deduction vs. induction

The one-step argument **A** is a splendid bit of reasoning; if the premisses are true, then the conclusion is *guaranteed* to be true too. Here's a similar case:

B   (1) Either Jill is in the library or she is in the coffee bar.
     (2) Jill isn't in the library.
So   (3) Jill is in the coffee bar.

Who knows whether the premises are true or not? But we can immediately see that necessarily, *if* the premises are true, then the conclusion will be true too: the inferential move is absolutely watertight. If premises B(1) and B(2) here are both true, then B(3) cannot conceivably fail to be true.

Now consider the following contrasting case. Wild philosophical scepticism apart, you are thoroughly confident that the cup of coffee you are drinking isn't going to kill you (and if you weren't really confident, you wouldn't be calmly sipping as you read this, would you?). What justifies your confidence? Well, you believe something like

C   (1) Cups of coffee that looked and tasted just fine haven't killed you in the past.

You also know that

     (2) This present cup of coffee looks and tastes just fine.

These premises will prompt you to conclude

     (3) This present cup of coffee won't kill you.

And the inference that moves from the premises C(1) and C(2) to the conclusion C(3) is, in the circumstances, surely perfectly reasonable: the facts recorded in C(1) and C(2) do give you excellent grounds for believing that C(3) is true. However – and here is the crucial contrast with the earlier 'Jack' and 'Jill' examples – it *isn't* the case that the truth of C(1) and C(2) *absolutely guarantees* C(3) to be true too.

Perhaps someone has slipped a slow-acting tasteless poison into the coffee, just to make the logical point that facts about how things have generally been in the past don't guarantee that the trend will continue in the future.

Fortunately for you, C(3) no doubt *is* true. The tasteless poison is a fantasy. Still, it is a *coherent* fantasy. It illustrates the point that your grounds C(1) and C(2) for the conclusion that the coffee is safe to drink are strictly speaking quite compatible with the falsity of that conclusion. Someone who agrees to C(1) and C(2) and yet goes on to deny C(3) might be saying something highly improbable given her premises, but she won't actually be contradicting herself. We can make sense of the idea of C(1) and C(2) being true and yet C(3) false.

In his little book *The Problems of Philosophy*, Bertrand Russell has a nice example which presses home the same basic point. Here is the chicken in the farmyard. Every day of its life, the farmer's wife has come into the hen yard in the morning to feed it. Here she comes again, bucket in hand. The chicken believes that it is about to be fed, on the best possible evidence. But tough luck: today is the day when it gets its neck wrung. Again, what's happened in the past is a very good guide to what will happen next (what else can we rely on?): but reasoning from past to future isn't absolutely watertight.

In summary, then, there is a crucial difference between the 'Jack' and 'Jill' examples on the one hand, and the 'coffee' example on the other. In the 'Jack' and 'Jill' cases, the premisses guarantee the conclusion. There is no conceivable way that A(1) and A(2) could be true and yet A(3) false: likewise if B(1) and B(2) are true then B(3) has to be true too. Not so with the 'coffee' case: it *is* conceivable that C(1) and C(2) are true while C(3) is false.

We'll need some terminology to mark this difference: so, as a first shot definition, we will say

> An inference step is *deductively valid* just if, given that its premisses are true, then its conclusion is absolutely guaranteed to be true as well.

Equivalently, when an inference is deductively valid, we'll say that the premisses *logically entail* the conclusion. So: arguments **A** and **B** each involve deductively valid inferential steps, but the coffee argument **C** doesn't. The latter argument instead involves reasoning from previously examined cases to a new case: this kind of extrapolation from the past to the future, or more generally from old cases to new cases, is standardly called *inductive*.

Note, the deductive/inductive distinction is *not* the distinction between good and bad reasoning. The 'coffee' argument is a perfectly decent one. It involves the sort of generally reliable reasoning to which we rightly trust our lives, day in, day out. The conclusion is highly likely to be true given the premisses. It is just that the inference in this case doesn't *guarantee* that the conclusion is true, even assuming the given premisses are true.

We'll say very little more about inductive inferences in this book. They are an important and difficult topic, but they are not *our* topic. Our concern is with arguments which aim to use deductively valid inferences, where the premisses are supposed to logically entail the conclusion. And our principal task will be to find various techniques for establishing whether a putatively valid inferential move really *is* deductively valid.

1.4   More examples

The 'Jack' and 'Jill' arguments were examples where the inferential moves are *obviously* deductively valid. Take the inference

D   (1)   All Republican voters support capital punishment.
　　　 (2)   Jo supports capital punishment.
　So (3)   Jo is a Republican voter.

That is equally obviously *invalid*. Even if D(1) and D(2) are true, D(3) doesn't follow. Maybe lots of people in addition to Republican voters support capital punishment, and Jo is one of them.

But questions of deductive validity needn't be *immediately* obvious, one way or the other. Suppose someone holds both of

E   (1)   Most Irish are Catholics.
　　　 (2)   Most Catholics oppose abortion.

On those grounds he concludes

> So (3)   At least some Irish oppose abortion.

What are we to make of *this* argument? I don't know whether the premises are true or not, and anyway that is hardly a matter for the logician (it is an empirical sociological matter to determine the distribution of religious affiliation among the Irish, and to find out how many Catholics support their church's official teaching about abortion). So let's leave aside the question of whether the premises are in fact correct. What we *can* seek to determine here – from our armchairs, so to speak – is whether the inferential move is valid: *if* the premises are true, then must the conclusion be true too?

Suppose the Irish are a tiny minority of the world total of Catholics. Then it could be that nearly all the other (non-Irish) Catholics oppose abortion, and hence most Catholics do, even though *none* of the Irish oppose abortion. In other words, E(1) and E(2) could be true, yet E(3) false. The truth of the premisses doesn't absolutely guarantee the truth of the conclusion, so the inference step can't be a deductively valid one.

Here's a last example:

> F   (1)   Some philosophy students admire all logicians.
>   (2)   No philosophy student admires any rotten lecturer.
> So (3)   No logician is a rotten lecturer.

Does this argument make a valid inferential move?

So far, all we can do to find the answer is to cast around experimentally, trying to work out somehow or other whether the truth of the premises would guarantee the truth of the conclusion – though, as with the 'Irish' argument, a little thought should give the right answer (which is revealed in §3.1). Still, it would be good to be able to proceed more methodically and find some *general* techniques for evaluating arguments like this. Ideally, we would like general techniques that work *mechanically*, that can be applied to settle questions of validity as automatically as we can settle arithmetical questions by calculation. But are such techniques for systematically evaluating arguments ever available?

Later we will explore some general techniques – though we will also have occasion to note that there are intriguing limitations on what can be done mechanically. First, however, we should say a little more about what makes any kind of systematic approach possible.

1.5 The systematic evaluation of arguments

Consider again our first sample argument with its deductively valid inference step:

> A   (1)   All philosophers are eccentric.
>   (2)   Jack is a philosopher.
> So (3)   Jack is eccentric.

And compare it with the following arguments:

A'   (1) All logicians are cool.
      (2) Russell is a logician.
   So (3) Russell is cool.

A''   (1) All robins have red breasts.
      (2) Tweety is a robin.
   So (3) Tweety has a red breast.

A''' (1) All existentialists write bosh.
      (2) Sartre is an existentialist.
   So (3) Sartre writes bosh.

(We can evidently keep on going *ad nauseam*, churning out arguments to the same A-pattern!)

The four displayed arguments – and, of course, any others cast in the same mould – are all equally good in the sense of all involving valid inference steps. In each case the truth of the premisses would logically guarantee the truth of the conclusion. And plainly, it is no mere accident that the arguments are all internally cogent. The validity of the inference move in the first argument hasn't anything especially to do with being a philosopher or being eccentric. Likewise the validity of the inference move in the 'robin' argument isn't due to special facts about robins. The four arguments all work for the same reason: the shared pattern of inference is a reliable one.

We can describe the general principle here as follows.

> Any inference step which moves from a pair of premisses, one of which says that everything of a certain kind has a given property, and the other of which says that a particular individual is of the former kind, to a conclusion which says that that individual has the latter property, is deductively valid.

A moment's reflection shows that this principle is a correct one. But we have hardly presented it in the most perspicuous way. How much easier to put it instead like this:

> Any inference step of the following type
>
> *All F are G*
> *n is F*
> *So, n is G*
>
> is deductively valid.

Here the italic letters '*n*', '*F*', '*G*' are used to exhibit the skeletal pattern of an argument (with '*n*' holding the place for a name, '*F*' and '*G*' for predicates, i.e. expressions which attribute properties or pick out kinds of thing). How we flesh out the skeletal pattern or *schema* doesn't matter: any sensible way of substituting for the *schematic variables* (as we'll call the likes of '*n*', '*F*', '*G*') will yield another argument with a valid inference step of the same type.

We said at the outset, logic is concerned with the systematic study of validity. We now get a first glimpse of how systematicity might be achieved – by noting that the same *patterns* or *forms* of inference can appear in many different particular arguments.

Some inference patterns like the one just shown are reliable. Others aren't. For example, consider this pattern of inference:

> *Most F are G*
> *Most G are H*
> *So, at least some F are H*

This is the pattern of inference the 'Irish' argument was relying on, and we saw that it doesn't work.

There will be much more on the idea of patterns of inference in later chapters; but first, a quick summary.

1.6 Summary

- We can evaluate a piece of reasoning in two distinct ways. We can ask whether the premises are actually true. And we can ask whether the truth of the premises actually supports the truth of the conclusion. Logic is concerned with the second dimension of evaluation: i.e. the province of logic is the question whether a given argument internally hangs together, whether its inferential moves are cogent ones.

- We are setting aside inductive arguments (and other kinds of non-conclusive reasoning). We will be concentrating on arguments involving inferences that purport to be *deductively valid*. In other words, we are going to be concerned with the study of inferences that aim to *logically entail* their conclusions – aim to be, as it were, absolutely watertight. (Imagine an argument as a hydraulic device: if truth is poured in at the top, then we want it to trickle down to the bottom. The arguments with non-conclusive, possibly leaky, inference moves are ones where putting truth in at the top doesn't guarantee getting any truth out at the bottom; the deductively valid inferences are those where *truth in* guarantees *truth out*.)

- Arguments typically come in families whose members share good or bad types of inferential move; looking at such general patterns of inference will be a step towards making logic a systematic study.

Exercises 1

By 'conclusion' we do *not* simply mean what concludes, what is stated at the end of, a passage of reasoning. We mean what the reasoning aims to establish – and that might in fact be stated at the outset of an argument. Likewise, 'premiss' does *not* mean (contrary to what the *Concise Oxford Dictionary* says!) 'a *previous* statement from which another is inferred'. Reasons supporting a certain claim might well be given after the target conclusion has been stated.

Indicate the premises and conclusions of the following arguments. Which of these arguments do you suppose involve deductively valid reasoning? Why? (We haven't developed any techniques for you to use yet: just improvise and answer the best you can!)

1. Whoever works hard at logic does well. Accordingly, if Russell works hard at logic, he does well.

2. Most politicians are corrupt. After all, most ordinary people are corrupt – and politicians are ordinary people.

3. It will snow tonight. For the snow clouds show up clearly on the weather satellite, heading this way.

4. Anyone who is well prepared for the exam, even if she doesn't get an A grade, will at least get a B. Jane is well prepared, so she will get at least a B grade.

5. John is taller than Mary; and Jane is shorter than Mary. So John is taller than Jane.

6. At eleven, Fred is always either in the library or in the coffee bar. And assuming he's in the coffee bar, he's drinking an espresso. Fred was not in the library when I looked at eleven. So he was drinking an espresso then.

7. The Democrats will win the election. For the polls put them 20 points ahead, and no party has ever overturned even a lead of 10 points with only a week to go to polling day.

8. Jekyll isn't the same person as Hyde. The reason is that no murderers are sane – but Hyde is a murderer, and Jekyll is certainly sane.

9. No experienced person is incompetent. Jenkins is always blundering. No competent person is always blundering. Therefore Jenkins is inexperienced.

10. Many politicians take bribes. Most politicians have extra-marital affairs. So many people who take bribes have extra-marital affairs.

11. (Lewis Carroll) Babies cannot manage crocodiles. Because babies are illogical. But illogical persons are despised. And nobody is despised who can manage a crocodile.

12. (Lewis Carroll again) No interesting poems are unpopular among people of real taste. No modern poetry is free from affectation. All your poems are on the subject of soap bubbles. No affected poetry is popular among people of real taste. Only a modern poem would be on the subject of soap bubbles. Therefore all your poems are uninteresting.

13. 'If we found by chance a watch or other piece of intricate mechanism we should infer that it had been made by someone. But all around us we do find intricate pieces of natural mechanism, and the processes of the universe are seen to move together in complex relations; we should therefore infer that these too have a maker.'

14. 'I can doubt that the physical world exists. I can even doubt whether my body really exists. I cannot doubt that I myself exist. So I am not my body.'

Validity and soundness

The first chapter briefly introduced the idea of an inferential move being deductively valid. This chapter explores the idea of validity in a little more depth, and also emphasizes the centrality of deductive reasoning.

2.1 Validity and possibility

We said that an inference step is valid just if, given the input premisses are true, then the output conclusion is absolutely guaranteed to be true as well. But what do we mean here by an 'absolute guarantee'?

Previously, we gave a number of other, equally informal, explanations of the idea of validity in terms of what is 'conceivable' or what it would be 'self-contradictory' to assert. Here now is another definition of validity: and this time, it is a version of the standard, 'classical', textbook definition of validity.

> An inference step from given premisses to a particular conclusion is *(classically) valid* if and only if there is no possible situation in which the premisses would be true and the conclusion false.

We'll take this as our official definition from now on. But it is only as clear as the notion of a 'possible situation'. It needs to be stressed, then, that 'possible' here is meant in the widest sense – a sense related to those ideas of what is 'conceivable', or 'involves no contradiction' that we met before.

Consider, for example, this inference:

Premiss: Jo jumped out of a twentieth floor window (without parachute, safety net, etc.) and fell freely and unimpeded onto a concrete pavement.

Conclusion: Jo was injured.

And let's grant that there is no situation that can really obtain in the actual world, with the laws of physics as they are, in which the premiss would be true and the conclusion false. In the world as it is, falling unimpeded onto concrete from twenty floors up will always produce serious (surely fatal) injury. Does that make the inference from the given premiss to the conclusion deductively valid?

No. Although it isn't, let's agree, *physically* possible to jump in those circumstances without being injured, it remains possible in a weaker sense: it is 'logically possible', if you like. We can coherently conceive of situations in which the laws of nature are different or are miraculously suspended, and someone jumping from twenty floors up will float delicately down like a feather. There is no internal logical contradiction in that idea. *And all that is required for the deductive invalidity of an inference is that it is possible in this very weak sense for the premisses to be true and the conclusion false.* As we put it before, an inference is invalid if there is no internal contradiction in the notion of the truth of the premisses being combined with the falsity of the conclusion.

This very weak notion of logical possibility (any consistent fantasy counts as possible in this sense) goes with a correspondingly strong notion of logical *im*possibility. If something is *logically* impossible, like being circular and square at the same time, then it is absolutely ruled out: the very idea is inconsistent, incoherent, logically absurd. For a classically valid inference, it is impossible in this very strong sense for the premisses to be true and the conclusion false.

It will be useful to introduce a further definition:

> A set of propositions is *logically consistent* if it is logically possible for the propositions all to be true together. Likewise, a set of propositions is *logically inconsistent* if it is logically impossible for the propositions to be true together.

If there is no possible situation in which a bunch of premisses would be true and a certain conclusion false, then the premisses and the *denial* of the conclusion form an inconsistent set of propositions. So another way of characterizing the notion of validity is this: an inference is classically valid if the premisses taken together with the denial of the conclusion form an inconsistent set.

To be sure, we have still only gestured at the notions of logical possibility and logical impossibility that are in play here. As we go along, we will need to sharpen up our understanding of these ideas. Still, we've said enough to get us started.

2.2   What's the use of deduction?

It needs to be stressed again that deductively valid inferences are *not* the only acceptable ones. Inferring that Jo is injured from the fact she fell twenty stories onto concrete is perfectly reasonable. Inductive reasoning like this is very often good enough to trust your life to: the premisses may render the conclusion a racing certainty. But such reasoning isn't deductively valid.

Now consider a more complex kind of inference: take the situation of the detective – call him Sherlock. In the ideally satisfying detective story, Sherlock assembles a series of clues and then solves the crime by an *inference to the best explanation.* In other words, the detective arrives at an hypothesis that neatly accommodates all the strange events and bizarre happenings, an hypothesis that strikes us (once revealed) as obviously the right explanation. Why is the bed bolted to the floor? Why is there a useless bell rope hanging by the bed? Why is there a ventilator near the top of the rope leading through into the next room? What is that strange music heard at the dead of night? All the pieces fall into place when Sherlock infers a dastardly plot to kill the sleeping heiress in her unmovable bed by means of a poisonous snake, trained to descend the rope in response to the music. But although this is an impressive 'deduction' in one everyday sense of the term, it is *not* deductively valid reasoning in the logician's sense. We may have a set of clues, and some hypothesis *H* may be the only plausible explanation we can find: but in the typical case it won't be a contradiction to suppose that, despite the way the evidence stacks up, our hypothesis *H* is false. That won't be a logically inconsistent supposition, only a very unlikely one. Hence, the detective's plausible 'deductions' are not (normally) valid deductions in the logician's sense.

This being so, we might begin to wonder: if our inductive reasoning about the future on the basis of the past is not deductive, and if inference to the best explanation is not deductive either, just how interesting *is* the idea of deductively valid reasoning?

To make the question even more worrisome, consider that paradigm of systematic rationality, *scientific reasoning*. We gather data, and try to find the best theory that fits; rather like the detective, we aim to come up with the best explanation of the actually observed data. But a useful theory goes well beyond merely summarizing the data. In fact, it is precisely because the theory goes beyond what is strictly given in the data that the theory is useful for making novel predictions. Yet since the excess content isn't *guaranteed* by the data, because it is (in a broad sense) an induction beyond the data, the theory cannot be validly deduced from observation statements. So again it might be asked: what's the interest of deductive validity, except maybe to mathematicians, if deductive inference doesn't feature even in the construction of scientific theories?

But that's too quick! It is true that we can't simply deduce a theory from the data it is based on. However, it just doesn't follow that deductive reasoning plays no essential part in scientific reasoning.

Here's a picture of what goes on in science. Inspired by patterns in the data, or by models of the underlying processes, or by analogies with other phenomena, etc., we *conjecture* that a certain theory is true. Then we use the conjectured theory (together with assumptions about 'initial conditions') to *deduce* a range of testable predictions. The first stage, the conjectural stage, may involve flair and imagination, rather than brute logic, as we form our hypotheses about the underlying processes. But at the second stage, having made our conjectures, we need to infer testable consequences; and this *does* involve deductive logic. For we need to examine what else must be true *if* the theory is true: we want to know what our hypothesized theory logically entails. Once we have deduced testable predictions, we seek to check them out. Often our predictions turn out to be false. We have to reject the theory. Or else we have to revise it as best we can to accommodate the new data, and then go on to deduce more testable consequences. The process is typically one of repeatedly improving and revising our hypotheses, deducing consequences which we can test, and then refining the hypotheses again in the light of test results.

This *hypothetico-deductive* model of science (which highlights the role of deductive reasoning *from* theory *to* predictions) no doubt itself needs a lot of development and amplification and refinement. But with science thus conceived, we can see why deduction is absolutely central to the enterprise after all.

2.3 The invalidity principle

Classical validity is defined in terms of possibility and necessity. An inference is valid if it is *impossible* for the premisses to be true and the conclusion false. Valid inferences are *necessarily* truth-preserving. So, to evaluate for validity, we usually have to consider not only what is the case in the actual world but also consider alternative ways things might have been – or, to use a helpful image, we need to consider what is the case 'in other possible worlds'. Take, for example, the following argument:

A   (1) No Welshman is a great poet.
    (2) Shakespeare is a Welshman.
So (3) Shakespeare is not a great poet.

The premisses and conclusion are all false. But this fact is quite compatible with the inferential move here being a valid one. Which indeed it is. For there is no possible situation in which the premisses would be true and the conclusion false. In other words, any possible world which *did* make the premisses true (Shakespeare being brought up some miles to the west, and the Welsh not going in for verse), would also make the conclusion true.

Take a couple more examples:

B   (1) No father is female.
    (2) Bill Clinton is a father.
So (3) Bill Clinton is not female.

C   (1) No one whose middle name is 'Winifrid' is a Democrat.
    (2) George W. Bush's middle name is 'Winifrid'.
So (3) George W. Bush is not a Democrat.

These again are valid inferences (cut to the same reliable pattern as the Shakespeare inference), but the inference step leads in argument B from true premisses to a true conclusion, and in argument C from false premisses to a true conclusion.

That last claim may for a moment seem odd: how can a truth be validly inferred from two falsehoods? Well, C(3) is true – George W. Bush is a Republican – but here in the actual world it has nothing to do with facts about his middle name. It just happens to be true quite independently, and someone who believed C(3) on the basis of the daft premisses C(1) and C(2) would be right merely by luck. Still, the point remains that in any world where those premisses *were* true, C(3) would also be true, and so the inference step is deductively valid.

There can of course be mixed cases too, with some true and some false premisses. So allowing for these, we have in summary:

- A valid inference can have actually true premisses and a true conclusion, (some or all) actually false premisses and a false conclusion, or (some or all) false premisses and a true conclusion.
- The only combination ruled out by the definition of validity is having all true premisses and a false conclusion. Validity is about the *necessary preservation of truth* – so a valid inference cannot take us from actually true premisses to a false conclusion.

That last point is worth stressing, and indeed we might usefully dignify it with a label:

> *The invalidity principle* An inference with actually true premisses and an actually false conclusion can't be deductively valid.

We'll see in §4.1 how this invalidity principle gets to do interesting work when combined with the observation that arguments come in families sharing the same kind of inference step. But here's an immediate (if rather trivial) application. Consider the following line of reasoning:

D   (1) No England cricket captain has written a world-famous philosophy text book.
     (2) All world-famous philosophy text books are written by people with some knowledge of philosophy.
     (3) Anyone who has taught philosophy at a university has some knowledge of philosophy.
So (4) No one who has taught philosophy at a university has captained England at cricket.

Here we have a three-premiss inference with a certain complexity, and it may take a moment's thought to figure out whether the inferential move is valid or not.

The question is immediately settled, however, once you put together the fact that the premisses are all *true* with the esoteric bit of information that the conclusion is actually *false* (as it happens, the one-time England captain Mike Brearley previously was a university lecturer in philosophy). Hence, by the invalidity principle, the inference move in D must be invalid: and that settles that!

2.4 Inferences and arguments

So far, we have spoken of *inferential steps in arguments* as being valid or invalid. For the moment, we'll largely stick to this way of talking. But it is more common simply to describe *arguments* as being valid or otherwise. In that usage, we say that a one-step argument, like those we've been looking at so far, is valid just if the inference step from the premiss(es) to the conclusion is valid.

This is absolutely standard shorthand. But it can mislead beginners. After all, saying that an argument is 'valid' can sound rather like an all-in endorsement. However, to repeat, to say that an argument is valid is only to commend the cogency of the *inferential move* between premisses and conclusion. A valid argument, i.e. one that is internally cogent like **A**, can still have premisses that are quite hopelessly false.

It is useful, then, to have a term for arguments that *do* deserve all-in endorsement – which both start from truths and proceed by deductively cogent inference steps. The usual term is '*sound*'. So:

> A (one-step) argument is *valid* just if the inference from the premisses to the conclusion is valid.
>
> A (one-step) argument is *sound* just if it has all true premisses and the inference from those premisses to the conclusion is valid.

A few authors annoyingly use 'sound' to mean what we mean by 'valid'. But everyone agrees there is a key distinction to be made between mere deductive cogency and the all-in virtue of having true premisses *and* making a cogent inference. There's just an irritating divergence over how this agreed distinction should be labelled. Our usage of the labels is the majority one.

We typically want arguments that aim at deductive success to be sound in our double-barrelled sense. But not always! Sometimes we argue from premisses we think are false (or even inconsistent) precisely to bring out some absurd consequence of them, in the hope that a disputant will agree that the consequence *is* absurd and so come to accept that the premisses *aren't* all true.

Note three immediate consequences of our definition:

(a)   any sound argument has a true conclusion;
(b)   no pair of sound arguments can have inconsistent conclusions;
(c)   no sound argument has inconsistent premisses.

First, a sound argument starts from actually true premisses and involves a necessarily truth-preserving inference move – so it must end up with an actually true conclusion. Second, since a pair of sound arguments will have a pair of actually true conclusions, that means that the conclusions are true together. If they *are* actually true together, then of course they *can* be true together. And if they can be true together then (by definition) the conclusions are *consistent* with each other. Finally, since a bunch of inconsistent premisses cannot all be true together, an argument starting from those premisses cannot fulfil the first half of the definition of soundness.

Replacing 'sound' by '(classically) valid' in (a) to (c) yields falsehoods, however. That is easy to see in the first two cases. For the third case, reflect again that sometimes we show that certain premisses are implicitly inconsistent by validly drawing some explicit contradiction from them: we'll return to explore this point in Chapter 5.

2.5 What sort of thing are premisses and conclusions?

The basic constituents of arguments are premisses and conclusions. But what sort of thing are they? Well, for a start, they are the sort of thing that can be true or false; and we'll follow one convention by calling the bearers of truth or falsity, whatever they should turn out to be, *propositions*. So in our usage, arguments are uncontentiously made up out of propositions: but what are *they*?

> *A terminological aside*   Propositions, such as premisses and conclusions, are assessed for truth/falsity. Inference moves are assessed for validity/invalidity. These dimensions of assessment, as stressed before, are fundamentally different: so we should mark the distinction very carefully. Despite the common misuse of the terms, we should *never* say that a proposition is 'valid' when we mean it is true, or that an argument is 'true' when we mean that it is valid (or sound).

You might suggest: propositions are just *declarative sentences* (like 'Jack kicks the ball' as opposed to questions like 'Does Jack kick the ball?', or imperatives like 'Jack, kick the ball!'). But it would be odd to identify premisses and conclusions with sentences in the sense of the actual printed inscriptions, particular physical arrays of blobs of ink on the page. Surely we want to say that the *same* premisses and conclusions can appear in the many different printed copies of this book (or indeed, when the book is read aloud). So the suggestion is presumably that propositions are declarative sentence *types*, which can have printed instances appearing in many copies (and can have many spoken instances too).

But even so understood, the suggestion is not without difficulties. Consider (a) an ambiguous sentence like the proverbial wartime headline 'Eighth Army push bottles up Germans' – are the Germans being bottled up by a push, or are the Eighth Army pushing bottles? The sentence says two different things, one of which may be true and the other false: so, same sentence, different things said, i.e. different propositions expressed. What about (b) sentences involving pronouns? For example, consider the argument 'No logician is irrational. Russell is a logician. So, he is rational.' We naturally read this as a valid argument. But the conclusion as written is in a sense incompletely expressed. Who is being referred to by 'he'? In context, we take it to be Russell. In another context, the pronoun 'he' could refer to someone else. And ripped out of all context, an occurrence of 'He is rational' wouldn't convey any determinate message – wouldn't express any particular proposition. So again, same sentence but expressing different propositions (or none). (c) There can be converse cases, where it seems natural to say that different sentences express the same proposition. After all, don't 'Snow is white' and 'La neige est blanche' say the same? Or within a language, how about 'Jack loves Jill' and 'Jill is loved by Jack'?

Some find these sorts of reasons for distinguishing sentences and propositions compelling, and conclude that propositions aren't sentences but are – as we've just been saying – what are *expressed* by declarative sentences. Others respond that cases of ambiguity and incompleteness only show that we need to identify propositions with *fully interpreted* sentences (i.e. sentences parsed as having a certain meaning, and with the reference of pronouns such as 'he', etc. determined); and they will say that cases of equivalent sentences (whether from different languages or the same one) are no more than that – i.e. we have equivalent but still distinct propositions.

We aren't going to try to settle that dispute. That may sound rather irresponsible. But it will turn out that for our purposes later in this book, we don't actually *need* to settle the nature of propositions here. So we fortunately can leave that tricky issue in 'philosophical logic' hanging as we proceed with our more formal explorations.

2.6 Summary

- Although inductive arguments from past to future, and inferences to the best explanation, are not deductive, the hypothetico-deductive picture shows how there can still be a central role for deductive inference in scientific and other enquiry.

- An *inference step* is (classically) valid if and only if there is no possible situation in which its premisses are true and the conclusion false.

- The relevant notion of a possible situation is (roughly) the notion of a coherently conceivable situation, one whose characterization involves no internal contradiction.

- A one-step *argument* is valid if and only if its inference step is valid.

- An argument which is valid and has true premisses is said to be *sound*.

- The constituents of arguments (premisses and conclusions) are *propositions*. But we needn't settle the nature of propositions.

Exercises 2

Which of the following claims are true, which false, and why?

1. If someone produces an invalid argument, his premisses and conclusion must make a logically inconsistent set.

2. If an argument has false premisses and a true conclusion, then the truth of the conclusion can't really be owed to the premisses: so the argument cannot really be valid.

3. Any inference with actually true premisses and a true conclusion must be truth-preserving and so valid.

4. If the conclusion of an argument is false and all its premisses are true, then the argument cannot be deductively valid.

5. You can make a valid inference invalid by adding extra premisses.

6. You can make a sound inference unsound by adding extra premisses.

7. You can make an invalid inference valid by adding extra premisses.

8. If a set of propositions is consistent, then adding a further true proposition to it can't make it inconsistent.

9. If a set of true propositions is consistent, then adding a further false proposition to it must make it inconsistent.

10. If a set of propositions is inconsistent, then if we remove some proposition *P*, we can validly infer that *P is false* from the remaining propositions in the set.

Patterns of inference

We noted in the first chapter how arguments can come in families which share the same pattern of inference step. We now explore this idea in more detail.

3.1 More patterns

Consider again the argument:

> **A**   (1)   No Welshman is a great poet.
>   (2)   Shakespeare is a Welshman.
>   So (3)   Shakespeare is not a great poet.

The inference step here is deductively valid. Likewise for the parallel 'Clinton' and 'Bush' arguments in §2.3. These are valid too:

> **A′**   (1)   No three-year old understands quantum mechanics.
>   (2)   Daisy is three years old.
>   So (3)   Daisy does not understand quantum mechanics.

> **A″**   (1)   No elephant ever forgets.
>   (2)   Jumbo is an elephant.
>   So (3)   Jumbo never forgets.

And we can improvise endless variations on the same theme.

The inference steps in these arguments aren't validated by any special facts about poets, presidents, three-year-olds or elephants. Rather, they are all valid for the same reason, namely the meaning of 'no' and 'not' and the way that these logical concepts distribute in the premises and conclusion. In fact, we have

> Any inference of the type
>
> > *No F is G*
> > *n is F*
> > *So, n is not G*
>
> is valid.

As before (§1.5), 'F' and 'G' here stand in for predicates – expressions that attribute properties like *being an elephant* or *understanding quantum mechanics*

– and '*n*' holds the place for a name. Note again that this sort of use of symbols is in principle dispensable. We *could* say, e.g.,

> Any one-step inference is valid which has one premiss which says that nothing of some given kind has a certain property, whose other premiss says that a certain individual is of the given kind, and whose conclusion says that the individual in question lacks the property in question.

But it is surely a lot more transparent to use the symbolic shorthand.

Let's be clear, then, that there is nothing essentially mathematical involved in this use of symbols like '*F*' and '*n*'. We are simply exploiting the fact that it is easier to talk about a pattern of inference by *displaying* the pattern using these 'schematic variables', instead of trying to *describe* it in cumbersome words.

Here's another, slightly trickier, example which we have met before (§1.4):

> B   (1)   Some philosophy students admire all logicians.
>   (2)   No philosophy student admires any rotten lecturer.
> So (3)   No logician is a rotten lecturer.

Do the premisses logically entail the conclusion?

Suppose the premisses are true. Then by B(1) there are some philosophy students (as it might be, Jack and Jill) who admire all logicians. We know from B(2) that Jack and Jill (since they are philosophy students) don't admire rotten lecturers. That is to say, people admired by Jack and Jill aren't rotten lecturers. So in particular B(3), logicians – who are all admired by Jack and Jill – aren't rotten lecturers. The inference is valid.

Having now done the work for the argument about rotten lecturers, and shown that it is valid, what about this following argument?

> B'   (1)   Some opera fans buy tickets for every new production of Wagner's *Siegfried*.
>   (2)   No opera fan buys tickets for any merely frivolous entertainment.
> So (3)   No new production of Wagner's *Siegfried* is a merely frivolous entertainment.

That again is valid; and a moment's reflection shows that it essentially involves the *same* type of inference move as before, and can be shown to be valid in just the same kind of way. We could put it symbolically like this:

> Any inference of the type
>
> > *Some F have relation R to every G*
> > *No F has relation R to any H*
> > *So, no G is H*
>
> is valid.

By talk of a *relation* we just mean whatever is expressed by a predicate with (as it were) more than one slot in it waiting to be filled by subject terms. For example the predicate '... loves ...' has two empty places waiting to be filled up by names like 'Romeo' and 'Juliet' (or by more complex terms like 'Someone in this room' and 'every philosopher'). Likewise, '... is married to ...' also expresses a two-place relation, as does '... is taller than ...' and (to return to our examples) '... admires ...' and '... buys tickets for ...'. So, with that notion of a relation in play, we can see that the arguments **B** and **B'** do share the same indicated pattern of inference.

And just try expressing that common pattern *without* using the symbols: it can be done, to be sure, but at what cost in perspicuity!

In summary, then, we have now displayed three examples of general patterns of valid inference. In §1.5 we noted the simple inference pattern

> *All F are G*
> *n is F*
> *So, n is G*

We have just noted the pattern

> *No F is G*
> *n is F*
> *So, n is not G*

and, rather more complicatedly,

> *Some F have relation R to every G*
> *No F have relation R to any H*
> *So, no G is H*

Any inference following one of these three patterns will be valid. To repeat, it is not the special subject matter of the instances that makes them valid. The way that 'all', 'every' and 'any', 'some', 'no' and 'not' distribute between the premisses and conclusion suffices for validity in these cases.

Here are some more examples of reliable inference patterns (with varying numbers of premises – there is nothing special about having exactly two):

> *No F is G*
> *So, no G is F*

> *All F are H*
> *No G is H*
> *So, no F is G*

> *All F are either G or H*
> *All G are K*
> *All H are K*
> *So, all F are K*

(Construct some instances, and convince yourself that your examples and other inferences of the same types will all be valid.)

In sum, evaluating a particular argument for validity very often goes with seeing a certain general pattern of inference as doing the work in the argument. And evaluating this inference pattern will, in one fell swoop, simultaneously give a verdict on a whole range of arguments making the same sort of inferential move. This point underlies the whole business of logic as a systematic study – though, as we will explain in §3.4 below, the point does need to be handled with a bit of care. But first …

3.2 Three simple points about inference patterns

Point One   Take again the inference pattern

> *All F are G*
> *n is F*
> *So, n is G*

Does the following inference count as an instance of the pattern?

> C   (1) All men are mortal.
> (2) Snoopy is a dog.
> So (3) Bill Clinton is female.

Of course not! True, when taken separately, C(1) is of the form *All Fs are G* (i.e. attributes a certain property to everything in some class), C(2) is a simple subject/predicate claim of the form *n is F*, and likewise C(3) has the form *n is G*. But when we describe an inference as having a pair of premises of the pattern

> *All F are G*
> *n is F*

and a conclusion

> *n is G*

we are indicating that the same predicate *F* is involved in the two premises. Likewise, the name *n* and predicate *G* that are in the premises should recur in the conclusion. The whole point of the schematic variables is to represent *patterns of recurrence* in the premises and conclusion. So in moving back from an abstract schema to a particular instance of the type, we must preserve the patterns of recurrence by being consistent in how we interpret the 'F's and 'G's and 'n's. A schematic variable is to be filled out the same way at each occurrence in the schema.

What about the following argument? Does this count as an instance of the displayed inference pattern?

> D   (1) All philosophers are philosophers.
> (2) Socrates is a philosopher.
> So (3) Socrates is a philosopher.

This time, instead of substituting at random, we have at least been consistent – though we have substituted in the *same* way for both 'F' and 'G'. We will allow this as a special case. The crucial thing, to repeat, is that we must hold the interpretation of each of 'F' and 'G' fixed throughout the argument. If an argument pattern is reliably valid whatever the substitutions for 'F' and 'G', etc., it will still be valid in the trivializing case where what is substituted for the 'F's and 'G's happens to be the same.

And indeed argument **D** *is* classically valid by our definition: it can't have true premises and a false conclusion. The inference is certainly no use at all as a means for *persuading* someone of the conclusion who does not already accept that very proposition! But there's an important distinction to make here: truth-preservation is one thing, persuasiveness is something else. And once we firmly make that distinction, counting **D** as deductively virtuous is acceptable. After all, an inference that covers no ground hasn't got a chance to go wrong! (We will return to expand this remark in Chapter 6.)

Point Two   Consider the two arguments:

E   (1) All men are mortal.
     (2) Socrates is a man.
 So (3) Socrates is mortal.

E′   (1) Socrates is a man.
     (2) All men are mortal.
 So (3) Socrates is mortal.

What's the relation between the inferences here? Validity, recall, is defined this way: an inference is valid just when there is no possible situation in which all the premises are true and the conclusion false. Assessments of validity for inference steps are therefore quite blind to the order in which the premises are 'fed in'. Hence for our purposes – concentrating, as logic does, on assessments of validity – the order in which the premises of an inference step are presented is irrelevant. So we'll take the two E arguments as involving the same inference step. And generalizing, we'll take two schemas

> *All F are G*
> *n is F*
> *So, n is G*
>
> *n is F*
> *All F are G*
> *So, n is G*

as representing the *same* basic pattern of inference.

Point Three   What is the relation between that pattern and these ones? –

> *All H are K*
> *m is H*
> *So, m is K*
>
> *All Φ are Ψ*
> *α is Φ*
> *So, α is Ψ*
>
> *All ① are ②*
> *◇ is ①*
> *So, ◇ is ②*

Again, all four schemas are just more possible ways of representing the *same* kind of inferential move. The choice of '*F*'s and '*G*'s as against '*H*'s and '*K*'s, or Greek letters, or some other place-holders, doesn't matter at all; what we are trying to reveal is a pattern of recurrence, and the '*F*'s or '*Φ*'s or whatever are just used to mark different places in the pattern. (It would be a rather bad mistake, then, to *identify* the pattern of inference with one or other of these various symbolic expressions used to *display* the pattern.)

3.3 Generality and topic neutrality

Our examples so far have all had a very abstract structure. But consider:

F   (1) Jill is a mother.
    So (2) Jill is female.

G   (1) Jack has a first cousin.
    So (2) Jack's parents are not both only children.

H   (1) Tom is taller than Dick.
    (2) Dick is taller than Harry.
    So (3) Tom is taller than Harry.

These arguments are all valid. If Jill is a mother, she is a female parent, and hence female. If Jack has a first cousin (where a first cousin, by definition, is a child of one of his aunts or uncles) he must have an aunt or uncle. So his parents cannot both have been only children. And if three people are lined up in descending order of height, then the first must be taller than the last.

Now, in each case we can expose some inferentially relevant structure. The first two arguments are not made valid by something peculiar to Jill or Jack. Any inference of these kinds is valid:

> *n is a mother*
> *So, n is female*
>
> *n has a first cousin*
> *So, n's parents are not both only children*

Likewise, any argument with the following pattern is valid:

> *k is taller than l*
> *l is taller than m*
> *So, k is taller than m*

Here, then, we have further examples of reliable patterns of inference; but in an obvious sense, these are less general patterns than before. In our earlier examples, we could display the pattern using, in addition to the symbols, only very general vocabulary like 'all' or 'none' or 'any' or 'not'. That vocabulary can be called *topic neutral*, for it is vocabulary that we might equally well use in discussing any topic, be it history, pure mathematics, philosophy or whatever. The logical structures that the skeletal forms displayed were correspondingly general. By contrast, our new examples of valid inference patterns involve concepts which belong to more specific areas of interest, concerning familial relations, or physical size.

These latest examples are none the worse for that; they are perfectly good kinds of valid inferences. But in what follows we are not going to be much interested in arguments whose validity depends on features of special-interest concepts like motherhood or tallness. We are going to be concentrating on the kinds of argument whose salient patterns can be laid out using only topic-neutral vocabulary like 'if' and 'or' and 'some' and 'all', and so can occur when we are reasoning about any subject matter at all.

3.4 Arguments instantiate many patterns

Consider this inference again:

F   (1) Jill is a mother.
So (2) Jill is female.

Then note that this very simple one-step valid argument in fact exemplifies a number of different patterns of reasoning.

First, and most abstractly, it is an argument with one premiss and a (distinct) conclusion. So it is an instance of the pattern (a) 'A; so, C' (rather than the repetitious 'A; so, A' or the two-premiss 'A, B; so C'). It need hardly be remarked that the very general pattern of inference (a) is quite unreliable; most of its instances are invalid.

Second, our inference has a premiss which attributes a certain property to a given individual, and a conclusion which attributes another property to the same individual. So it is an instance of the slightly more specific pattern (b) 'n is F; so, n is G' – though again this is a totally unreliable inferential move in the general case.

To move to a third, more detailed, level of description, the argument is also an instance of the type (c) 'n is a mother; hence n is female'. And this more specific pattern of reasoning plainly *is* deductively reliable in the sense that all of its instances are valid.

Finally, we could take the inference to be the one and only exemplar of the type (d) 'Jill is a mother; so Jill is female' – which is (so to speak) a pattern with *all* the details filled in. A reliable pattern again (of course!).

All this illustrates a very important general moral:

> There is no such thing as *the* pattern of inference that a given argument can be seen as instantiating.

The valid argument F exemplifies a number of different patterns, at different levels of generality, some risky. Generalizing, while valid inferences are typically instances of reliable general patterns, they will also be instances of other, 'too general', unreliable patterns. Or to put it the other way around, a generally *un*reliable pattern of inference can have specific instances that happen to be valid (being valid for some other reason than exemplifying the unreliable pattern).

This point becomes important in the next chapter.

   Finally, a terminological remark. It is quite common to say that a type of inference is itself 'valid' when all its instances are valid: otherwise the type is said to
be 'invalid'. This usage shouldn't cause confusion when you encounter it, so long
as you remember that a generally invalid inference *type* (like '*A*; so, *C*') can have
valid *instances* (like **F**).

3.5 'Logical form'

Consider the following arguments:

   I   (1) All dogs have four legs.
          (2) Fido is a dog.
     So (3) Fido has four legs.

   I′   (1′) Every dog has four legs.
          (2) Fido is a dog.
     So (3) Fido has four legs.

   I″   (1) Any dog has four legs.
          (2) Fido is a dog.
     So (3) Fido has four legs.

In an obvious sense, these arguments exemplify different patterns of inference,
involving respectively 'all', 'every' and 'any'.

   However it is very tempting to say that the differences here are superficial and
the underlying logical structure is intuitively the same. In each case, we have one
premiss which says that everything/anything/all things of a certain general kind
has/have some property (those are arguably just different ways of expressing the
same proposition); another premiss which says that a given individual belongs to
the general kind in question; and a conclusion which says that the relevant individual has the mentioned property.

   It is tempting to put it this way: the various versions of **I** may differ in surface
form, but they share the same *logical form*. At the level of the propositions
expressed, the inferences are all instances of a single pattern

> *All/every/any F are G*
> *n is F*
> *So, n is G*

We might also want to say that arguments couched in e.g. Aristotle's Greek or in
a medieval logician's Latin can again have this same inferential structure (even
though *they* don't involve the words 'all', 'any' or 'every' at all).

   Another simple illustration. At the beginning of the chapter, we cheerfully
counted the inference in

   A″   (1) No elephant ever forgets.
          (2) Jumbo is an elephant.
     So (3) Jumbo never forgets.

as belonging to the family of inferences of the type

> No F *is* G
> n *is* F
> So, n *is not* G

despite the absence of the words 'is not' from the conclusion. It is again very nat-ural to say that the different surface idiom provides only the lightest of disguises for the logical form that argument **A″** shares with others in the family.

So the suggestion is now emerging that the schematic representations of forms of inference that we've been introducing are perhaps best thought of, not as just displaying surface patterns in the *sentences* used in an argument, but rather as indicating underlying logical patterns in the *propositions* expressed. But again we are not going to try to really develop and explore this tempting suggestion. That's because, as with the sentence/proposition distinction itself, we will find that we can largely sidestep issues about *this* idea of logical form. (For more dis-cussion, see e.g. §§12.3, 13.6.)

3.6 'Arguments are reliable in virtue of their form'

We should, however, pause over one point. Talk of 'logical form' is associated with a popular slogan: *Valid arguments are valid in virtue of their form.* We will conclude this chapter with some deflationary remarks.

We have seen that valid arguments typically exemplify a number of different inferential patterns, some good, some bad. It is therefore more than a bit mis-leading to say baldly that an argument is valid in virtue of its form (as if it has only got one). The idea behind the popular slogan is presumably that for every valid argument there is *some* relevant inferential 'form' which captures what makes it valid.

But that idea is empty if we are allowed to take an argument as the one and only instance of its own entirely specific inference pattern. Of course a valid argument always exemplifies some reliable pattern if you count the argument as a one-off pattern for itself! So, to avoid triviality, the claim had better be some-thing like this: every valid argument belongs to a family of valid arguments cut to the same pattern, a family which also has other members, and the argument is valid in virtue of the inference pattern it shares with the other members of the family.

Well, that's true of all our examples of valid arguments so far. Is it true in every case? Take for example the following one-step argument:

> J   (1) Everyone is female.
> So (2) Any siblings are sisters.

This is valid. Any situation in which the premiss is true – for instance, a plague having killed off all the men – must be a situation without brothers. But what more general but still reliable form does the inference exemplify? Certainly, at the surface level, there is no way of replacing various parts of the argument with place-holding letters so that you end up representing a schematic pattern whose instances are all valid. (You might reply: 'Still, "sister" means *female sibling* so at the level of propositions expressed the argument *is* of the valid general form

Everyone is F.
So, anyone who is G is F and G.'

This riposte, however, takes us back to troublesome questions about propositions and their constituents. For a start, what makes it the case that the concept of a sister is to be analysed into the constituents *female* and *sibling*, rather than the concept of a sibling being the complex one, to be analysed as *brother-or-sister*? Let's not open up these murky issues!)

As we will see later, e.g. in §13.6, there *are* good uses for crisply defined technical notions of being 'valid in virtue of logical form'. These notions will sharpen up the idea of owing validity to the way in which all-purpose, topic-neutral, logical concepts like 'and' and 'not' or 'all' and 'some' distribute between the premisses and conclusion. However, these crisp notions are put to work in contexts where we want to *contrast* arguments which are valid in virtue of 'logical form' (in some narrow, technical sense) with other kinds of valid arguments like examples F and J (which depend for their validity on something other than the presence of topic-neutral logical notions like 'not' and 'all'). Hence many valid arguments are *not* valid-in-virtue-of-form in the technical senses to be later explained.

In fact, it is very unclear whether there is any useful notion of being valid-in-virtue-of-form which is applicable to *all* valid inferences.

3.7 Summary

- Arguments with valid inference steps are often – but not always – valid in virtue of the way that various key logical concepts (like 'all', 'no' and 'most') are distributed in a patterned way between the premisses and conclusion.

- Inferential patterns can be conveniently represented using schematic variables, whose use is governed by the obvious convention that a given variable represents the same name or predicate whenever it appears within an argument schema.

- Intuitively, inferential patterns can be shared even when the surface linguistic form differs (e.g. it is natural to say that various arguments beginning 'Every man ...', 'Any man ...', 'Each man ...' may yet exemplify the same inferential move).

- There is strictly no such thing as the unique form exemplified by an argument. Valid arguments will typically be instances of reliable forms; but they can also be instances of other, more general, unreliable forms.

- Our main concern henceforth will be with rather general, 'topic-neutral', patterns of inference.

Exercises 3

A   Which of the following types of inference step are valid (i.e. are such that all their instances are valid)? If you suspect an inference-type is invalid, find an instance which obviously fails because it has plainly true premisses and a false conclusion.

1.   Some F are G; no G is H; so, some F are not H.
2.   Some F are G; some F are H; so, some G are H.
3.   All F are G; some F are H; so, some H are G.
4.   No F is G; some G are H; so, some H are not F.
5.   No F is G; no G is H; so, some F are not H.

(Arguments of these kinds, with two premisses and a conclusion, with each proposition being of one of the kinds 'All ... are ...', 'No ... is ...' or 'Some ... are/are not ...', and each predicate occurring twice, are the traditional *syllogisms* first discussed by Aristotle.)

B   What of the following patterns of argument? Are these valid?
1.   All F are G; so, nothing that is not G is F.
2.   All F are G; so, at least one thing is F and G.
3.   All F are G; no G are H; some J are H; so, some J are not F.
4.   There is an odd number of F, there is an odd number of G; so there is an even number of things which are either F or G.
5. m is F; n is F; so, there are at least two F.
6.   All F are G; no G are H; so, all H are H.

C   Consider the following argument:

Dogs have four legs.
Fido is a dog.
So Fido has four legs.

Is it of the same inferential type as the I-family arguments in §3.5?

The counterexample technique

In this chapter we exploit the fact that many arguments come in families which share an inferential structure (Chapter 3) and put that together with the invalidity principle (Chapter 2) to give a technique for showing that various *invalid* arguments are indeed invalid.

4.1 The technique illustrated

Recall the following argument (§1.4):

> **A** (1) Most Irish are Catholics.
> (2) Most Catholics oppose abortion.
> So (3) At least some Irish oppose abortion.

We quickly persuaded ourselves that the inference is invalid by imagining how the premisses could be true and conclusion false. But let's now proceed a bit more methodically. Notice first that this argument exemplifies the argumentative pattern

> **M** *Most F are G*
> *Most G are H*
> *So, at least some F are H*

And the 'Irish' argument surely stands or falls with this general pattern of inference – for if this pattern isn't reliable, what *else* can the argument depend on?

So, next ask: *is* this inference pattern **M** in fact a reliable one? Are inferences of this shape always valid? Consider:

> **A'** (1) Most chess masters are men.
> (2) Most men are no good at chess.
> So (3) At least some chess masters are no good at chess.

The premisses of this argument are true. Chess is a minority activity, and still (at least at the upper levels of play) a predominantly male one. But the conclusion is crazily false. So we have true premisses and a plainly false conclusion: hence by the invalidity principle, argument **A'** is invalid. We have shown, then, that the inference pattern is *not* reliable: an inference of this type can have true premisses and a false conclusion. There is no special feature to save the 'Irish' argument – the cases are exactly parallel – so that too shares in the general unreliability.

We need to be absolutely clear about what is going on here. We can't just say 'The inference pattern **M** is unreliable, so any instance (such as **A**) is invalid'. For as we stressed in §3.4, a generally unreliable pattern of inference may have special instances that happen to be valid. Still, when an unreliable pattern does happen to have a valid instance, there must be something *else* about the argument (i.e. something other than exhibiting the unreliable pattern) which makes it valid. So faced with an instance of an unreliable pattern, *if there are no other redeeming features of the argument*, then we can conclude that the argument isn't valid. That's the case with **A**. The inference pattern **M** is unreliable. There's nothing else that the 'Irish' argument can be depending on for its validity. In other words there's nothing to make it any better than the invalid argument **A'**. They really are parallel cases. So **A** is invalid.

4.2 More illustrations

Here's another example of the same technique for criticizing an argument. Consider the inference:

> **B**   (1) Some philosophers are great logicians.
>   (2) Some philosophers are Germans.
>   So (3) Some great logicians are Germans.

This has true premises and a true conclusion; so we can't yet apply the invalidity principle. But the argument exemplifies the pattern

> **N**   *Some F are G*
>   *Some F are H*
>   *So, some G are H*

Argument **B** is valid if inference pattern **N** is reliable: and equally, it is invalid if **N** is unreliable – for argument **B** has no special feature that could make it valid for some other reason.

So is the pattern **N** reliable? Take this instance:

> **B'**   (1) Some humans are adults.
>   (2) Some humans are babies.
>   So (3) Some adults are babies.

True premisses, false conclusion: so by the invalidity principle, the inference in **B'** is invalid. **N** *is* unreliable. And hence any argument like **B** that depends on this pattern is unreliable.

Another quick example: what are we to make of the following argument?

> **C**   (1) Most philosophy professors have read Spinoza's *Ethics*.
>   (2) Few football fans have read Spinoza's *Ethics*.
>   So (3) Few philosophy professors are football fans.

This has one true and one false premiss (most philosophy lecturers have never waded through Spinoza!). Maybe the conclusion is true as well. So we can't apply the invalidity principle directly. But the argument evidently appeals to the inference move

O   *Most F are G*
             Few H are G
             So, few F are H

Here's another instance of O:

C′   (1)   Most philosophy professors have got a doctorate in philosophy.
       (2)   Few owners of televisions have got a doctorate in philosophy.
   So (3)   Few philosophy professors own televisions.

The premises are true and the conclusion false (*lots* of philosophy professors are worldly enough to have televisions!). So we have a 'counterexample' which shows the unreliability of the argument form O: this form of inference doesn't guarantee to take you from truths to truths. The Spinoza argument C relies on this form of reasoning – there is nothing to make it any better than argument C′; so it is a bad argument.

4.3 The technique described

We have been using a two-stage technique for showing arguments to be invalid. Let's set down the general recipe, and then use it in some more examples.

The counterexample technique

Stage 1   Locate the pattern of inference that the argument is actually relying on (i.e. the pattern of inference that needs to be valid if the argument is to work).

Stage 2   Show that the pattern is not a reliable one by finding a *counterexample*, i.e. find an argument with this pattern which is evidently invalid, e.g. because it has actually true premises and a false conclusion.

There is absolutely nothing mysterious or difficult or novel about this counterexample technique; we use it all the time in everyday evaluation of arguments. We could also call it the *'But you might as well argue …'* gambit. Someone says that Mrs Jones must be a secret alcoholic, because she has been seen going to the Cheapo Booze Emporium and everyone knows that that is where all the local alcoholics hang out. You reply, for example: 'But you might as well argue that Hillary Clinton is a Republican Senator, because she's been seen going into the Senate, and everyone knows that that's where all the Republican Senators hang out'.

The gossip used the reasoning:

D   (1)   All the local alcoholics use the Cheapo Booze Emporium.
       (2)   Mrs Jones uses the Cheapo Booze Emporium.
   So (3)   Mrs Jones is an alcoholic.

That has the form

> P   *All (the) F are G*
>   *n is G*
>   *So, n is F*

You then point out that this line of reasoning is unreliable by coming up with a simple counterexample:

> D′   (1) All the Republican Senators frequent the Senate.
>   (2) Hillary Clinton frequents the Senate.
> So   (3) Hillary Clinton is a Republican Senator.

In sum, the counterexample technique is simple, powerful and natural. And its rationale is clear. Certain patterns of inference are guaranteed to be truth-preserving, other kinds of inferential move can lead from true premises to a false conclusion. Now, given an argument, we want to know whether the pattern of inference it is crucially relying on is a safe one, that is bound to lead from true premises to a true conclusion. Very obviously, the pattern will not be a safe one if you can find an instance of it with actually true premises and a false conclusion – in other words, if we can find a 'real life' counterexample to the reliability of the inferential pattern.

And in fact, now we've got the basic point, we can see that telling counterexamples don't *have* to be constructed from 'real life' situations. *An uncontroversially coherent but merely imaginable counterexample will do just as well.* Why? Because that's still enough to show that the inferential pattern in question is unreliable: it doesn't necessarily preserve truth in all possible situations. Using 'real life' counterexamples just has the advantage that you don't get into any disputes about what is coherently imaginable.

4.4 More examples

Consider the following quotation (in fact, the opening words of Aristotle's *Nicomachean Ethics*):

> Every art and every enquiry, and similarly every action and pursuit, is thought to aim at some good; and for this reason the good has rightly been declared to be that at which all things aim.

On the face of it – though many scholars rightly challenge this interpretation – this gives us an initial premiss, roughly

> E   (1) Every practice aims at some good.

And then a conclusion is drawn (note the inference marker 'for this reason')

>   (2) There is some good ('*the* good') at which all things aim.

This argument has a very embarrassing similarity to the following one:

> E′   (1) Every assassin's bullet is aimed at some victim.
> So   (2) There is some victim at whom every assassin's bullet is aimed.

And that is obviously bogus. Every bullet has its target (let's suppose), without there being a single target shared by them all. Likewise, every practice may aim at some valuable end or other without there being a single good which encompasses them all.

Drawing out some logical structure, Aristotle's argument appears to rely on the inference form

Q   *Every F has relation R to some G*
          So, there is some G such that every F has relation R to it

and then the 'assassin' argument is a counterexample (an inference of the same pattern, which is evidently invalid).

Expressions like 'every' and 'some' are standardly termed *quantifiers* by logicians (see Chapter 21); and the invalidity of Q shows that in general we can't shift around the order of quantifiers in a sentence. It is perhaps worth noting that a number of supposed arguments for the existence of God commit the same *quantifier-shift fallacy*. Thus, even if you grant that

E˜   (1) Every causal chain has an uncaused first link,

it does not follow that

       (2) There is something (God) that is the uncaused first link of every causal chain.

Again, even if you suppose that

E°   (1) Every ecological system has a designer,

(though that's a premiss which Darwin exploded), it still does not follow that

       (2) There is one designer (God) of every ecological system.

There may, consistently with the premiss, be no one Master Designer of all the ecosystems; perhaps each system was produced by a competing designer.

4.5 Countering the counterexample technique?

Did Aristotle really use the terrible argument E? Many, perhaps most, scholars would deny that he did, and offer alternative readings.

Quite generally, faced with a play of the 'But you might as well argue ...' gambit, the defender of the argumentative text up for criticism may always try to counter that the intended form of argument has been misidentified, and that the supposedly parallel counterexample isn't really parallel at all. And in complex cases, as with Aristotle's text which we have ripped out of context, it may well not be immediately obvious quite what the intended argument is and what pattern of inference is being relied on.

However, faced with an apparent counterexample, the proponent of the original argument must at least meet the challenge of saying *why* the supposed counterexample fails to make the case, and must tell us what the principle of inference used in the argument *really* is. If Aristotle isn't wanting to rely on the fallacious inference-type Q then what *is* he up to? His defender has to come up with an alternative suggestion here.

This kind of interpretative exchange can be very instructive. More generally, there's a considerable skill involved in taking a real-life passage and trying to tease out what is supposed to be going on argumentatively, and knocking it into its best shape. Rarely does ordinary prose serve up fully explicit arguments neatly packaged into premises and conclusions all set out just waiting for the propositions to be numbered off! Which is why there are whole books on 'informal logic' which aim to foster the skill of extracting nuggets of hard argument from the relaxed prose presentations.

The focus of this book, however, is different. Here we will assume that *that* kind of interpretative work is done, and arguments are already neatly packaged into premises and conclusions. Our concern is with the assessment of arguments once we have sifted them from the prose.

4.6 Summary

- The counterexample technique to show invalidity is applicable to a target argument when (1) we can find a pattern of inference that the argument is depending on, and (2) we can show that the pattern is not a reliable one by finding a *counterexample* to its reliability, i.e. find an argument exemplifying this pattern which has (or evidently could have) true premises and a false conclusion.

- This technique is familiar in everyday evaluation of arguments as the 'But you might as well argue …' gambit.

- But in applications of this gambit to real-world arguments, step (1) – i.e. locating the relevant form of inference that is being relied upon – can be a contentious business.

Exercises 4

Some of the following arguments are invalid. Which? Use the counterexample technique to prove invalid the ones that are.

1. Many ordinary people are corrupt, and politicians are ordinary people. So, some politicians are corrupt.

2. Many great pianists admire Glenn Gould. Few, if any, unmusical people admire Glenn Gould. So few, if any, great pianists are unmusical.

3. Everyone who admires Bach loves the *Goldberg Variations*; some who admire Chopin do not love the *Goldberg Variations*; so some admirers of Chopin do not admire Bach.

4. Some nerds are trainspotters. Some nerds wear parkas. So some trainspotters wear parkas.

5. Anyone who is good at logic is good at assessing philosophical arguments. Anyone who is mathematically competent is good at logic. Anyone who is good at assessing philosophical arguments admires Bertrand Russell. Hence no-one who admires Bertrand Russell lacks mathematical competence.

6. (Lewis Carroll) Everyone who is not a lunatic can do logic. No lunatics are

fit to serve on a jury. None of your cousins can do logic. Therefore none of your cousins is fit to serve on a jury.

7. Most logicians are philosophers; few philosophers are unwise; so at least some logicians are wise.

8. Few Sicilians approve of abortion; many atheists approve of abortion; so few atheists are Sicilians.

9. All logicians are rational; no existentialists are logicians; so if Sartre is an existentialist, he isn't rational.

10. If Sartre is an existentialist, he isn't rational; so if he is irrational, he is an existentialist.

5

Proofs

The counterexample technique is a way of demonstrating the *invalidity* of some inferences. This chapter introduces a way of showing the *validity* of some other inferences.

5.1 Two sample proofs

It can be hard to pin down the deductive structure that is being relied on in a passage of argument. But suppose we *have* agreed what type of inference move is at stake. Then, if we *can* find a counterexample which shows the inference move is not reliable, that's that. The argument is shown to be invalid.

Suppose, however, that we have tried for some time to find a counterexample and failed. What does that show? Perhaps nothing very much. It could be that the argument is invalid but we have just been too dim to think up a suitable counterexample. Or it could be that the argument is in fact valid after all. But if failure to find a counterexample doesn't settle the matter, how *can* we demonstrate that a challenged inference really is a valid one?

Take this charmingly daft example from Lewis Carroll:

> A   Babies are illogical; nobody is despised who can manage a crocodile; illogical persons are despised; so babies cannot manage a crocodile.

This three-premiss inference is in fact valid. Maybe, just for a moment, that isn't obvious. So consider the following two-step argument:

> A′ (1) Babies are illogical.   (premiss)
> (2) Nobody is despised who can manage a crocodile. (premiss)
> (3) Illogical persons are despised.   (premiss)
> (4) Babies are despised.   (from 1, 3)
> (5) Babies cannot manage a crocodile.   (from 2, 4)

Here, we have inserted an extra step between the original premises and the target conclusion. The inference from the premises to the intermediate step, and then the inference from there to the conclusion, are both evidently valid. So we can indeed get from the initial premises to the conclusion by a truth-preserving route.

Here's a second quick example, equally daft (due to Richard Jeffrey):

B   Everyone loves a lover; Romeo loves Juliet; so everyone loves Juliet.

Take the first premiss to mean 'everyone loves anyone who is a lover' (where a lover is a person who loves someone). Then this inference too is, slightly surprisingly, deductively valid!

For consider the following multi-step argument:

B′	(1)	Everyone loves a lover.	(premiss)
	(2)	Romeo loves Juliet.	(premiss)
	(3)	Romeo is a lover.	(from 2)
	(4)	Everyone loves Romeo.	(from 1, 3)
	(5)	Juliet loves Romeo.	(from 4)
	(6)	Juliet is a lover.	(from 5)
	(7)	Everyone loves Juliet.	(from 1, 6)

We have again indicated on the right the 'provenance' of each new proposition in the argument. And by inspection we can again see that each small inference step is truth-preserving. As the argument grows, then, we are adding new propositions which must be true if the original premisses are true. These new propositions then serve in turn as inputs (as premisses) to further valid inferences. Everything is chained together so that, if the original premisses are true, everything later, and hence the final conclusion, must be true too. Hence the inference B that jumps straight across the intermediate stages must be valid.

These examples illustrate how we can establish an inferential leap as valid – namely by breaking down a big leap into smaller steps, each one of which is clearly valid. And as a first-shot definition, we'll say:

> When a chain of argument leads from initial premisses to a final conclusion via intermediate inferential steps, each one of which is clearly valid, then we will say that the argument constitutes a *proof* of the conclusion from those premisses.

More or less informal proofs, defending inferences by reducing them to a sequence of smaller steps, occur in everyday contexts. Some systems of formal logic – so-called 'natural deduction' systems – aim to regiment such everyday proofs into tidy deductive structures governed by strict rules. Our examples of proofs in this chapter fall somewhere between loose everyday argumentation and the rigorous presentations of natural deduction systems. For want of a better label, these are *semi-formal* proofs.

5.2 Fully annotated proofs

In our two examples so far, we indicated at each step which earlier propositions the inference depended on. But we could do even better, and also indicate what *type* of inference move is being invoked at each stage.

Take another example from Lewis Carroll (an inexhaustible source!):

C   Anyone who understands human nature is clever; every true poet can stir the heart; Shakespeare wrote *Hamlet*; no one who does not understand human nature can stir the heart; no one other than a true poet wrote *Hamlet*; so Shakespeare is clever.

The big inferential leap from those *five* premisses to the conclusion is in fact valid. (Try to prove that it is, before reading on!)

There is a number of ways of demonstrating this. Suppose we grant that the following two patterns of inference are deductively reliable ones (which they are):

U   *Any/every F is G; n is F; so n is G*
V   *No one who isn't F is G; so any G is F*

Then we can argue as follows, using these two patterns:

C′   (1)   Anyone who understands human nature   (premiss)
           is clever.
     (2)   Every true poet can stir the heart.   (premiss)
     (3)   Shakespeare wrote *Hamlet*.   (premiss)
     (4)   No one who does not understand human
           nature can stir the heart.   (premiss)
     (5)   No one other than a true poet wrote *Hamlet*.   (premiss)
     (6)   Anyone who wrote *Hamlet* is a true poet.   (from 5, by V)
     (7)   Shakespeare is a true poet.   (from 3, 6, by U)
     (8)   Shakespeare can stir the heart.   (from 2, 7, by U)
     (9)   Anyone who can stir the heart   (from 4, by V)
           understands human nature.
    (10)   Shakespeare understands human nature.   (from 8, 9, by U)
    (11)   Shakespeare is clever.   (from 1, 10, by U)

That's all very, very laborious, to be sure: but now we have the provenance of every move fully documented. Since each of the inference moves is clearly valid, and together they get us from the initial premisses to the final conclusion, we can see that the original inference C must be valid.

Here, then, we glimpse an argumentative ideal – the ideal of setting out all the needed premisses and inferential moves systematically, with all the commitments of the argument explicitly acknowledged. Call that a 'fully annotated proof'.

Of course, setting out a fully annotated proof gives a critic a particularly clear target to fire at. If the critic can rebut one of the premisses or come up with a counterexample to the general reliability of any one of the inference patterns that is explicitly called on, then the proof (as presented) is sabotaged. On the other hand, *if* the premisses are agreed and *if* the inference moves are uncontentiously valid, *then* the proof will indeed establish the conclusion beyond further dispute.

5.3 Enthymemes

Plato writes (*Phaedrus* 245c)

D   Every soul is immortal. For that which is ever moving is immortal.

Plainly the premiss that whatever is ever-moving is immortal doesn't by itself logically entail the desired conclusion. We need some additional assumption. The intended extra assumption in this case is obvious: Plato is assuming that every soul is ever-moving (whatever exactly that means!).

Another quick example. Consider the argument

E   The constants of nature have to take values in an extremely narrow range (have to be 'fine-tuned') to permit the evolution of intelligent life. So the universe was intelligently designed.

For this to work as a deductively valid inference, we presumably need to add two more premisses. One is uncontroversial: intelligent life *has* evolved. The other is much more problematic: intelligent design is needed in order for the universe to be fine-tuned for the evolution of intelligent life. Only with some such additions can we get a cogent argument (exercise: construct a fully annotated argument with versions of these additional premisses).

Arguments like **D** and **E** with missing premisses left to be understood (that is, with 'suppressed premisses') are traditionally called *enthymemes*. And the exercise of trying to regiment informal reasoning into semi-formal annotated proofs can often help us in spotting an enthymeme and finding a suitable repair. Note too that the same exercise can also help identify *redundancies*, when it turns out that a premiss given at the outset needn't actually be used in getting to the conclusion.

5.4 Reductio arguments

Suppose someone couldn't see why in the 'crocodile argument' **A′** it was correct to infer that babies cannot manage a crocodile (given that nobody is despised who can manage a crocodile, and babies are despised). How might we convince them? We might amplify that step, as follows:

F   (1)   Nobody is despised who can wrestle a crocodile.   (premiss)
     (2)   Babies are despised.   (premiss)
                 Suppose temporarily, for the sake of argument,
     (3)   │   Babies *can* manage a crocodile.   (supposition)
     (4)   │   Babies are not despised.   (from 1, 3)
     (5)   │   *Contradiction!*   (from 2, 4)
                 │ *So the supposition that leads to this absurdity must be wrong.*
     (6)   Babies cannot manage a crocodile   (RAA)

Here we are arguing 'indirectly'. We want to establish (6). But instead of aiming directly for the conclusion, we branch off by temporarily supposing the exact opposite is true, i.e. we suppose (3). This supposition is very quickly shown to lead to something that contradicts an earlier claim. Hence the supposition (3) is 'reduced to absurdity': its opposite must therefore be true after all. The label '(RAA)' indicates that the argument terminates with a *reductio ad absurdum* inference.

Another quick example. We'll demonstrate the validity of the inference 'No girl loves any sexist pig; Caroline is a girl who loves whoever loves her; Henry loves Caroline; hence Henry is not a sexist pig'.

G   (1) No girl loves any sexist pig.   (premiss)

        (2) Caroline is a girl who loves whoever loves her.   (premiss)

        (3) Henry loves Caroline.   (premiss)

        (4) Caroline is a girl who loves Henry.   (from 2, 3)

        (5) Caroline is a girl.   (from 4)

        (6) Caroline loves Henry.   (from 4)

                | *Suppose temporarily, for the sake of argument,*

        (7) | Henry is a sexist pig.   (supposition)

        (8) | No girl loves Henry.   (from 1, 7)

        (9) | Caroline does not love Henry.   (from 5, 8)

      (10) | *Contradiction!*   (from 6, 9)

                | *So the supposition that leads to this absurdity must be wrong.*

      (11) Henry is not a sexist pig   (RAA)

And we are done.

Let's say – at least for now (compare §12.1) – that a contradiction is a pair of propositions, *C* together with its exact opposite *not-C* (or what comes to the same thing, a single proposition of the type *C and not-C*). The principle underlying the last inference move in **F** and **G** can then be summarized as follows:

> *Reductio ad absurdum* If the propositions $A_1, A_2, ..., A_n$ plus the temporary supposition S logically entail a contradiction then, keeping $A_1, A_2, ..., A_n$ as premisses, we can validly infer that *not-S*.

Why does this principle hold? Suppose that a bunch of premisses plus *S* do entail a contradiction. Then there can be no situation in which those other premisses plus *S* are all true together (or else this would be a situation in which the entailed contradiction would also have to be true, and it can't be). So any situation in which the other premisses are all true is one in which *S* has to be false. Hence those premisses logically entail it isn't the case that *S*, i.e. they entail *not-S*.

It is worth noting that the *reductio* principle continues to apply in the special case where the number of original premisses is *zero*. In other words, if the supposition *S* by itself leads to a contradiction, then it can't be true.

As an illustration of the last point, here's a third – and rather more interesting – example of an RAA argument. We'll prove that the square root of 2 is an irrational number, i.e. $\sqrt{2}$ is not a fraction m/n, where m, n are integers.

If $\sqrt{2}$ *is* a fraction m/n, then we can put it in lowest terms (that is to say, divide out any common factors from m and n). So we'll start by making the following initial assumption:

H | *Suppose temporarily, for the sake of argument,*

    (1) | $\sqrt{2} = m/n$ where m and n are integers with no common factor.

We now show that this leads to contradiction, so it can't be true:

(2)	$2 = m^2/n^2$	(squaring 1)
(3)	$m^2 = 2n^2$	(rearranging 2)
(4)	m^2 is even.	(from 3)
(5)	m is even.	(from 4)

That last move depends on the easily checked point that odd integers always have odd squares: so if m^2 is even, m cannot be odd. Continuing:

(6)	$m = 2r$, for some integer r.	(from 5)
(7)	$(2r)^2 = 2n^2$	(from 3, 6)
(8)	$n^2 = 2r^2$	(rearranging 7)
(9)	n^2 is even.	(from 8)
(10)	n is even.	(from 9)
(11)	m and n are both even.	(from 5, 10)
(12)	m and n have a common factor (i.e. 2).	(from 11)
(13)	*Contradiction!*	(from 1, 12)
	So the supposition (1) that leads to (13) must be wrong.	
(14)	$\sqrt{2} \neq m/n$ where m and n are integers with no common factor.	(RAA)

Despite the relative simplicity of the argument, the result is deep. (It shows that there can be 'incommensurable' quantities, i.e. quantities which lack a common measure. Take a right-angled isosceles triangle. Then there can't be a unit length that goes into one of the equal sides exactly n times and into the hypotenuse exactly m times.)

Finally, two quick remarks. The first concerns the way we have displayed these arguments involving RAA. When we are giving proofs where temporary suppositions are made *en route*, we need to have some way of clearly indicating which steps of the argument are being made while that temporary supposition is in play. A fairly standard way of doing this is to *indent the argument to the right* while the supposition is in play; and then to go back left when the supposition is 'discharged' and is finished with – we've also used vertical lines to help signal the indentation. That gives us a neat visual display of the argument's structure.

Second, allowing suppositions to be made temporarily complicates the range of permissible structures for arguments. However, well-formed proofs will still march steadily on in the sense that *each new step of the proof that isn't a new assumption will depend on what has gone before*. It is just that now what goes before can not only be previous *propositions* but also whole indented *subproofs*. For example, the conclusion in **G** is derived from the subproof from lines (7) through to (10); the conclusion in **H** is derived from the subproof from (1) to (12).

5.5 Limitations

The styles of direct and indirect proof we've been illustrating can be used informally and can also be developed into something much more rigorous. However, there is a certain limitation to the technique of trying to warrant inferences by finding step-by-step proofs. If we are lucky/clever, and spot a proof that takes us from the given premises to the target conclusion, that will show that a putative inference is indeed valid. But we may be unlucky/dim, and not spot the proof. In the general case, even if a proof exists, proof-discovery requires ingenuity, and failing to find a proof doesn't show that the original inference from premises to conclusion is invalid. (Compare the counterexample technique: if we are lucky/clever, deploying that technique will show us that an invalid inference is invalid. But we may miss the counterexample.)

It would be nice to have less hit-and-miss techniques for establishing validity and invalidity. Ideally, we would like a *mechanical* general technique for working out whether an inference is or is not valid that doesn't need any imaginative searching for proofs or counterexamples. Can we ever (as it were) *calculate* the validity or invalidity of inferences in as mechanical a way as we calculate how much five pints of beer will cost, given the price per pint?

Profound results (mostly discovered in the 1930s) show that, in general, a mechanically applicable calculus will *not* in fact be available. But in a limited class of cases, we can pull off the trick. In the next part of this book, starting in Chapter 7, we will be exhibiting one such calculus.

5.6 Summary

- To establish validity we can use many-step deductions, i.e. *proofs*, which chain together little inference steps that are obviously valid and that lead from the initial premises to the desired conclusion.

- Some common methods of proof, however, are 'indirect', like reductio ad absurdum.

- In the general case, even if a proof exists, proof-discovery requires ingenuity.

Exercises 5

Which of the following arguments are valid? Where an argument is valid, provide a proof. Some of the examples are enthymemes that need repair.

1. No philosopher is illogical. Jones keeps making argumentative blunders. No logical person keeps making argumentative blunders. All existentialists are philosophers. So, Jones is not an existentialist.

2. Jane has a first cousin. Jane's father had no siblings. So, if Jane's mother had no sisters, she had a brother.

3. Every event is causally determined. No action should be punished if the agent isn't responsible for it. Agents are only responsible for actions they can avoid doing. Hence no action should be punished.

4. Something is an elementary particle only if it has no parts. Nothing which has no parts can disintegrate. An object that cannot be destroyed must continue to exist. So an elementary particle cannot cease to exist.

5. No experienced person is incompetent. Jenkins is always blundering. No

competent person is always blundering. So, Jenkins is inexperienced.

6.   Only logicians are good philosophers. No existentialists are logicians. Some existentialists are French philosophers. So, some French philosophers are not good philosophers.

7.   Either the butler or the cook committed the murder. The victim died from poison if the cook did the murder. The butler did the murder only if the victim was stabbed. The victim didn't die from poison. So, the victim was stabbed.

8.   Promise-breakers are untrustworthy. Beer-drinkers are very communicative. A man who keeps his promises is honest. No one who doesn't drink beer runs a bar. One can always trust a very communicative person. So, no one who keeps a bar is dishonest.

9.   When I do an example without grumbling, it is one that I can understand. No easy logic example ever makes my head ache. This logic example is not arranged in regular order, like the examples I am used to. I can't understand these examples that are not arranged in regular order, like the examples I am used to. I never grumble at an example, unless it gives me a headache. So, this logic example is difficult.

Validity and arguments

Before turning to explore some formal techniques for evaluating arguments, we should pause to say a little more about the 'classical' conception of validity for individual inference steps, and also say more about what makes for cogent multi-step arguments.

6.1 Classical validity again

In Chapter 1, we introduced the intuitive idea of an inference's being deductively valid – i.e. absolutely guaranteeing its conclusion, given the premises. Then in Chapter 2, we gave the classic definition of validity, which is *supposed* to capture the intuitive idea. But does our definition really do the trick?

A reminder, and a quiz. We said that an inference is ('classically') valid if and only if there is no possible situation in which the premises are true and the conclusion false. So consider: *according to that definition*, which of the following inferences are classically valid?

    A Jo is married; so Jo is married.

    B Socrates is a man; Jo is married; all men are mortal; so Socrates is mortal.

    C Jack is married; Jack is single; so today is Tuesday.

    D Elephants are large; so either Jill is married or she is not.

And then ask: which of the inferences are *intuitively* deductively compelling? Do your answers match?

(1) Trivially, there is no possible way that **A**'s premiss 'Jo is married' could be true and its conclusion 'Jo is married' be false; so the first inference is indeed valid by the classical definition.

What is the intuitive verdict about this argument? Of course, the inference is no use at all as a means for *persuading* someone who does not already accept the the conclusion as correct. Still, truth-preservation is one thing, being useful for persuading someone to accept a conclusion is something else (after all, a valid argument leading to an unwelcome conclusion can well serve to persuade its recipient of the falsehood of one of the premisses, rather than the truth of the conclusion). Once we make the distinction between truth-preservation and use-fulness, there is no reason not to count the 'Jo' argument as deductively valid. As we put it in §3.2, an inference that covers no ground hasn't got a chance to go wrong!

Objection: 'It was said at the very outset that arguments aim to give *reasons* for their conclusions – and the premiss of **A** doesn't give an independent reason for its conclusion, so how can it be a good argument?' Reply: Fair comment. We've just agreed that **A** *isn't* a persuasive argument, in the sense that it can't give anyone who rejects the conclusion a reason for changing their minds. Still, to repeat, that doesn't mean that the inferential relation between premiss and conclusion isn't absolutely secure. And the intuitive idea of validity is primarily introduced to mark absolutely secure inferences: **A** just illustrates the limiting case where the security comes from unadventurously staying in the same place.

(2) Argument **B** with a redundant and irrelevant premiss also comes out as valid by the classical definition. There can be no situation in which Socrates is a man, and all men are mortal, yet Socrates is not mortal. So there can be no situation in which Socrates is a man, all men are mortal, and (as it happens) Jo is married, yet Socrates is not mortal.

What's the intuitive verdict on **B**? Well, introducing redundant premisses may be pointless (or worse, may be misleading). But **B**'s conclusion is still guaranteed, if all the premisses – including the ones that really matter – are true.

Generalizing, suppose $A_1, A_2, ..., A_n$ logically entail C. Then $A_1, A_2, ..., A_n, B$ logically entail C, for any additional premiss B (both intuitively, and by our official definition).

Deductive inference contrasts interestingly with inductive inference in this respect. Recall the 'coffee' argument in §1.3:

E   (1) Cups of coffee that looked and tasted just fine haven't killed you in the past.
     (2) This present cup of coffee looks and tastes just fine.
So   (3) This present cup of coffee won't kill you.

The premisses here probabilistically support the conclusion. But add the further premiss

(2′) The coffee contains a tasteless poison,

and the conclusion now looks improbable. Add the further premiss

(2″) The biscuit you just ate contained the antidote to that poison,

and (3) is supported again (for it is still unlikely the coffee will kill you for any other reason). This swinging to and fro of degree of inductive support as further relevant information is noted is one reason why the logic of inductive arguments is difficult to analyse and codify. By contrast, a deductively cogent argument can't be spoilt by adding further premisses.

(3) So far so good. We can readily live with the results that completely trivial arguments and arguments with redundant premisses count as deductively valid

by the classical definition. But now note that there is no possible situation in which Jack is both married and single. That is to say, there is no possible situation where the premises of argument C are both true. So, there is no possible situation where the premises of that argument are both true and the conclusion false. So, this time much less appealingly, the inference in C also comes out as valid *by the classical definition.*

The point again generalizes. Suppose that a bunch of propositions $A_1, A_2, ..., A_n$ is logically inconsistent, i.e. there is no possible situation in which $A_1, A_2, ..., A_n$ are all true together. Then there is also no situation in which $A_1, A_2, ..., A_n$ are all true and C is false, whatever C might be. Hence by our definition of classical validity, we can validly infer *any proposition we choose* from a bunch of inconsistent premises.

That all looks decidedly counter-intuitive. Surely the propositions 'Jack is married' and 'Jack is single' can't really entail anything about what day of the week it is.

(4) What about the 'Jill' argument? There is no possible situation in which the conclusion is false; it is inevitably true that either Jill is married or she isn't. So, there is no possible situation where the conclusion is false and the premiss about elephants is true. Hence the inference in argument D is also valid by the classical definition.

The point again generalizes: If C is necessarily true, i.e. there is no possible situation in which C is false, then there is also no situation in which $A_1, A_2, ..., A_n$ are all true, and C is false, whatever $A_1, A_2, ..., A_n$ might be. Hence by our definition, we can validly infer a necessarily true conclusion *from any premises we choose*, including elephantine ones.

Which again looks absurd. What to do?

6.2 Sticking with the classical definition

We have just noted a couple of decidedly counter-intuitive results of the classical definition of validity. Take first the idea that contradictory premises entail anything. Then one possible response runs along the following lines:

> We said that the premises of a valid argument *guarantee* the conclusion: how can contradictory premises guarantee anything? And intuitively, the conclusion of an inferentially cogent inference ought to have *something* to do with the premises. So argument C commits a fallacy of irrelevance. Our official definition of validity therefore needs to be tightened up by introducing some kind of relevance-requirement in order to rule out such examples as counting as 'valid'. Time to go back to the drawing board.

Another response runs:

> The verdict on C is harmless because the premises can never be true together; so we can never *use* this type of inference in a *sound* argument for an irrelevant conclusion. To be sure, allowing the 'Jack' argument to count as valid looks surprising, even counter-intuitive. But maybe we shouldn't

slavishly follow *all* our pre-theoretical intuitions, especially our intuitions about unusual cases. The aim of an official definition of validity is to give a tidy 'rational reconstruction' of our ordinary notion which smoothly captures the uncontentious cases; the classical definition does that particularly neatly. So we should bite the bullet and accept the mild oddity of the verdict on **C**.

Likewise there are two responses to the observation that the classical definition warrants argument **D**. We could either say this just goes to show that our definition allows more gross fallacies of irrelevance. Or we could bite the bullet again and deem this result to be a harmless oddity (we already need to know that a conclusion *C* is necessarily true before we are in a position to say that it is classically entailed by arbitrary premises – and if we already know that *C* is necessary, then extra valid arguments for it like **D** can't extend our knowledge).

The majority of modern logicians opt for the second responses. They hold that the cost of trying to construct a plausible notion of 'relevant' inference turns out to be too high in terms of complexity to make it worth abandoning the neat simplicity of the idea of classical validity. But there is a significant minority who insist that we really should be going back to the drawing board, and hold that there are well-motivated and workable systems of 'relevant logic'. We won't be able to explore their arguments and their alternative logical systems here – and in any case, these variant logics can really only be understood and appreciated in contrast to classical systems. We just have to note frankly that the classical definition of validity is not beyond challenge.

Still, our basic aim is to model certain forms of good reasoning; and models can be highly revealing even if some of their core constructs idealize. The idea of a 'classically valid argument' – like e.g. the idea of an 'ideal gas' – turns out at least to be a *highly* useful and illuminating idealization: and this can be so even if it would be wrong to claim that it is in the end the uniquely best and most accurate idealization. Anyway, we are going to stick to the classical definition in our introductory discussions in this book.

6.3 Multi-step arguments again

We characterized deductive validity as, in the first place, a property of individual *inference steps*. We then fell in with the habit of calling a *one-step* argument valid if it involves a valid inference step from initial premises to final conclusion. In this section, we'll see the significance of that qualification 'one-step'.

Consider the following mini-argument:

F   (1) All philosophers are logicians.
    So (2) All logicians are philosophers.

The inference is obviously fallacious. It just doesn't follow from the assumption that all philosophers are logicians that only philosophers are logicians. (You might as well argue 'All women are human beings, hence all human beings are women'.) Here's another really bad inference:

G   (1)   All existentialists are philosophers.
    (2)   All logicians are philosophers.
So (3)   Hence, all existentialists are logicians.

It plainly doesn't follow from the claims that the existentialists and logicians are both among the philosophers that *any* of the existentialists are logicians, let alone that *all* of them are. (You might as well argue 'All women are human beings, all men are human beings, hence all women are men'.)

But now imagine someone who chains this pair of rotten inference steps together into a two-step argument as follows. He starts from the premisses

H   (1)   All existentialists are philosophers.   (premiss)
    (2)   All philosophers are logicians.   (premiss)

He then cheerfully makes the same fallacious inference as in F, and infers that

    (3)   All logicians are philosophers.   (from 2!!)

Then he compounds the sin by committing the same inferential howler as in G, and supposes that it follows from H(1) and the newly inferred H(3) that

    (4)   All existentialists are logicians.   (from 1, 3!!)

Here our reasoner has got from the initial premisses H(1) and H(2) to his final conclusion H(4) by two quite *terrible* moves: it would therefore be extremely odd to dignify his supposed 'proof' as deductively cogent.

Note, however, that in this case the two howlers by luck happen to cancel each other out; there *are* no possible circumstances in which the initial premisses H(1) and H(2) are both true and yet the final conclusion H(4) is false. If the existentialists are all philosophers, and all philosophers are logicians, then the existentialists must of course be logicians. Hence, the inferential jump from initial premisses to final conclusion is in fact *valid*.

So if we were to say (without any qualification) that an argument is valid if the initial premisses absolutely guarantee the final conclusion, then we'd have to count the two-step argument H as 'valid'. Which is, to say the least, a very unhappy way of describing the situation, given that H involves a couple of nasty fallacies!

The previous chapter reminds us that many arguments involve more than one inference step. To be sure, we don't very often set them out as semi-formal proofs; but multi-step arguments are common enough in any form of extended enquiry – indeed our initial informal definition of an argument (§1.1) was as a *chain* of reasoning. The point of multi-step chains is to show that you can legitimately infer some (perhaps highly unobvious) conclusion by breaking down the big inferential leap from initial premisses to final conclusion into a series of smaller, more evident steps. But if an argument goes astray along the way, then it will be mere good luck if it happens to get back on track to a conclusion that does genuinely follow via some other route from the starting premisses.

The moral here is a simple one: we need to make a distinction. Suppose we have a step-by-step argument from premisses $A_1, A_2, ..., A_n$ to conclusion C.

Then it is one thing to say that the overall, big-leap, inference from A_1, A_2, ..., A_n to C is valid. It is another thing to say that the argument we give from A_1, A_2, ..., A_n to C is deductively cogent. Of course, in the case where the argument has only one step, then the argument is deductively cogent if and only if the overall inference is valid. But as example **H** shows, in the case of multi-step arguments, this two-way link breaks down. The inferential leap from the premisses to conclusion might still be valid, even if some step-by-step argument we give to defend the inference is not deductively cogent.

You might think that we are making much ado about very little here. For doesn't this just show that we must define a genuine proof – a deductively virtuous multi-step argument, where luck doesn't enter into it – to be an argument *whose individual inferential steps are all valid?* (In fact, that was our first-shot definition in §5.1.)

But things aren't quite that easy. Not only must individual steps be valid, but they must be chained together in the right kind of way.

To illustrate, consider this one-step inference:

I (1) Socrates is a philosopher.
(2) All philosophers have snub noses.
So (3) Socrates is a philosopher and all philosophers have snub noses.

That's trivially valid (if you are given *A* and *B*, you can infer *A-and-B*). Here is another equally trivial but valid inference:

J (1) Socrates is a philosopher and all philosophers have snub noses.
So (2) All philosophers have snub noses.

From *A-and-B*, you can infer *B*. And thirdly, this too is plainly valid:

K (1) Socrates is a philosopher.
(2) All philosophers have snub noses.
So (3) Socrates has a snub nose.

Taken separately, then, those three little inferences are quite unproblematic. But now imagine someone chains them together, not in the usual way (where each step depends on something that has gone before) but in the following tangle:

L (1) Socrates is a philosopher. (premiss)
(2) Socrates is a philosopher and all philosophers have snub noses. (from 1, 3: as in I)
(3) All philosophers have snub noses. (from 2: as in J)
So (4) Socrates has a snub nose. (from 1, 3: as in K)

By separately valid steps we seem to have deduced the shape of Socrates' nose just from the premiss that he is a philosopher! What has gone wrong?

The answer is obvious enough. In the middle of the argument we have gone round in a circle. L(2) is derived from L(3), and then L(3) is derived from L(2). Circular arguments can't take us anywhere.

So that suggests we need an improved account of what makes for a deductively cogent proof: the individual inferential steps must all be valid, *and* the steps must be chained together in a non-circular way, with steps only depending on what's gone before in the argument.

That's an improvement, but more still needs to be said. For recall, we need to allow for multi-step arguments like the 'reductio' arguments we met in §5.4 – i.e. arguments where we can introduce a new temporary supposition that doesn't follow from what's gone before (and then later 'discharge' the assumption by an inference step that relies, not on prior propositions, but on a prior 'sub-proof'). Then there will be more complications when we introduce 'downward-branching tree proofs' (see §20.1). Recognizing such various argument-styles as all being in good order complicates the task of framing a definition of what makes for a deductively cogent proof in general. Fortunately, for our purposes in this book, we don't need to come up with a general story here: so we won't attempt to do so.

6.4 Summary

- Our classical definition of validity counts as valid any inference which has inconsistent premisses, or has a necessarily true conclusion. Despite being counter-intuitive, these are upshots we will learn to live with.

- It is unhappy to say that a multi-step argument is 'valid' just if there is no way the initial premisses can be true and the conclusion false. For we don't want to endorse arguments where there are inferential steps which are invalid but where it happens that the fallacies cancel each other out.

Interlude

Logic, formal and informal

What has been covered so far?

- We have explained, at least in an introductory way, the notion of a classically valid inference-step.

- We have explained the corresponding notion of a deductively valid one-step argument.

- We have distinguished deductive validity from other virtues that an argument might have (like being a good inductive argument).

- We have noted how different arguments may share the same pattern of inference. And we exploited this fact when we developed the counterexample technique for demonstrating invalidity.

- We have seen some simple examples of direct and indirect multi-step proofs, where we show that a conclusion really can be validly inferred from certain premises by filling in the gap between premises and conclusion with evidently valid intermediate inference steps.

What haven't we done so far?

- We still haven't given a very sharp characterization of the notion of deductive validity (we invoked an intuitive notion of 'possible situation' without really spelling out what that came to).

- Nor have we given a really clear story about what counts as a 'pattern' of inference.

- Nor have we given a sharp characterization of what counts as a legitimate 'well-built' form of multi-step proof.

- Indeed, we even set aside the question of the nature of the basic constituents of informal arguments, namely propositions (so in a sense, we don't really know the first thing about valid arguments, i.e. what they are made of).

At this point, then, we could go in a number of directions. We could remain content with an informal level of discussion, and first explore more techniques for teasing out the inferences involved in real-life passages of extended prose argumentation. Then we could explore a variety of ways of evaluating a wide range of everyday deductions. We could also turn our attention to various inductive, probabilistic modes of argument, and explore the sorts of fallacies we can fall into when trying to argue inductively. And that kind of wide-ranging study in *informal logic* (as it is often called) would be a highly profitable enterprise. But this book takes a different direction. We are going to presuppose that we are already moderately competent at filleting out the bones of an argument from everyday prose; and we are going to be focusing on just a couple of broad classes of deductive inferences, with the aim of constructing rigorous, comprehensive, *formal* theories about them.

We are going, in short, for depth rather than breadth. Our approach will give us a powerful set of tools for dealing with certain kinds of real-life arguments. But that isn't the only reason for taking the formal route. For just as in-depth theories about genes or electrons tell us what genes and electrons really are, we might hope that these rigorous theories about some core types of argument will tell us more about what validity for such arguments amounts to. Moreover, in the course of these discussions we are going to be developing a range of other ideas which belong in the toolkit of any analytic philosopher.

I'd claim, in fact, that the now well understood theory of 'first-order quantification theory' – the main branch of formal logic which we are introducing in this book – is one of the great intellectual achievements of twentieth-century logicians. It opens the door into a rich and fascinating field: and although, in this introductory book, there will only be occasional glimpses through the door, what follows will take us right to the threshold.

There are, however, different possible routes to the threshold, different (though ultimately equivalent) ways of starting to develop formal logical systems. Once upon a time, so-called axiomatic systems were the fashion. More recently, most students were introduced to serious logic via some sort of 'natural deduction' system. More recently still, systems using 'tree' proofs have become popular, being particularly easy to understand and manipulate by those without a mathematical bent. Given world enough and time, we'd develop all three main approaches, and discuss their pros and cons. But done in enough detail, that survey would take too long; and done quickly, it would just confuse beginners. We have to choose. So this book chooses to introduce formal logic by trees. What that involves will become clear soon enough.

7

Three propositional connectives

In the next part of the book, we will be concentrating on a very restricted class of inferences, whose logical structure involves the so-called propositional connectives, 'and', 'or' and 'not'. Such arguments may not be very exciting in themselves: but the exercise of developing systematic techniques for evaluating them is the simplest way of introducing a number of the centrally important concepts and strategies of formal logic. This present chapter explains the advantages of introducing a *formalized language* in which to express arguments involving the propositional connectives.

7.1 'And', 'or' and 'not'

Consider the following argument:

    A   (1) Either we are out of petrol or the carburettor is blocked.
          (2) We are not out of petrol.
  So (3) The carburettor is blocked.

That argument is valid; so is this one:

    A'   (1) Either Jill is in the library or she is in the coffee bar.
          (2) Jill isn't in the library.
  So (3) Jill is in the coffee bar.

Evidently, we can generalize here. Our two arguments have the form

    Either A or B
    Not A
    So, B

And any inference sharing the same form will be valid.

Three quick comments about this. First, we are here using 'A' and 'B' to stand in for whole propositions, such as 'We are out of petrol' or 'Jill is in the coffee bar'. We can call these (schematic) *propositional variables*. We have used such propositional variables before (e.g. in §§3.3 and 5.4): contrast the use of 'F' and 'G' which schematically represent predicates, and 'n' which holds the place of a naming or subject term.

Second, what about our representation of the form of the second premiss in each case? We have written *Not A*. But the premiss 'We are not out of petrol' does not *begin* with 'not'. Still, inserting the 'not' into 'We are out of petrol' has the result of denying what was asserted by the original sentence. And we can very naturally represent the proposition which says exactly the opposite of *A* by *Not A*.

Third, note that in representing the structure of the second argument we have ignored the surface difference between using the name 'Jill' and using the pronoun 'she' to refer to Jill. You might say that we are trying to represent logical structure, not just surface form (see §3.5).

Here's another pair of valid arguments:

B   (1)   It's not the case that Jack played lots of football and also did well in his exams.
          (2)   Jack played lots of football.
    So (3)   Jack did not do well in his exams.

B′   (1)   It's not true that Tony supports the policy and George does too.
        (2)   Tony supports the policy.
   So (3)   George does not support it.

These two evidently share the same reliable inferential form, which we might represent as follows:

> *Not (A and B)*
> *A*
> *So, not B*

In each case, the first premiss says that something is not the case, where the something in question is a combination of two propositions linked by 'and'. Why use brackets in representing this structure? Well, we certainly do want to distinguish

> *Not (A and B)*

– which joins together *A* and *B*, and then says that this combination doesn't hold – from

> *(Not-A and B)*

which denies *A*, but then adds that *B does* hold. For suppose we replace

> A: The President orders an invasion.
> B: There will be devastation.

Then *Not (A and B)* represents the hopeful thought that it isn't true that the President will order an invasion with consequent devastation; while *(Not-A and B)* is the distinctly less cheering claim that although the President will not order an invasion, there will still be devastation all the same.

We have seen, then, two examples of valid inference patterns where the relevant logical materials are simply 'and' 'or' and 'not'. Here is a patently invalid inference:

C   (1) It is not the case that Alice is clever and Bob is clever too.
   So (2) Alice is not clever.

Suppose Alice is clever but Bob is a dunce: then the premiss is true and the conclusion false in that situation. More generally; from *Not (A and B)* you can't in general infer *Not A* – for maybe *B* is to blame for the falsity of (*A and B*).

These initial examples are extremely trite; but they give us some first cases of arguments whose validity or invalidity depends on the distribution among the premisses and conclusion of the three *propositional connectives* 'and' 'or' and 'not'.

Talk of 'connectives' here is obviously apt for 'and' and 'or'. Thus 'and' connects two propositions *A*, *B* to give us the more complex proposition *A and B*. Likewise 'or' connects two propositions *A*, *B* to give us the more complex proposition *A or B*. It is a bit stretched to call 'not' a *connective*, since it operates on just *one* proposition and (so to speak) turns it on its head. Still, 'not' is like 'and' and 'or' in giving us a way of building more complex propositions from simpler ones (and we will soon discover some closer similarities). So we will follow convention and count 'not' as an honorary connective.

We are going to explore, then, arguments involving these three propositional connectives; and later we will present a mechanical way of demonstrating validity for such arguments. But before getting to that point, a fair amount of preliminary but important ground-clearing needs to be done.

7.2 Quirks of the vernacular

Let's remark briefly on some of the complexities and quirks of 'and', 'or' and 'not' as they function in vernacular English.

(1) We can use 'and' to conjoin a pair of matched expressions belonging to almost any grammatical category – as in 'Jack is fit and tanned', 'Jill won quickly and easily', 'Jack and Jill went up the hill', or 'Jo smoked and coughed', where 'and' conjoins respectively pairs of adjectives, adverbs, proper names and verbs. But set aside all those uses, and consider only the cases where 'and' joins two whole sentences to form a new one. In most such cases what we get is a sentence which expresses the bare conjunction of the propositions expressed by the original sentences:

> The *conjunction* of two propositions *A* and *B* is true when *A* and *B* are both true, and is false otherwise.

However, inserting an 'and' between two claims in English doesn't *always* express their bare conjunction. Take the two claims

   (a) There is not a word of truth in the reports that Jill loves Jack.
   (b) Jack loves Jill.

Perhaps both these are true: Jack's love is sadly unrequited. But you could hardly assert this just by sticking an 'and' between (a) and (b). For that yields

(c)   There is not a word of truth in the reports that Jill loves Jack and Jack loves Jill,

which is naturally heard as *denying* (b) rather than as asserting it! Mechanically gluing together two claims with 'and' isn't an infallible way of arriving at their conjunction.

There is another kind of complication: compare

(d)   Eve became pregnant and she married Adam.

(e)   Eve married Adam and she became pregnant.

These are naturally taken as conveying different messages about temporal order. Does 'and' in English therefore sometimes mean *more* than bare conjunction and connote, e.g., temporal succession as well? It might appear so.

But perhaps appearances deceive. For delete the 'and' from (d), take what is left as two separate sentences, and the implication of temporal succession still remains because of our default narrative practice of telling a story in the order the events happened. Since the implication remains even without the use of the connective 'and', there is perhaps no need to treat temporal succession as one possible meaning of the connective itself.

However, we won't pursue that debate further. The point is that there is at least an issue here about whether 'and' sometimes means more than pure conjunction.

(2) Turning next to 'or', we can again use this connective to combine a pair of matched expressions belonging to almost any grammatical category – as in 'Jack is cycling or driving', 'Jack or Jill went up the hill', and so forth. But set aside all those uses, and consider only the cases where 'or' is used to join two whole propositional clauses. The obvious complication here is that sentences of the kind *A or B* can often be heard as meaning *A or B but not both* ('Either you will hand your logic exercises in by midday or you will have to leave the class'; 'Either a peace treaty will be signed this week or the war will drag on another year'). But about equally often, such sentences can be heard as meaning *A or B or perhaps both* (of a student with bad grades, I say 'Either he didn't try, or he really isn't any good at logic').

Those two readings of 'or' are called the 'exclusive' and 'inclusive' readings respectively. So:

> The *exclusive disjunction* of two propositions A and B is true when exactly one of A and B is true, and is false otherwise.
>
> The *inclusive disjunction* of two propositions A and B is true when at least one of A and B is true, and is false otherwise.

If we are going to consider the logic of 'or', then we will need for a start to have some explicit way of signalling when we are dealing with exclusive disjunctions and when we are dealing with inclusive cases.

(3) The complexities in the behaviour of 'not' are different again. What concerns us here is the operation which yields the negation of a proposition, which says the

exact opposite of the original proposition:

> The *negation* of a proposition *A* is true exactly when *A* is false and is false exactly when *A* is true.

In English we can often negate a claim by inserting the word 'not' at an appropriate place and then making any required grammatical adjustments. Thus the negation of 'Jo is married' is 'Jo is not married', and the negation of 'Jack smokes' is 'Jack does not smoke'.

But contrast: inserting a 'not' into the truth 'Some students are good logicians' at the only grammatically permissible place in fact gives us another truth, namely 'Some students are not good logicians'. Hence, inserting 'not' isn't always sufficient to produce the negation of a proposition. Nor is it always necessary (for example, replacing 'some' by 'no' in 'Some students are good logicians' *does* give us the negation).

It might be supposed that English does have a cumbersome but sure-fire way of forming negations, namely prefixing a sentence with the phrase 'It is not the case that'. For example, 'It is not the case that some students are good logicians' is indeed the negation of 'Some students are good logicians'. And this works in nearly every case. But not always. Compare

(f) Jack loves Jill and it is not the case that Jill loves Jack.

(g) It is not the case that Jack loves Jill and it is not the case that Jill loves Jack.

Here, (g) results from prefixing 'It is not the case that' to (f). Yet on the natural readings, *both* (f) and (g) are false if Jack and Jill love each other; so these two sentences don't express propositions which are real contradictories. In fact, (g) is naturally read as having the form (*Not-A and Not-B*) – while the real negation of (f) would need to have the form *Not* (*A and Not-B*).

(4) Finally, let us note that English sentences which involve more than one connective can be ambiguous. Consider

(h) Either Jack took Jill to the party or he took Jo and he had some fun.

Does this say that Jack went to the party with Jo and had fun, or else he went with Jill. Or does it say, differently, that he took Jill or Jo to the party and either way enjoyed himself?

Logicians talk about the *scope* of connectives (and other logical operators). For example, in the natural reading of (g), the scope or 'coverage' of the first 'It is not the case that' is just the next clause, 'Jack loves Jill'. In (h), however, there is an ambiguity of scope, i.e. an ambiguity in how to group together the clauses. Thus what is the scope of the 'and'? Does it (first reading) just conjoin 'he took Jo' and 'he had some fun'? Or is the scope or 'coverage' wider (second reading), so that the whole of 'either Jack took Jill to the party or he took Jo' is being conjoined with 'he had some fun'?

7.3 Formalization

Even the handful of examples in the previous section is enough to establish that natural-language connectives can behave in some rather quirky and ambiguous ways. The same goes for other logically salient constructions (as we will see later, for example, when discussing expressions like 'all' and 'any' in §21.1). So to treat the logic of arguments couched in ordinary English really requires taking on two tasks simultaneously; we need *both* to clarify the many vagaries of English *and* to deal with the logical relations of the messages variously expressible in the language.

Given our interest is in logic rather than linguistics, is there a way to divide the labour?

Here is a very inviting, though radical, strategy. We can bulldoze past many of the problems generated by the quirkiness of natural language if, instead of using the messy vernacular to express arguments, we use some much more austere and regimented *formal language* which is designed to be logically well-behaved and quite free of ambiguities and shifting meanings. We then assess arguments involving (for instance) the English 'and', 'or' and 'not' by a two-step process. We start by (1) rendering the vernacular argument into a well-behaved artificial language; and then (2) we investigate the validity or otherwise of the reformulated argument. Step (1) of this two-step process may raise issues of interpretation: but as logicians we can largely focus on step (2). For even given the premises and conclusion translated into a perspicuous and unambiguous notation, the central question of how to assess the argument for validity remains.

This two-step strategy has its roots as far back as Aristotle who used Greek letters as symbols in presenting syllogisms. But the strategy came of age in the nineteenth century with the German logician Gottlob Frege (who wrote works such as the *Begriffsschrift* which presented a 'concept-script' designed for the perspicuous representation of propositions). Frege was soon followed by, among others, Bertrand Russell and A. N. Whitehead early in the last century: their monumental *Principia Mathematica* aims to derive core chunks of classical mathematics from a tiny handful of axioms. And in order to enforce total precision, the derivations are couched not in the informal mix of English and symbols that mathematicians usually use, but in an artificial symbolic language.

In the rest of this book, we are going to follow this two-step strategy, which really underlies the whole project of modern formal logic. The defence of this strategy is its abundant fruitfulness, and that can only become fully clear as we go along. At this stage, therefore, you'll have to take it on trust that the (small) effort required in mastering some simple formal languages will pay off handsomely.

7.4 The design brief for PL

For our immediate purposes, then, we want to design an appropriate language for regimenting arguments involving conjunction, disjunction and negation, a language which must avoid the quirks and ambiguities of ordinary language.

PL (the label is supposed to suggest 'Propositional Logic') is a set of symbolic devices designed to do this job in an entirely straightforward and simple way. We can usefully carve up its design brief into three parts.

(1) First, we want symbols for the three connectives.

- So we give **PL** a connective '∧' which is stipulated *invariably* to signify bare conjunction, no more and no less: i.e. (A ∧ B) is always to be true just when A and B are both true.

- We add a connective '∨' which is stipulated *invariably* to signify *inclusive* disjunction, no more and no less (we will see later how to deal with exclusive disjunction in **PL**). So (A ∨ B) is always to be true just when A is true or B is true or both. '∨' is often read *vel*, which is Latin for inclusive 'or'.

- Finally we add a negation prefix '¬' which always works as the English 'It is not the case that' usually works; i.e. ¬A *invariably* expresses the strict negation of A, and is true just when A is false.

By all means, read '∧', '∨' and '¬' aloud as respectively 'and', 'or' and 'not' – but remember too that these new symbols shouldn't be thought of as mere abbreviations of their ordinary-language readings. For mere abbreviations would inherit the ambiguities of their originals, and then what would be the point of introducing them? These **PL** connectives are better thought of as cleaned-up *replacements* for the vernacular connectives, which unambiguously express their originals' core meanings.

(2) Next, we want to avoid any structural scope ambiguities as in 'Either Jack took Jill to the party or he took Jo and he had a really good time'. So ambiguous expressions of the type A ∨ B ∧ C must be banned. How should we achieve this? Compare arithmetic, where again we can form potentially ambiguous expressions like '1 + 2 × 3' (is the answer 7 or 9?). Here we can disambiguate by using brackets, and write either '(1 + (2 × 3))' and '((1 + 2) × 3)', which both have unique values. We'll use the same simple device in **PL**. That is to say, we'll insist on using brackets, and write either (A ∨ (B ∧ C)) or ((A ∨ B) ∧ C). So, more generally:

- *every* occurrence of '∧' and '∨' is to come with a pair of brackets to indicate the *scope* of the connective, i.e. to demarcate exactly the limits of what they are connecting;

- but since the negation sign '¬' only combines with a single sentence, we don't need to use brackets with that.

The convention, then, is that (roughly speaking) the negation sign binds tightly to the **PL** clause that immediately follows it. So:

- in a sentence of the form (A ∨ (¬B ∧ C)), just B is being negated;

- in (A ∨ ¬(B ∧ C)), all the bracketed conjunction (B ∧ C) is negated;

- and in ¬(A ∨ (B ∧ C)), the whole of (A ∨ (B ∧ C)) is negated.

(3) Finally, what do these three connectives '∧', '∨' and '¬' (ultimately) connect? Some simple, connective-free, sentences of the language. So we'll need to give **PL** a base class of items which express propositional messages: we can think of these as the *atoms* of an artificial language. Since for present purposes we are not interested in further analysing what is conveyed by these atomic sentences – we are concentrating on the logical powers of 'and', 'or' and 'not' when they connect whole sentences – we might as well waste as little ink as possible and just use single letters for them. So, for maximum brevity, the single letters 'P', 'Q', 'R', 'S' can serve as atoms for the language. (Though, since we could well want to use more than four atoms, we'll also allow ourselves to cook up as many more as we need by adding 'primes' to yield 'P'', 'P''', 'Q'''''', etc.)

And that's all we need. So, by way of interim summary: **PL** has a base class of symbols which express whole 'atomic' propositional messages, and which might as well be as simple as possible. More complex, 'molecular', symbolic formulae are then built up using '∧', '∨' and '¬', bracketing very carefully as we go. And since **PL** is purpose-designed for expressing arguments whose relevant logical materials are just conjunction, disjunction and negation and no more, that is *all* the structure which we need to give the language.

7.5 Some simple examples

Suppose we adopt the following translation key:

'P' means *Jack loves Jill.*
'Q' means *Jill loves Jack.*
'R' means *Jo loves Jill.*
'S' means *Jack is wise.*

Let's render the following into **PL** (even with our very quick explanations so far, these should not cause too much trouble: so try your hand at the translations before reading on).

1. Jack doesn't love Jill.
2. Jack is wise and he loves Jill.
3. Either Jack loves Jill or Jo does.
4. Jack and Jill love each other.
5. Neither Jack loves Jill nor does Jo.
6. It is not the case that Jack loves Jill and it is not the case the Jill loves Jack.
7. Either Jack is not wise or both he and Jo love Jill.
8. It isn't the case that either Jack loves Jill or Jill loves Jack.

The first four are indeed done very easily:

1′. ¬P
2′. (S ∧ P)
3′. (P ∨ R)
4′. (P ∧ Q)

Just three comments. First remember that we are insisting that *whenever* we introduce an occurrence of '∧' or '∨', there needs to be a pair of matching brackets to indicate the scope of the connectives. To be sure, the brackets in (2′) to (4′) are strictly speaking redundant; if there is only one connective in a sentence, there is no possibility of a scope ambiguity needing to be resolved by bracketing. No matter, our bracketing policy will be strict. Second, note that rendering the English as best we can into **PL** isn't a matter of mere phrase-for-phrase transliteration or mechanical coding. For example, in rendering (2) we have to assume that the 'he' refers to Jack; likewise we need to read (3) as saying the same as 'either Jack loves Jill or Jo loves Jill' – and assume too that the disjunction here is inclusive. Third, note that it is customary *not* to conclude sentences of our mini-language **PL** by full stops.

To continue. We don't have a 'neither …, nor …' connective in **PL**. But we can translate (5) in two equivalent ways:

5′. ¬(P ∨ R)
5″. (¬P ∧ ¬R)

(Convince yourself that these renditions *do* come to the same.)

The natural reading of (6), as remarked in §7.2, treats it as the conjunction of 'It is not the case that Jack loves Jill' and 'It is not the case the Jill loves Jack'; so translating into **PL** we get:

6′. (¬P ∧ ¬Q)

And the next proposition is to be rendered as follows:

7′. (¬S ∨ (P ∧ R))

(Check you see why the brackets need to be placed as they are.) Finally, we have

8′. ¬(P ∨ Q)

So far, so simple. But now we've got going, we can translate ever more complicated sentences. For example, consider

9.   Either Jack and Jill love each other or it isn't the case that either Jack loves Jill or Jill loves Jack.

This is naturally read as the disjunction of (4) and (8); so we translate it by disjoining the translations of (4) and (8), thus:

9′. ((P ∧ Q) ∨ ¬(P ∨ Q))

If we want to translate the negation of (9), we write

10′. ¬((P ∧ Q) ∨ ¬(P ∨ Q))

And so it goes.

In the next chapter we will begin characterizing **PL** rather more carefully. This won't involve any major new ideas; the aim will be to be maximally clear and careful about what is going on. But before plunging into those details, let's just summarize again *why* we are engaging in this enterprise.

7.6 Summary

- There is a class of inferences whose validity (or invalidity) depends on the way conjunction, disjunction, and negation feature in the premisses and conclusion.

- 'And', 'or' and 'not' in English are used to express conjunction, disjunction, and negation; but there are many quirks and ambiguities in the way ordinary language expresses these logical operations.

- So to avoid the quirks of the vernacular, we are going to use a special-purpose language **PL** to express arguments whose relevant logical materials are the connectives 'and', 'or' and 'not'. The design brief for **PL** is for the simplest language to do this without any ambiguities or obscurities.

Exercises 7

A   The strict negation of a proposition can usually be expressed by prefixing the operator 'It is not the case that'. Can you think of any exceptions in addition to the kind described in §7.2? Can you think of further cases which suggest that 'and' in English might sometimes mean more than bare conjunction?

B   What is the most natural way in English of expressing the strict negations of the following?

1. No one loves Jack.
2. Only unmarried men love Jill.
3. Everyone who loves Jack admires Jill.
4. Someone loves both Jack and Jill.
5. Jill always arrives on time.
6. Whoever did that ought to be prosecuted.
7. Whenever it rains, it pours.
8. No one may smoke.

C   Two propositions are *contraries* if they cannot be true together; they are *contradictories* if one is true exactly when the other is false. (Example: 'All philosophers are wise' and 'No philosophers are wise' are contraries – they can't both be true. But maybe they are both false, so they are not contradictories.) Give examples of propositions which are contraries but not contradictories of **B** 1–7.

D   Using the same translation manual as in §7.5, render the following into **PL**:

1. Jack is unwise and loves Jill.
2. It isn't true that Jack doesn't love Jill.
3. Jack loves Jill and Jo doesn't.
4. Jack doesn't love Jill, neither is he wise.
5. Either Jack loves Jill or Jill loves Jack.
6. Either Jack loves Jill or Jill loves Jack, but not both.
7. Either Jack is unwise or he loves Jill and Jo loves Jill.

The syntax of PL

In the previous chapter, we explained why it is advantageous to introduce an artificial language **PL** in which to express arguments involving conjunction, disjunction and negation. In this chapter, we'll describe the structure of our new language much more carefully.

8.1 Syntactic rules for PL

The grammar or *syntax* of **PL** determines which strings of symbols count as potentially meaningful sentences of the language. In other words, to use the standard terminology, the syntax defines what counts as a *well-formed formula* of the language – henceforth '*wff*' for short.

First, then, we specify which symbols form the basic building blocks for individual wffs:

The *basic alphabet* of **PL** is: P, Q, R, S, ', ∧, ∨, ¬, (,).

Later, we will want to string wffs together into arguments, and for those purposes it is useful to have three more symbols, a comma (to separate wffs in lists of premises), an inference marker (the **PL** analogue of 'therefore'), and an absurdity marker (meaning 'Contradiction!'). So let's add them to the official alphabet straight away. Hence,

> The *full alphabet* of **PL** is the comma, plus: P, Q, R, S, ', ∧, ∨, ¬, (,), ∴, ✱.

Now we must specify which strings of symbols from the basic alphabet form *atomic wffs*. As explained, we will set the atomic wffs to be as simple as possible:

P, Q, R, S, P', Q', R', S', P″, Q″, ...

But how are we supposed to continue the sequence here? We want to be absolutely precise: there should be no obscurity or vagueness about what does or doesn't belong to the set of atomic wffs, so an open-ended list won't really do. On the other hand, we can't just list a few more atoms and then stop, since we don't want to put an arbitrary finite bound on the size of this set (if there were e.g. just 14 atoms, we couldn't use **PL** to encode an argument that involved 15 different propositional components). So, more formally, we will define the set of atomic wffs as follows:

> (A1)   'P', 'Q', 'R' and 'S' are atomic wffs.
> (A2)   Any atomic wff followed immediately by a prime ''' is an atomic wff.
> (A3)   Nothing is an atomic wff other than what can be shown to be so by the rules (A1) and (A2).

By (A1) and (A2), 'P'' is an atomic wff; and then repeatedly applying (A2) again shows that 'P''', 'P'''', etc. are atomic wffs. (By the way, a rule like this, which can be reapplied to the results of its own previous application is standardly called *recursive*.) So (A1) and (A2) generate all the atomic wffs we want. But, as far as those first two rules go, Julius Caesar (for example) could count as an atomic wff too! To delimit the class of atomic wffs, we therefore need to rule him out by adding the 'extremal' clause (A3). It is then a mechanical task to determine whether a given string of symbols is an atomic wff according to our definition (an idiot computer program can check an input string of symbols to see whether the first symbol is 'P' or 'Q' etc. and any following ones are primes).

Having fixed the class of atomic wffs, we next need to give the rules for building up more complex wffs out of simpler ones to arrive at the full class of wffs. For the record, we should state explicitly that the atomic wffs do count as belonging to the full class of wffs, atomic or molecular: so,

(W1)   Any atomic wff is a wff.

Second, we can build up wffs using the one-place ('monadic') connective '¬':

(W2)   The result of writing a negation sign followed by a wff is also a wff.

The two-place (or 'binary' or 'dyadic') connectives '∧' and '∨' take two wffs to yield another. We insisted that to avoid ambiguity, the two-place connectives always come with a pair of brackets to indicate their 'scope'. We'll build that requirement in as follows:

(W3)   The result of writing a left-hand bracket followed by a wff followed by a conjunction sign followed by a wff followed by a right-hand bracket is another wff.

(W4)   The result of writing a left-hand bracket followed by a wff followed by a disjunction sign followed by a wff followed by a right-hand bracket is another wff.

Finally, we again need an extremal clause which tells us that these initial four rules essentially wrap things up (we need again to rule out Julius Caesar as a possible wff):

(W5)   Nothing is a wff other than whatever can be constructed by using rules W1 to W4.

We can helpfully and perhaps more perspicuously summarize these rules using an obvious shorthand:

> (W1) Any atomic wff is a wff.
> (W2) If *A* is a wff, so is ¬*A*.
> (W3) If *A* and *B* are wffs, so is (*A* ∧ *B*).
> (W4) If *A* and *B* are wffs, so is (*A* ∨ *B*).
> (W5) Nothing else is a wff.

Here '*A*' and '*B*' are schematic variables standing in for whole wffs (wffs of any complexity, by the way). For example, the wff *A* might be '(Q ∧ P)', and the wff *B* might be '¬(Q ∨ S')'. Then rule W3 tells us that '((Q ∧ P) ∧ ¬(Q ∨ S'))' is a wff.

It is again a mechanical task to determine whether a given string of symbols is a wff according to these recursive W-rules, though it is not quite so simple to write a program that will do the trick. (Exercise for enthusiasts: design the flow chart for an appropriate program. One answer is implicit in §8.3 below.)

And that's it. The A-rules plus the W-rules, elementary as they are, completely define the syntax of **PL**.

8.2 Construction trees

Let's use our rules to check carefully (if very laboriously) that the sequence of **PL** symbols

¬(P ∧ ¬(¬Q ∨ R))

is indeed a wff. We can set out a fully annotated proof in the style of §5.3, listing on the right the principles of construction that are being appealed to at each stage:

A	(1)	'Q' is a wff.	(by A1 and W1)
	(2)	'¬Q' is a wff.	(from 1 by W2)
	(3)	'R' is a wff.	(by A1 and W1)
	(4)	'(¬Q ∨ R)' is a wff.	(from 2, 3 by W4)
	(5)	'¬(¬Q ∨ R)' is a wff.	(from 4 by W2)
	(6)	'P' is a wff.	(by A1 and W1)
	(7)	'(P ∧ ¬(¬Q ∨ R))' is a wff.	(from 5, 6 by W3)
	(8)	'¬(P ∧ ¬(¬Q ∨ R))' is a wff.	(from 7 by W2)

Here's another example. To show

((P ∨ (R ∧ ¬S)) ∨ ¬(Q ∧ ¬P))

is a wff, we can set out the following equally laborious reasoning:

B	(1)	'R' is a wff.	(by A1 and W1)
	(2)	'S' is a wff.	(by A1 and W1)
	(3)	'¬S' is a wff.	(from 2 by W2)
	(4)	'(R ∧ ¬S)' is a wff.	(from 1, 3 by W3)
	(5)	'P' is a wff.	(by A1 and W1)
	(6)	'(P ∨ (R ∧ ¬S))' is a wff.	(from 4, 5 by W4)
	(7)	'Q' is a wff.	(by A1 and W1)
	(8)	'¬P' is a wff.	(from 5 by W2)

(9) '(Q ∧ ¬P)' is a wff.   (from 7, 8 by W3)
(10) '¬(Q ∧ ¬P)' is a wff.   (from 9 by W2)
(11) '((P ∨ (R ∧ ¬S)) ∨ ¬(Q ∧ ¬P))' is a wff.   (from 6, 10 by W4)

Any wff at all will have a 'constructional history' that can be set out in this sort of way because, by rule W5, the wffs are just the strings of symbols that can be systematically built up using the building rules W1–W4.

However, these linear proofs aren't the most revealing way of representing the constructional history of a wff. We can rather naturally reorganize the proof **A** into this form

A′

'Q' is a wff
——————————
'¬Q' is a wff   'R' is a wff
——————————————————————
'(¬Q ∨ R)' is a wff

'P' is a wff   '¬(¬Q ∨ R)' is a wff
——————————————————————————
'(P ∧ ¬(¬Q ∨ R))' is a wff
——————————————————
'¬(P ∧ ¬(¬Q ∨ R))' is a wff

This presents the same derivation in a *tree* structure. At the tips of the branches we note that various atoms are wffs. And as we move down the tree, we derive further facts about wffs in accordance with the rule W2 (which keeps us on the same branch), or in accordance with the rules W3 and W4 (when we join two branches).

Don't get fazed by the rather formal look of these trees: they just give us neatly perspicuous renditions of some perfectly common-or-garden bits of reasoning *about* wffs, essentially being carried out in ordinary English!

Here's proof **B** redone in tree form:

B′

'S' is a wff   'P' is a wff
'R' is a wff   '¬S' is a wff   'Q' is a wff   '¬P' is a wff
————————————   ————————————
'P' is a wff   '(R ∧ ¬S)' is a wff   '(Q ∧ ¬P)' is a wff
——————————————————   ————————————————
'(P ∨ (R ∧ ¬S))' is a wff   '¬(Q ∧ ¬P)' is a wff
——
'((P ∨ (R ∧ ¬S)) ∨ ¬(Q ∧ ¬P))' is a wff

Now, we could, if we like, make things look even neater by not bothering to write down the frame "'…' is a wff" which appears at each stage in **A′** and **B′**, and just take that as understood. This pruning would give us the following more austere trees:

A″

Q
————
¬Q   R
——————————————
(¬Q ∨ R)
P   ¬(¬Q ∨ R)
————————————————————
(P ∧ ¬(¬Q ∨ R))
————————————
¬(P ∧ ¬(¬Q ∨ R))

$$\frac{\frac{\frac{}{R}\quad\frac{\dfrac{S}{\neg S}}{(R\ \wedge\ \neg S)}}{(P\ \vee\ (R\ \wedge\ \neg S))}\qquad\frac{\dfrac{Q}{(Q\ \wedge\ \neg P)}\quad\dfrac{P}{\neg P}}{\neg(Q\ \wedge\ \neg P)}}{((P\ \vee\ (R\ \wedge\ \neg S))\ \vee\ \neg(Q\ \wedge\ \neg P))}$$

Such trees are often called *construction trees* for obvious reasons.

(Warning! Be careful reading these abbreviated construction trees. They do represent inferences; but of course, at the top of **A**″ we are not daftly inferring ¬Q from Q. Unpacking the shorthand here, the inference here, to repeat, is the perfectly valid one from the proposition that 'Q' is a wff to the proposition that '¬Q' is a wff.)

If we adopt the convention that, when branches join, the wffs above the inference line appear in the same left-to-right order in which they appear as components in the more complex wff below the line, then we have the following important result: *a PL wff has a unique construction tree.* (We won't pause to establish this; it is one of those results which should seem obvious but is a bit of a fuss to prove. Another exercise, then, for enthusiasts.)

8.3 Main connectives

Because construction trees are unique, we can properly talk of *the* construction tree for a wff. That means that the following definition is in good shape:

> The *main connective* of a non-atomic wff is the connective that is introduced at the final stage of its construction tree.

In our two example wffs above, the main connectives are respectively '¬' and '∨' – for don't forget, '¬' counts as a connective! Obviously, atomic wffs don't have a main connective, because they don't have *any* connective.

It is important to fix the main connective of a wff, because that determines the wff's logical powers. To explain what that means, note that the following pattern of argument is evidently reliable (compare §7.1):

$(A \vee B)$, $\neg A$ therefore B

And that of course remains reliable however complex A and B might be. But to apply this fact to warrant a particular inference in **PL** we need to know whether the relevant wff in the argument really *does* have the structure $(A \vee B)$, really is overall a disjunction, and that's a matter of its *main* connective's being the '∨'.

In the case of most wffs you'll actually encounter in practice, there's no problem in spotting the main connective. But in fact, it is easy to find the main connective even in *very* complex wffs. Here's how:

- If the first symbol is '¬', that is the main connective – an initial negation sign must have been the last connective introduced in building up the wff.

- Otherwise a wff must have the form $(A \wedge B)$ or $(A \vee B)$, where A and B are themselves wffs. Now, it is a trivial consequence of our rules that any wff A has to have the same number of left-hand and right-hand brackets (the number may be zero, of course!). So, to the left of the main connective, there is altogether one more left-hand than right-hand bracket (namely the opening one before the wff *A*). So, count from the left, find the first '∧' or '∨' which is preceded by exactly one more left-hand than right-hand bracket, and *that* has to be the main connective.

For an example, take the long wff

$$(((P \lor (R \land \neg S)) \lor \neg(Q \land \neg P)) \lor \neg(P \land \neg(\neg Q \lor R)))$$

Since this doesn't start with '¬' we apply the second rule; and a simple bracket-count then shows that the main connective is the third '∨'.

This trick for finding main connectives also gives us a trick for growing construction trees from the bottom up. Start with the given wff. Find the main connective. If it is '¬', write the wff again the next line up, without its initial negation sign. If it is '∧' or '∨', so the wff has the form (*A* ∧ *B*) or (*A* ∨ *B*), split the tree and write *A* (the part before the main connective, minus the opening bracket) on the left branch, and *B* (the part after the main connective, minus the closing bracket) on the right branch. Repeat the process on the new wffs at the top of the branches, and keep on going until we get to atomic wffs which cannot be further disassembled. (Exercise: apply this to the two wffs in §8.2.)

We can apply this process, in fact, to *any* string of symbols, wff or otherwise. If all goes well, and we end up with a completed construction tree with atoms at the tips of the branches, that shows the string is indeed a wff. If the process breaks down, and we get garbage appearing on the tree where a wff should be, then we know that the original string isn't a wff.

Here's a couple of quick examples. First, take the long string

$$((P \land S) \lor \neg(((R \land S) \lor \neg P) \lor \neg\neg Q) \lor \neg(S \land (\neg Q \lor R)))$$

Applying the rules for disassembling this into a construction tree, the counting trick yields

$$\frac{(P \land S) \qquad\qquad \neg(((R \land S) \lor \neg P) \lor \neg\neg Q) \lor \neg(S \land (\neg Q \lor R))}{((P \land S) \lor \neg(((R \land S) \lor \neg P) \lor \neg\neg Q) \lor \neg(S \land (\neg Q \lor R)))}$$

The left-hand side is unproblematic. Tackling the right-hand side, we next strip off the negation sign, and then apply the counting trick to get

$$\frac{\qquad P \qquad S \quad}{\dfrac{(P \land S)}{\dfrac{((R \land S) \lor \neg P) \qquad\qquad \neg\neg Q) \lor \neg(S \land (\neg Q \lor R)}{\dfrac{(((R \land S) \lor \neg P) \lor \neg\neg Q) \lor \neg(S \land (\neg Q \lor R))}{\dfrac{\neg(((R \land S) \lor \neg P) \lor \neg\neg Q) \lor \neg(S \land (\neg Q \lor R))}{((P \land S) \lor \neg(((R \land S) \lor \neg P) \lor \neg\neg Q) \lor \neg(S \land (\neg Q \lor R)))}}}}}$$

And now we are obviously in trouble on the right-most branch: we are not going to be able to complete a well-formed tree, so the original string we started from cannot be a wff.

Suppose we'd started instead with the following horrendous string:

$$(((((R \wedge S) \vee \neg P) \vee \neg\neg Q) \vee \neg(S \wedge (\neg Q \vee R))) \wedge \neg(P \wedge \neg(\neg Q \vee R)))$$

(Don't panic, you aren't really going to encounter this sort of monstrosity in practice: we're just making the point that our technique works irrespective of complexity!) The counting trick tells us to first split the string like this:

$$\frac{(((((R \wedge S) \vee \neg P) \vee \neg\neg Q) \vee \neg(S \wedge (\neg Q \vee R))) \qquad \neg(P \wedge \neg(\neg Q \vee R))}{(((((R \wedge S) \vee \neg P) \vee \neg\neg Q) \vee \neg(S \wedge (\neg Q \vee R))) \wedge \neg(P \wedge \neg(\neg Q \vee R)))}$$

The rules for bracket-counting and wff-splitting applied on the left, and for negation-stripping applied on the right, tell us to continue the tree upwards thus:

$$\frac{\dfrac{((((R \wedge S) \vee \neg P) \vee \neg\neg Q) \qquad \neg(S \wedge (\neg Q \vee R)) \qquad (P \wedge \neg(\neg Q \vee R))}{(((((R \wedge S) \vee \neg P) \vee \neg\neg Q) \vee \neg(S \wedge (\neg Q \vee R))) \qquad \neg(P \wedge \neg(\neg Q \vee R))}}{(((((R \wedge S) \vee \neg P) \vee \neg\neg Q) \vee \neg(S \wedge (\neg Q \vee R))) \wedge \neg(P \wedge \neg(\neg Q \vee R)))}$$

Another cycle of bracket-counting and negation-stripping and we can go another level up the tree, as follows:

$$\frac{\dfrac{\dfrac{((R \wedge S) \vee \neg P) \qquad \neg\neg Q \qquad (S \wedge (\neg Q \vee R)) \qquad P \qquad \neg(\neg Q \vee R)}{(((R \wedge S) \vee \neg P) \vee \neg\neg Q) \qquad \neg(S \wedge (\neg Q \vee R)) \qquad (P \wedge \neg(\neg Q \vee R))}}{(((((R \wedge S) \vee \neg P) \vee \neg\neg Q) \vee \neg(S \wedge (\neg Q \vee R))) \qquad \neg(P \wedge \neg(\neg Q \vee R))}}{(((((R \wedge S) \vee \neg P) \vee \neg\neg Q) \vee \neg(S \wedge (\neg Q \vee R))) \wedge \neg(P \wedge \neg(\neg Q \vee R)))}$$

By this stage it is clear that the tree is going to finish satisfactorily with atoms at the tip of every branch (complete the tree to show that this claim is true). So the original horrendous string is a genuine wff.

8.4 Subformulae and scope

Here are two more useful syntactic notions which again are neatly introduced using the idea of a construction tree:

> A wff *S* is a *subformula* of a wff *A* if *S* appears anywhere on the construction tree for *A*.
>
> The *scope* of a connective in *A* is the wff on *A*'s construction tree where the connective is introduced.

So, a subformula is a string of symbols occurring within a wff which could equally stand alone as a wff in its own right.

Returning to example **B** above, in the wff

$$((P \vee (R \wedge \neg S)) \vee \neg(Q \wedge \neg P))$$

the strings 'P', '¬P', '(Q ∧ ¬P)' and '¬(Q ∧ ¬P)' are sample subformulae, and the string '¬S)) ∨ ¬(Q' isn't.

Note that our definition allows a wff *A* trivially to count as a subformula of itself, since it appears on its own construction tree. The subformulae of *A* which are shorter than *A* itself can be called the *proper* subformulae of *A*.

Our other definition sharpens up the intuitive idea of scope. By our definition, the scope of the first '∨' in our sample wff **B** is '(P ∨ (R ∧ ¬S))', the scope of the second '∨' is the whole wff, and the scope of the second negation is '¬(Q ∧ ¬P)'. A wff, or the occurrence of another connective, is then 'in the scope of' a given connective in the *intuitive* sense if it is literally *in*, i.e. is part of, the scope as we've just technically defined it.

For example, compare the wffs '(P ∨ (Q ∧ R))' and '((P ∨ Q) ∧ R)'. Intuitively '∧' is in the scope of '∨' in the first of these, and vice versa in the second. And that tallies with our definition.

Again, note that a connective counts as being in its own scope by our definition: but that's a harmless wrinkle.

8.5 Bracketing styles

So much, then, for the official story. Now a quick suggestion about how to make life easier – or at least, how to make long wffs more readable: mark up pairs of brackets which belong together, to reveal more clearly how the wff is constructed. You can do this in handwriting by using different sized brackets, or different coloured inks, to signal matching pairs. Here, in this book, we can make use of the different available styles for printing brackets. So, for example, take again the wff:

$$(((P ∨ (R ∧ ¬S)) ∨ ¬(Q ∧ ¬P)) ∨ ¬(P ∧ ¬(¬Q ∨ R)))$$

Then, if we bend some of the matching brackets into curly ones '{', '}' and straighten other pairs into square ones '[', ']', we get

$$[\{(P ∨ (R ∧ ¬S)) ∨ ¬(Q ∧ ¬P)\} ∨ ¬\{P ∧ ¬(¬Q ∨ R)\}]$$

and the string instantly looks rather more readable (and much more obviously a wff). But we will use this dodge quite sparingly in what follows, since it is good to get used to reading less complex wffs without this assistance.

8.6 A final brief remark on symbolism

Languages like **PL** have been studied and used for a century. In that time, a fair variety of symbolism has been used. But things have settled down, and the symbolism used in this book is pretty much the standard one. The only divergences between modern books you are likely to see are these:

- some authors use '&' rather than '∧';
- some authors use '~' rather than '¬';
- authors differ in which letters count as atoms, and some provide an unending list of atoms by using numerical subscripts (thus, 'P_0', 'P_1', 'P_2', ...).

In addition, some allow a few special conventions for leaving out brackets in such a way that no ambiguities will arise (e.g. you are allowed to omit outermost brackets). None of these variations should cause you any problems if you encounter them elsewhere.

You will find, however, that many books from the outset add to **PL** a fourth connective '⊃' (or '→'). It will become clear soon enough why we aren't following suit: this additional connective will get introduced in due course, at a more appropriate place (Chapter 14).

8.7 Summary

- We have described the official syntax of **PL**, which first defines the class of atomic wffs, and then gives rules for constructing legitimate complex wffs out of simpler ones. The constructional history of a wff can be set out as a unique construction tree.

- The main connective of a (non-atomic) wff is the final connective added to its construction tree. There are simple rules for finding the main connective of even very complex wffs, and hence simple rules for 'disassembling' a complex string and recovering its construction tree, if it has one.

- A subformula of a wff is a substring that would make a stand-alone wff. The scope of a connective is the subformula where it is introduced in the construction tree.

Exercises 8

A   Show the following expressions are wffs by producing construction trees for them. Which is the main connective of each wff? Also, list all the sub-formulae of the last two examples.

     1. $((P \lor P) \land R)$
     2. $(\neg(R \land S) \lor \neg Q)$
     3. $\neg\neg((P \land Q) \lor (\neg P \lor \neg Q))$
     4. $((((P \lor P) \land R) \land Q) \lor (\neg(R \land S) \lor \neg Q))$
     5. $(\neg(\neg(P \land Q) \land \neg(P \land R)) \lor \neg(P \land (Q \lor R)))$
     6. $(\neg((R \lor \neg Q) \land \neg S) \land (\neg(\neg P \land Q) \land S))$

What is the scope of each connective in the wffs (3) and (4)? What wffs are in the scope of each '$\land$' in (5) and (6)?

B   Which of the following expressions are wffs of **PL**?

     1. $((P \lor Q) \land \neg R))$
     2. $((P \lor (Q \land \neg R) \lor ((Q \land \neg R) \lor P))$
     3. $\neg(\neg P \lor (Q \land (R \lor \neg P))$
     4. $((P \land (Q \lor R)) \land (Q \lor R))$
     5. $(((P \land (Q \land \neg R)) \lor \neg\neg\neg(R \land Q)) \lor (P \land R))$
     6. $((P \land (Q \land \neg R)) \lor \neg\neg\neg((R \land Q) \lor (P \land R)))$
     7. $(\neg(P \lor \neg(Q \land R)) \lor (P \land Q \land R)$
     8. $\neg\neg((\neg P \land (S \lor \neg S)) \land ((Q \land (P \land \neg R)) \lor \neg S))$

Repair the defective expressions by adding/removing the minimum number of brackets needed to do the job. Show the results are indeed wffs by producing construction trees in each case. What is the main connective in each case?

The semantics of PL

Having described the syntax of **PL**, we now turn to *semantics*, i.e. to questions of meaning and truth. And while our syntactic story was thoroughly conventional, now there are more options: the stories in various text books are subtly but importantly different. Introductory clarity requires that we fix on one approach and stick to it. Our line follows through the implications of the two-stage strategy for assessing arguments. The proposal, recall, is that we first render a vernacular argument into a well-behaved artificial language; and then we investigate the validity or otherwise of the reformulated argument. But if the renditions of arguments into **PL** are still to be *arguments*, with contentful premisses and conclusions, that means that the atoms of **PL** cannot be left as mere uninterpreted symbols: they must be given meaning.

9.1 Interpreting atomic wffs

We interpret the wffs of **PL**, then, in two stages. First, we assign content to the atomic wffs; and then we explain how to work out the meanings of molecular wffs, given we know what their atomic components mean.

We have to specify the content of atoms by a bald list that fixes which propositions are expressed by which atoms – as follows, perhaps:

> 'P' means *Alun loves Bethan.*
> 'Q' means *Bethan loves Caradoc.*
> 'R' means *Alun has a dog.*
> 'S' means *The dog is in the kennel.*
> 'P'' means *The cat is on the mat.*
> 'Q'' means ....

But how long should this list be? We want to be able to render arguments on all kinds of topics into **PL** and then assess them: does that mean we need a single *huge* English-to-**PL** dictionary to enable all those translations?

No. Remember that our central concern is with the way that certain arguments depend for their validity or invalidity on the use of conjunction, disjunction and negation (rather than on the meaning of the atoms). So the translation of atoms is a step to be got out of the way as simply as possible, before revealing how the propositional connectives are distributed among premisses and conclusions. *We therefore treat the atoms of PL as available to code different messages on different occasions of use.* On any particular occasion, we fix on a code-book that tells us how to interpret the atoms we need. We've just described the beginnings of one code-book. We described another in §7.5. And if, for example, we want to express argument **A** of §7.1 – 'Either we are out of petrol or the carburettor is blocked; we are not out of petrol; so the carburettor is blocked' – we can stipulate that on this occasion:

'P' means *We are out of petrol.*
'Q' means *The carburettor is blocked.*

The argument then gets translated

$(P \lor Q), \neg P \therefore Q$

(Now you see what the comma, and the familiar symbol '∴' for *therefore*, are going to be needed for!) If encoding that argument is our only current concern, then we don't need to interpret any of the myriad other atoms of **PL** which are potentially available to code propositions.

Similarly, if we want to express argument **B′** of §7.1 – 'It's not true that Tony supports the policy and George does too; Tony supports the policy; so George does not support it' – then we can stipulate that on *this* occasion

'P' means *Tony supports the policy.*
'Q' means *George supports the policy.*

So the whole argument gets translated

$\neg(P \land Q), P \therefore \neg Q$

In summary, then, the plan is to let the interpretation of atoms (enough atoms for the purposes at hand) vary between different applications of **PL** to encode arguments. What stays permanently fixed as the reading of the atoms changes is the interpretation of the connectives.

9.2 Interpreting molecular wffs

As already explained in Chapter 7, the three connectives get the following constant interpretations.

- '∧' invariably signifies bare conjunction, no more and no less. So a conjunctive wff of the form $(A \land B)$ – i.e. a wff whose main connective is '∧' – means just that A holds and B does too.

- Similarly, the vel sign '∨' invariably signifies bare inclusive disjunction, no more and no less. In other words, a disjunctive wff of the form $(A \lor B)$ – i.e. a wff whose main connective is '∨' – expresses no more than that A holds, or B holds, or both hold.

- The basic rule for interpreting wffs starting with a negation sign is even simpler: a wff of the form $\neg A$ expresses the contradictory of A, i.e. that proposition which denies exactly what plain A asserts.

Given the meanings of the atoms in a wff, it is then easy to work out the corresponding meanings of complex wffs (we've already implicitly assumed an ability to do this in the translation examples in §7.5). The only problems will be in hitting on reasonably natural vernacular versions that unambiguously express the relevant message.

For example, suppose we set:

'P' means *Jack took Jill to the party.*
'Q' means *Jack took Jo to the party.*
'R' means *Jack had some real fun.*

Then it won't do to interpret

(P ∨ (Q ∧ R))

as

Either Jack took Jill to the party or he took Jo and he had some real fun,

for that just reintroduces the very kind of ambiguity which the careful bracketing in **PL** is there to eliminate (see §7.2, point (4)). An English rendition that expresses the correct interpretation unambiguously would be e.g.

Either Jack took Jill to the party or else he both took Jo and also had some real fun.

9.3 Valuations

On the first set of interpretations above, 'P' means *Alun loves Bethan.* So 'P' says something true just so long as Alun really does love Bethan. Likewise, 'Q' means *Bethan loves Caradoc.* So 'Q' says something true just so long as Bethan loves Caradoc. And so on. Now, it is up to us under what conditions 'P' and 'Q' say something true, for that's a matter of the meanings we assign. But having assigned the meanings, and so fixed the conditions which have to hold for the various atomic wffs to be true, it is up to the world whether the relevant conditions actually obtain. If Alun does love Bethan, then 'P' is true; if not, 'P' is false. Likewise if Bethan indeed loves Caradoc, then 'Q' is true; if not, then 'Q' is false. In this way, our readings of 'P', 'Q' etc. combine with the facts about Welsh life to determine a *truth-valuation* of those interpreted atomic wffs, i.e. an assignment of a *truth-value*, one of *true* or *false*, to each of them.

But you might ask: does an interpretation for 'P' combined with the facts *always* fix a determinate truth-value for the atomic wff? Here are four sorts of case to think about:

• Perhaps 'P' still means *Alun loves Bethan*, but Alun's feelings are mixed, on the borderline of love. So maybe it is neither definitely true that he loves Bethan nor definitely false.

• Perhaps you are told 'P' means *Mr Brixintingle loves Bethan*. But the name 'Mr Brixintingle' has just been made up on the spur of the moment and denotes no one. We might be inclined to say that in this case 'P' doesn't express a complete proposition – there is, so to speak, a gap in it – and hence 'P' doesn't get to the starting line for being assessed for truth or falsehood. Arguably, it is neither true nor false.

- Perhaps 'P' is interpreted as one of those paradoxical claims which says of itself that it is false. So if 'P' is false, it is true; and if it is true it is false. Contradiction! What should we say about such so-called 'liar sentences'? Perhaps here too we have a case in which 'P' is neither true nor false.

- Here's a more sophisticated worry. It can be tempting to think of the truth of some mathematical propositions as consisting in their provability, and their falsehood as consisting in their refutability. Or at least that view can be tempting if you are puzzled by the alternative idea that mathematical propositions are made true by how things are in a mysterious Platonic realm inhabited by abstract mathematical entities. But then, on the truth-as-provability view, if 'P' is interpreted as a mathematical proposition, we again may not be able to say that it is determinately true or determinately false (for there needn't be a proof one way or the other).

Well, these are difficult issues. But here we will just have to ignore questions of vagueness – their proper logical treatment is a mightily vexed topic that it would take another book to explore: our concern will have to be with arguments where vagueness isn't an issue. We will also have to largely ignore the question of 'empty names', and set aside the issue about paradoxical 'liar'-type sentences. The issue of the so-called 'intuitionistic' interpretation of mathematical truth is also far too big a question to take on here.

So, for our purposes, we just have to stipulate:

> We will henceforth take any atom of **PL** to be so interpreted that it is appropriate to treat it as either determinately true or determinately false.

Then the 'classical two-valued logic' we are about to explore can apply.

9.4 Evaluating complex wffs

Given a truth-valuation of some *atomic* wffs, we can then go on to work out the truth-values of more complex, *molecular*, wffs built up from those atoms.

For example, we stipulated that a conjunctive wff of the type $(A \land B)$ expresses a proposition which is true just when A and B both obtain. In other words, if A and B are both true, then $(A \land B)$ must be true too; otherwise it is false.

It is convenient at this point to introduce some shorthand: let's abbreviate 'A is true' (or in the standard phrase, 'A has the truth-value *true*') by '$A \Rightarrow T$', and abbreviate 'B is false' (or equivalently, 'B has the truth-value *false*') by '$B \Rightarrow F$'. Then the rule for conjunctions is:

(P1) For any wffs, A, B: if $A \Rightarrow T$ and $B \Rightarrow T$ then $(A \land B) \Rightarrow T$; otherwise $(A \land B) \Rightarrow F$.

Likewise, we stipulated that a disjunctive wff of the type $(A \lor B)$ expresses a proposition which is true when one of A and B, or maybe both, obtain. Hence, however complex A and B might be, if at least one of A and B is true, then $(A \lor B)$ must be true too; otherwise it is false. Or put another way,

  (P2) For any wffs, A, B: if $A \Rightarrow$ F and $B \Rightarrow$ F then $(A \lor B) \Rightarrow$ F; otherwise $(A \lor B) \Rightarrow$ T.

And thirdly, negation simply flips values: a wff of the form $\neg A$ takes the value *true* in just those situations when plain A is valued *false*. In other words

  (P3) For any wff A: if $A \Rightarrow$ T then $\neg A \Rightarrow$ F; otherwise $\neg A \Rightarrow$ T.

These three rules can be summed up and displayed in an even clearer tabular form – read off the lines of each table horizontally in the obvious way:

Truth tables for the connectives

A	B	$(A \land B)$
T	T	T
T	F	F
F	T	F
F	F	F

A	B	$(A \lor B)$
T	T	T
T	F	T
F	T	T
F	F	F

A	$\neg A$
T	F
F	T

Suppose, then, that we are given the truth-values of some atoms. We can then apply and re-apply these *truth tables* for the connectives, and so work out the value of *any* wff built up from the given atoms using the connectives. A valuation of the atoms determines the valuation of all the wffs constructed from them. We'll look at some initial examples of how this works in the next section.

But first let's pause to emphasize again the difference between giving an *interpretation* of some atoms (i.e. giving their propositional content) and giving a *truth-valuation*. Assigning an interpretation, in our sense, involves specifying the situations in which 'P', 'Q' etc. *would* be true; a valuation specifies whether the wffs *are* true or not. You may know the content of 'P', and so know that it says something true just so long as Alun loves Bethan, without knowing its truth-value (for you may not be well acquainted with Welsh affairs). Conversely you may be told that 'Q'' is true, i.e. that it corresponds to a situation which actually obtains, without thereby coming to know *which* situation 'Q'' corresponds to.

In what follows, for reasons that will become clear, we say a lot more about possible valuations than about the particular contents of wffs. Indeed, many logic books in effect ignore the contents of wffs and immediately focus just on what we are calling valuations (perhaps confusingly, they then call the valuations 'interpretations'). But if wffs aren't given content, they don't convey determinate propositions, and apparent 'arguments' couched in **PL** will not be genuine *arguments* with meaningful premises and conclusions at all. Our approach, where we allow **PL** atoms to express propositional messages (albeit different messages on different occasions of use) at least means that we still count as studying real arguments when we move from English to **PL** examples (rather as if we were moving from English to e.g. Polish examples).

9.5 Calculating truth-values

Take for example the wff

$$¬(P ∧ ¬(¬Q ∨ R))$$

and suppose that the interpretations of the atoms and the facts of the matter conspire to generate assignments of truth-values to the atoms as follows:

$$‘P’ ⇒ T, ‘Q’ ⇒ F, ‘R’ ⇒ T$$

Then we can easily work out that the whole wff is true. The key point to bear in mind when evaluating bracketed expressions is that we need to do our working 'from the inside out', just as in arithmetic. Here's the derivation in its tedious detail:

A	(1)	'Q' ⇒ F	(premiss)
	(2)	'¬Q' ⇒ T	(from 1 by table for '¬')
	(3)	'R' ⇒ T	(premiss)
	(4)	'(¬Q ∨ R)' ⇒ T	(from 2, 3 by table for '∨')
	(5)	'¬(¬Q ∨ R)' ⇒ F	(from 4 by table for '¬')
	(6)	'P' ⇒ T	(premiss)
	(7)	'(P ∧ ¬(¬Q ∨ R))' ⇒ F	(from 5, 6 by table for '∧')
	(8)	'¬(P ∧ ¬(¬Q ∨ R))' ⇒ T	(from 7 by table for '¬')

It is perhaps more perspicuous, though, to lay out the reasoning in a tree form:

Note the direct parallelism between these derivations (in linear and tree form) and the corresponding derivations **A** and **A′** in §8.2 which showed that the wff in question is genuinely well-formed. This is no accident! We have expressly designed **PL** so that the *syntactic* structure of a wff, the way that the symbols are put together into coherent strings, is a perfect guide to its *semantic* structure and the way the truth-value of the whole depends on the truth-values of the atoms.

Here's another example. Take the wff

$$((P ∨ (R ∧ ¬S)) ∧ ¬(Q ∧ ¬P))$$

and suppose 'P' ⇒ F, 'Q' ⇒ F, 'R' ⇒ T, 'S' ⇒ F. Then we have (in linear, step by step, detail)

B	(1)	'R' ⇒ T	(premiss)
	(2)	'S' ⇒ F	(premiss)

(3) '¬S' ⟹ T	(from 2 by table for '¬')
(4) '(R ∧ ¬S)' ⟹ T	(from 1, 3 by table for '∧')
(5) 'P' ⟹ F	(premiss)
(6) '(P ∨ (R ∧ ¬S))' ⟹ T	(from 4, 5 by table for '∨')
(7) 'Q' ⟹ F	(premiss)
(8) '¬P' ⟹ T	(from 5 by table for '¬')
(9) '(Q ∧ ¬P)' ⟹ F	(from 7, 8 by table for '∧')
(10) '¬(Q ∧ ¬P)' ⟹ T	(from 9 by table for '¬')
(11) '((P ∨ (R ∧ ¬S)) ∧ ¬(Q ∧ ¬P))' ⟹ T	(from 6, 10 by W3)

Or putting the same derivation in tree form

B′

$$
\begin{array}{cc}
\text{'P' ⟹ F} \quad \dfrac{\text{'R' ⟹ T} \quad \dfrac{\text{'S' ⟹ F}}{\text{'¬S' ⟹ T}}}{\text{'(R ∧ ¬S)' ⟹ T}} \quad & \quad \dfrac{\text{'Q' ⟹ F} \quad \dfrac{\text{'P' ⟹ F}}{\text{'¬P' ⟹ T}}}{\text{'(Q ∧ ¬P)' ⟹ F}}
\end{array}
$$

'(P ∨ (R ∧ ¬S))' ⟹ T   '¬(Q ∧ ¬P)' ⟹ T

'((P ∨ (R ∧ ¬S)) ∧ ¬(Q ∧ ¬P))' ⟹ T

Again, notice the exact parallelism with **B** and **B′** in §8.2 above, where we showed that the wff in question is indeed a wff.

9.6 Three points about valuations

(1) Note that we can do these calculations of the truth-values of complex wffs even when we've forgotten what their constituent atoms are supposed to mean. *That's because the truth-values of wffs in PL depend only on the truth-values of their parts, and not on the interpretation of those parts.* More on this very important point in Chapter 11.

(2) The fact that we evaluate a wff by in effect working down its construction tree means that the values of atoms which *aren't* on the construction tree can't matter for the value of the final wff. In other words, the truth-value of a wff depends *only* on the assignment of values to the atoms it actually contains.

(3) A valuation of the atoms in a wff determines the value of that wff uniquely. That's because a complex wff has a unique construction tree; and once values are assigned to the atoms at the tips of the branches, the values of the increasingly complex subformulae encountered as you work down the tree are uniquely fixed at each stage by applying the truth-table for the connective introduced at that stage.

Suppose we *hadn't* adopted our strict policy requiring brackets to be used with the (two-place) connectives. Suppose, in other words, that we allowed expressions like 'P ∨ Q ∧ R' to count as well formed. Then this expression would be *structurally ambiguous* in the sense that it wouldn't have a unique construction tree. It could be built up in either of the following ways:

$$
\dfrac{\dfrac{\text{'P' is a wff} \quad \text{'Q' is a wff}}{\text{'P ∨ Q' is a wff}} \quad \text{'R' is a wff}}{\text{'P ∨ Q ∧ R' is a wff}}
\qquad
\dfrac{\text{'P' is a wff} \quad \dfrac{\text{'Q' is a wff} \quad \text{'R' is a wff}}{\text{'Q ∧ R' is a wff}}}{\text{'P ∨ Q ∧ R' is a wff}}
$$

Imagine that 'P' ⇒ T, 'Q' ⇒ T, 'R' ⇒ F. The wff would then evaluate in *both* these ways:

$$\frac{\displaystyle \frac{\text{'P'} \Rightarrow T \qquad \text{'Q'} \Rightarrow T}{\text{'P} \vee \text{Q'} \Rightarrow T} \qquad \text{'R'} \Rightarrow F}{\text{'P} \vee \text{Q} \wedge \text{R'} \Rightarrow F} \qquad\qquad \frac{\text{'P'} \Rightarrow T \qquad \displaystyle\frac{\text{'Q'} \Rightarrow T \qquad \text{'R'} \Rightarrow F}{\text{'Q} \wedge \text{R'} \Rightarrow F}}{\text{'P} \vee \text{Q} \wedge \text{R'} \Rightarrow T}$$

Here the final valuations differ because, on the left, the wff is being seen as con-joining a truth and a falsehood ('∧' has wider scope than '∨'), and on the right as disjoining a different truth and falsehood ('∨' has wider scope). The *syntactic* structural ambiguity leads to a *semantic* ambiguity: fixing the truth-values of the atoms does not uniquely fix the truth-value of the whole unbracketed expression 'P ∨ Q ∧ R'.

Moral: leaving out brackets from **PL** formulae introduces the risk of semantic ambiguity. Don't do it!

9.7 Short working

And that's it as far as the basic semantics of **PL** is concerned. This final section merely introduces a way of presenting truth-value calculations rather more snappily.

Go back to example **A** again, and consider the following table of calculations:

P	Q	R	¬ (P ∧ ¬ (¬Q ∨ R))
T	F	T	T̲ T F F FTF T T
			8 6 7 5 2 1 4 3

On the left, we display the initial assignment of values to the atomic subformu-lae. On the right, we've copied the same values directly under the corresponding atoms and then, as we calculate the values of more and more complex parts of the final formula, we write the value *under the connective whose impact is being calculated* at each stage. For example, when we are evaluating '(¬Q ∨ R)' by considering how '∨' operates on the truth-values of '¬Q' and 'R', we write the result of evaluating that disjunction under the '∨'. (To aid comparison with the calculation in §9.5, we have numbered the stages at which values are calculated by the corresponding line number. For clarity, the final result is underlined.)

Here, in the same short form is the second calculation from §9.5:

P	Q	R	S	((P ∨ (R ∧ ¬S)) ∧ ¬(Q ∧ ¬P))
F	F	T	F	F T T T TF T̲ T F F TF
				5 6 1 4 32 11 10 7 9 8 5

This time, the numbers correspond to the lines of the derivation **B**.

For a third example, take the wff

((S ∧ ¬(Q ∧ ¬P)) ∨ (R ∨ ¬¬Q))

and suppose 'P' ⇒ T, 'Q' ⇒ F, 'R' ⇒ F, 'S' ⇒ F. This time, let's do the working directly in short form. Looking at the wff, and bracket counting, we see that the main connective is the first '∨'; so evaluating this wff involves evaluating the two disjuncts '(S ∧ ¬(Q ∧ ¬P))' and '(R ∨ ¬¬Q)'. It doesn't matter, of course, which half of the disjunction you tackle first, and the numbers here just indicate the order of steps in one way of doing the calculation (and we've used different shapes of brackets to highlight the structure of wff):

P	Q	R	S	({S ∧ ¬ (Q ∧ ¬P)} ∨ {R ∨ ¬¬Q})
T	F	F	F	F F T F F F T F F F F T F
				1 7 6 2 5 4 3 13 8 12 11 10 9

For a fourth example, let's evaluate the same wff for 'P' ⇒ F, 'Q' ⇒ T, 'R' ⇒ T, 'S' ⇒ T (and henceforth let's not repeat the values of atoms on the right; and we'll just number off the *remaining* steps of the calculation):

P	Q	R	S	({S ∧ ¬ (Q ∧ ¬P)} ∨ {R ∨ ¬¬Q})
F	T	T	T	F F T T T T T F
				4 3 2 1 8 7 6 5

Finally, let's take the same valuation of atoms but this time evaluate the new displayed wff (again, we've not repeated the values of the atoms on the right):

P	Q	R	S	¬[{P ∧ (¬S ∨ Q)} ∧ ¬{¬R ∧ ¬P}]
F	T	T	T	T F F T F T F F T
				9 3 1 2 8 7 4 6 5

Once more, we work from the inside outwards, evaluating the two subformulae in curly brackets, '{P ∧ (¬S ∨ Q)}' (step 3), and '{¬R ∧ ¬P}' (step 6) which immediately yields the value of '¬{¬R ∧ ¬P}' (step 7). That gives us the value of the square-bracketed conjunction (step 8), and so the value of the whole wff.

We will introduce some more ways of cutting down working in Chapter 11: but we've explained enough to enable you to tackle some first examples. Of course, you don't actually need to write down the step numbers: that's inessential commentary. Practice quickly makes perfect.

9.8 Summary

- Atomic wffs in **PL** get interpreted, on a particular occasion of use, by being directly assigned meanings (we do this for as many atoms as we need for the purposes at hand). The connectives get meanings – expressing bare conjunction, inclusive disjunction, and strict negation – which remain constant across different uses of **PL**.

- The factual meaning of an atom fixes its 'truth-conditions' (i.e. tells us which situation has to obtain for the atom to be true). The factual meaning and the state of the world combine to fix the *truth-value* of (interpreted) atoms. Or at least, they do so given that we are setting aside cases of vagueness, paradoxical sentences, etc.

- The standard interpretation of the connectives means that fixing the truth or falsity of each of the wffs A and B will also fix the truth or falsity of the whole conjunction (A ∧ B), the disjunction (A ∨ B), and the negation ¬A.

- This means that a valuation of the atoms in a wff of **PL** (i.e. an assignment of the value *true* or *false* to each of the relevant atoms) will in fact determine a valuation for the whole wff, however complex it is; and this valuation will be unique because of the uniqueness of its constructional history.

Exercises 9

A   Suppose 'P' means *Plato is a great philosopher*; 'Q' means *Quine is a great philosopher*; 'R' means *Russell is a great philosopher*. Translate the following sentences into **PL**.

     1.   Either Quine is a great philosopher or Russell is.
     2.   Neither Plato nor Quine is a great philosopher.
     3.   Plato, Russell and Quine are great philosophers.
     4.   Not both Quine and Russell are great philosophers.
     5.   Quine is a great philosopher and Russell isn't.
     6.   Either Quine and Russell are great philosophers, or Plato is.
     7.   It isn't the case the Quine is and Russell isn't a great philosopher.

B   Suppose 'P' means *Fred is a fool*; 'Q' means *Fred knows some logic*; 'R' means *Fred is a rocket scientist*. Translate the following sentences into **PL** as best you can. (What do you think is lost in the translations, given that **PL** only has the 'colourless' connectives '∧', '∨' and '¬'?)

     1.   Fred is a rocket scientist, but he knows no logic.
     2.   Fred's a fool, even though he knows some logic.
     3.   Although Fred's a rocket scientist, he's a fool and even knows no logic.
     4.   Fred's a fool, yet he's a rocket scientist who knows some logic.
     5.   Fred is not a rocket scientist who knows some logic.
     6.   Fred is a fool despite the fact that he knows some logic.
     7.   Fred knows some logic unless he is a fool.

C   Confirm that all the following strings are wffs by producing construction trees. Suppose that 'P' and 'R' are both true and 'Q' false. Evaluate the wffs by working down the trees. Then do the working again in the short form.

     1. $\neg(P \wedge R)$
     2. $\neg\,\neg\,\neg\,\neg(Q \vee \neg R)$
     3. $(\neg(P \vee \neg R) \wedge Q)$
     4. $((R \vee \neg Q) \wedge (Q \vee P))$
     5. $\neg(P \vee ((Q \wedge \neg P) \vee R))$
     6. $\neg(\neg P \vee \neg(Q \wedge \neg R))$
     7. $(\neg(P \wedge \neg Q) \wedge \neg\,\neg R)$
     8. $(((P \vee \neg Q) \wedge (Q \vee R)) \vee \neg\,\neg(Q \vee \neg R))$
     9. $(\neg(\neg P \vee \neg(Q \wedge \neg R)) \vee \neg\,\neg(Q \vee \neg P))$
     10. $\neg((\neg(P \wedge \neg Q) \wedge \neg\,\neg R) \wedge \neg(\neg(P \vee \neg R) \wedge Q))$

'A's and 'B's, 'P's and 'Q's

Before putting **PL** to use in the following chapters, we should pause to make it absolutely clear how the 'A's and 'B's, the 'P's and 'Q's, are being used, and why it is very important to differentiate between them. We should also explain the rules governing the use of quotation marks. By all means, skip this chapter for the moment if you are impatient to get to see our work on **PL** being put to use: but return later, because it is important to get clear on these points.

10.1 Styles of variable: our conventions

We have been using two styles of symbols, both of which may sensibly be called 'propositional variables' – namely, italicized letters from early in the alphabet, and roman letters (from a different font) from much later in the alphabet. Why two styles? What's the difference?

- 'A', 'B', etc. are handy *augmentations of English*: they are devices which we are employing in order to help us speak snappily about patterns of inference, forms of propositions, etc. For example, in §7.1, we used them to enable us to talk briskly and clearly, in a slightly expanded English, about patterns of argument involving conjunction and disjunction and couched in English. In §9.4, we used the same variables in order to talk equally briskly and clearly – still in the same augmented English – about wffs in **PL**.

  As we noted before, such variables, used in presenting patterns in sentences and arguments, are often called *schematic* variables (the *Concise Oxford Dictionary*'s entry for 'schematic' includes 'representing objects by symbols'). Their use is always dispensable. At the expense of some long-windedness, we could always use plain unadorned English prose instead. Schematic variables do the work that could instead be done by expressions like 'a first proposition', or 'the second wff' and more cumbersome variants. And hence, like those ordinary-language devices, these variables are used to *talk about* propositions or wffs but do not themselves express propositions.

- On the other hand, 'P', 'Q', etc. do not belong to English (augmented or otherwise) but instead are sentences of a new artificial language, **PL**. And unlike the schematic variables, these new symbols *do* themselves potentially express propositions; they encode whole messages, as it might be about Alun's loving Bethan, or George's supporting the policy, or whatever.

We noted that, on different occasions when we translate into **PL**, we may use 'P', 'Q', etc. to express different messages. We can choose different code-books for associating **PL** atoms with specific messages, depending on circumstances (what stays fixed from case to case is the meaning of the connectives). In *this* sense, the atomic letters can reasonably be thought of as a kind of 'variable', meaning that the atoms are open to various interpretations. But the contrast remains: schematic variables like '*A*', '*B*', etc. are used *in augmented English* to talk about propositions and arguments; 'P', 'Q' etc. *belong to* **PL** and express propositions on all kinds of subject matter, depending on the code-book or interpretation-manual currently in force.

Sometimes the point is put like this. 'P', 'Q' and so on belong to the language **PL** which is – for the moment – the object of our logical studies. In short, these symbols belong to the *object-language*. And we must in general distinguish the language which is the object of investigation from the *metalanguage,* i.e. the language in which we conduct the investigation and discuss what is going on in the object-language. In this book, the metalanguage is English augmented with variables like '*A*', '*B*', etc. (so those schematic variables are often also called *metalinguistic variables*). In a Spanish translation of this book, the object language will again be **PL**, but then the metalanguage will be Spanish. (Of course, we may often want to discuss facts about the logical features of English sentences in English – and then the metalanguage subsumes the object language: but in the general case they are distinct.)

> In this book we adopt the convention that the elements of formal languages like **PL** will always be printed in sans-serif type, like 'P' and 'Q'. Symbols printed in *serif italics*, like '*A*', '*n*', '*F*', etc., are always augmentations of our English metalanguage.

An alternative convention widely adopted in other books is to use Greek letters as metalinguistic variables.

10.2 Basic quotation conventions

Now for some quick remarks about our related usage of quotation marks. We start by warming up with a few general observations.

If we want to mention a certain man, we may use his name – as it might be

Socrates.

But sometimes we want to discuss not the man but his *name*. There are various ways of picking out that philosopher's name, e.g. laboriously

The name consisting of the nineteenth letter of the alphabet, followed by the fifteenth letter, followed by the third letter, ...

Or indirectly, e.g.

The name which is displayed on the fourth line of this section.

Most simply, we can use quotation marks, thus:

'Socrates'

with the convention that the whole expression *including* the quotation marks is to be construed as standing for the word displayed *inside* the quotes.

So, compare

(1)   Socrates is snub-nosed.
(2)   Socrates contains the first letter of the alphabet.
(3)   'Socrates' is snub-nosed.
(4)   'Socrates' contains the first letter of the alphabet.

The first is true; the second false (people don't contain letters of the alphabet). The third is false too (names don't have noses to be snub or otherwise); the fourth is true. Again, consider the sentence

(5)   Charles is called Chuck by his friends and Dad by his children.

Observing our convention of using quotation marks when we are talking about an expression, this strictly speaking needs to be marked up

(5')   Charles is called 'Chuck' by his friends and 'Dad' by his children.

And how could we insert quotes in the following to yield a truth? –

(6)   The first expression in our third example is Socrates

That would need to be marked up

(6')   The first expression in our third example is ' 'Socrates' ',

for to refer to the expression at the beginning of (3), which already involves quotation marks, we need to put the whole expression in quotes – hence the double ration. For clarity's sake, to make it plain that there are *two* pairs of quotation marks here, we could use different styles of quotation marks like this:

(6")   The first expression in our third example is " 'Socrates' ".

Here's another series of examples. In **PL**, we can *use* the disjunction sign to express various claims. Thus, on one interpretation of **PL** which we met before,

(P ∨ S)

says that either Alun loves Bethan or he has a dog. But sometimes we want to talk about the disjunction sign itself, i.e. to *mention* it in English. We can do that in various ways. We could, as we have just done twice, use the description

the disjunction sign;

or we can call it

the *vel* sign.

The *vel* sign itself is no part of English, though the expression 'vel' *is* used as part of logician's English: it is the name they use for the **PL** connective. We could also use quotation marks again, thus:

'∨'

And this composite expression *is* available for use *in English* to refer to the disjunction sign in **PL**, as when we say that '∨' is a two-place connective. (Compare: the word

soleil

is no part of English. However we can use the expression

'soleil'

in an English sentence – as when we say that the French word 'soleil' means *sun*.)

Similarly for the atomic letters of **PL**. With a translation manual in place, we can *use* them as sentences of **PL** (to talk about Welsh affairs, or whatever it might be). But we also want to *mention* the atoms, i.e. talk about them in English, our metalanguage. Again the easiest way of doing that is to use quotation marks to form expressions (augmenting English!) to refer to expressions in **PL**. Thus we may write, *in English*,

(7)   The atomic sentences 'P' and 'S' are constituents of the molecular
        sentence '(P ∨ S)'.

Likewise, we may write, still in English,

(8)   'P' takes the value *true*.

So compare

(9)   'P' is a sentence of **PL**
(10)   'P' is a sentence of English
(11)   ''P'' is an expression which can occur in **PL** sentences
(12)   ''P'' is an expression which can occur in English sentences

The first is true, but the second is false; the third is false too (since **PL** doesn't contain quotation names of anything, only atomic sentences and their molecular compounds). The last, however, is true; for example, (10) provides an instance of an English sentence where the quotation of 'P' occurs.

In summary: it is important to grasp both the distinction between object-language and metalanguage 'variables' (signalled in this book by the use of different fonts), and also the distinction between the use and mention of expressions (standardly signalled by quotation marks). And because the distinctions matter and ignoring them can lead to confusion, it is important to fix on official conventions for marking them clearly.

However, having said that, sometimes we can forgivably relax the conventions when doing so promotes readability, and there is no danger of confusion – after all, we are in charge, and we don't need to be slaves to our own conventions! For example, strictly we should write e.g.

'P' takes the value *true*, 'Q' takes the value *false*.

Or (as we did in the last chapter)

'P' ⇒ T, 'Q' ⇒ F

But from now on we will typically take '⇒' to come with unwritten quotation marks affixed, and write more casually

P ⇒ T, Q ⇒ F

and so forth. That's a bit tidier, cutting down on a rash of quotation marks. But it must just be remembered these are still metalinguistic claims: the expression '⇒ T' does *not* belong to **PL**, but is shorthand added to English.

10.3 A more complex convention

This section notes a wrinkle, and straightens it out.

In §8.1, we stated the wff-building rule W3 in abbreviated form as

If *A* and *B* are wffs, so is (*A* ∧ *B*).

The intended reading of this rule would probably have been entirely obvious, even if we hadn't given a non-abbreviated version of the rule at the same time. But our shorthand version does mix augmented English (the schematic variables '*A*' and '*B*') with expressions from **PL** (the conjunction sign, of course, but also the two brackets as well). And we can't untangle things by writing

If *A* and *B* are wffs, so is '(*A* ∧ *B*)',

for what is inside the quotes here is *not* a wff of **PL** (remember, '*A*' and '*B*' are not symbols of *that* language).

If we want to be really accurate, then we will need to introduce a new brand of quotation marks ('corner quotes', often called 'Quine quotes' after their inventor). So we write

If *A* and *B* are wffs, so is ⌜(*A* ∧ *B*)⌝ .

Here, '⌜(*A* ∧ *B*)⌝' is short for

the result of writing '(' followed by *A* followed by '∧' followed by *B* followed by ')'.

For example, suppose *A* is the wff '(P ∨ S)', and *B* is the wff 'Q'. Then the conjunction of those two wffs is indeed ⌜(*A* ∧ *B*)⌝ , i.e. it is the result of writing '(' followed by '(P ∨ S)' followed by '∧' followed by 'Q' followed by ')', which is '((P ∨ S) ∧ Q)', just as we want.

Here's another sort of case where we can use Quine-quotes. As we'll note in §12.2,

Whatever truth-value *A* has, ¬(*A* ∧ ¬*A*) is true.

Again, it is entirely clear what is meant here. But again '¬(*A* ∧ ¬*A*)' mixes object-language and meta-language expressions. More carefully, then, we should say: given a wff *A*, then whatever its value, the result of writing '¬' followed '(' followed by *A* followed by '∧' followed by '¬' followed by *A* followed by ')' is true. Or, abbreviating that,

Whatever truth-value *A* has, ⌜¬(*A* ∧ ¬*A*)⌝ is true.

Generalizing, the basic convention governing Quine-quotes can be summed up in the slogan *object-language symbols inside them are mentioned while meta-language variables are used.*

However, the explicit use of these Quine-quotes is perhaps unnecessarily punctilious, and we won't follow this practice in this book. After all, to repeat, we are in charge of our conventions. So we will simply take our version of W3,

If A and B are wffs, so is $(A \land B)$,

and similarly the claim

Whatever truth-value A has, $\neg(A \land \neg A)$ is true,

as acceptable just as they stand (read such claims, if you like, as if they have invisible Quine-quotes).

10.4 Summary

- In the general case, we need to distinguish the object language under discussion (whether it is **PL** or some natural language) from the metalanguage that it is being discussed in (here, slightly augmented English).

- Italicized letters like 'A', 'n', 'F' belong to augmented English, are dispensable devices to enable us to speak briskly about e.g. patterns of argument, or formation rules for wffs. Contrast 'P', 'Q', etc., which belong to the artificial language **PL**.

- When we want to *mention* an expression, rather than use it in the ordinary way, quotation marks give us a standard way of doing this.

Exercises 10

Insert quotation marks into the following where necessary to make them accord to the customary conventions and come out *true*.

1. ∧ means much the same as and.
2. P can be interpreted as meaning that grass is green.
3. P is a subformula of (Q ∧ ¬P).
4. If (Q ∧ ¬P) is a subformula of A so is P.
5. The first word in this sentence is the.
6. This is not a verb, but is is.
7. George Orwell is the same person as Eric Blair.
8. George Orwell was Eric Blair's pen-name.
9. The Evening Star and The Morning Star denote the same planet.
10. Sappho is a Greek poet.
11. Sappho is the name of a Greek poet.
12. If we want to refer not to Sappho but her name, we need to use the expression Sappho.

Truth functions

We noted in §9.6 that if we assign truth-values to the atoms in a wff of **PL**, then this will uniquely determine the truth-value of the whole wff. This feature of the language is called *truth-functionality*: it is the topic of the present chapter.

11.1 Truth-functional vs. other connectives

Let's start with a definition.

> A way of forming a complex sentence out of one or more constituent sentences is *truth-functional* if fixing the truth-values of the constituent sentences is always enough (whatever those values are) to determine the truth-value of the whole complex sentence.

Note that this definition applies quite generally, to ordinary-language constructions as well as to constructions in formal languages like **PL**.

Some ordinary-language devices do indeed seem to be truth-functional in this sense (e.g. many uses of 'and' as a propositional connective). For a contrasting case, consider the two-place connective 'because'. Suppose

(1)   The bridge fell down,

and

(2)   There was a storm yesterday,

are both true. That plainly doesn't settle the truth-value of

(3)   The bridge fell down *because* there was a storm yesterday.

The joint truth of (1) and (2) is compatible with the truth of (3); it's equally compatible with the falsehood of (3). Hence fixing the truth-values of the constituent sentences (1) and (2) isn't always enough to determine the truth-value of the complex sentence (3). Hence 'because' is not truth-functional.

Of course, if A is false or B is false, *A because B* will be false too. For example, (3) must be false if the bridge in fact didn't fall down, or if there wasn't a storm. But for full truth-functionality, as we defined it, the truth-values of the constituents must *always* determine the truth-value of the whole, whatever the values are. As we've just seen, that doesn't hold for 'because'.

The same goes for many other constructions which embed sentences inside more complex ones. For example, consider:

> It is widely believed that *A*.
> It is more likely that *A* than that *B*.
> It is necessarily true that *A*.

In each case, merely settling the truth-values of *A* and *B* won't settle (or at least, won't invariably settle) the truth-value of the whole sentence.

It is worth pausing on the last of these. Compare, e.g.

> (4)   The President of the USA in 2008 was male.
> (5)   If someone is a brother, then he is male.

Both are true. But the first might not have been so: we can readily imagine an alternative historical scenario which leads to a woman President occupying the White House in 2008. The second, however, is necessarily true: just in virtue of what it is to be a brother (i.e. to be a male with at least one sibling), we cannot coherently conceive of anyone who is a brother but not male. So despite the fact that (4) and (5) both take the truth-value *true*, the following differ in truth-value:

> (4′)   It is necessarily true that the President of the USA in 2008 was male.
> (5′)   It is necessarily true that if someone is a brother, then he is male.

In short, the truth-value of 'It is necessarily true that *A*' is sensitive to the particular *content* of the embedded proposition *A*, and not just to *A*'s truth-value.

However, every wff of pure **PL** (no matter its degree of complexity) *is* a truth-functional combination of the atoms it contains. That is to say, fixing the truth-values of atoms always uniquely fixes the truth-value of the whole wff. As we noted in §9.6, the basic valuation rules for the three connectives in **PL** plus our fussiness about bracketing secures this result.

11.2 A very brief word about 'functions'

Why is the term 'truth-functional' apt?

You will be familiar with the idea of an arithmetical function which takes as input one or more numbers, and spits out a unique value. For example, 'the square of x' denotes a function mapping a given number x to a definite value, the result of multiplying x by itself. '$x + y$' denotes another function, this time mapping two numbers to their sum.

We can generalize this idea of a function to cover any sort of mapping that relates one or more objects to a single determinate 'value'. For example, 'the biological mother of x' denotes a function mapping people to other people. 'The age in years of x' denotes another function defined over people, this time correlating each person with a number. 'The distance in metres between a and b' maps two spatially located objects to another number.

By contrast, 'teacher of x', 'house belonging to x', and 'daughter of x' don't express functions; they don't express ways of mapping people to unique 'values'. Some people have more than one teacher, more than one house and more than one daughter – or have none.

Given our general notion of a function, what is expressed by a truth-functional mode of sentence combination is indeed a truth-*function*. For by definition, a complex sentence which is a truth-functional combination of simpler embedded sentences *will* always map the truth-values of its constituents sentences to a single determinate truth-value for the combination as a whole.

11.3 Full truth-tables

Consider again the wff '¬(P ∧ ¬(¬Q ∨ R))'. In §9.5, we specified one particular valuation for its atoms. Because this is a truth-functional mode of combining the atoms, settling on that valuation sufficed to fix the value of the whole wff. But that was just one possible valuation out of many. There are in all *eight* possible valuations of three atoms. There are two ways we can assign a value to 'P', each of these can combine with either value for 'Q', and each of these combinations can combine with either value for 'R', and 2 × 2 × 2 = 8.

Running through these eight valuations in turn, we can mechanically compute the truth-value of our sample wff in every case. We can set out the working in a neat tabular form. Putting in absolutely all the details, we have:

P	Q	R	¬	(P	∧	¬	(¬	Q	∨	R))
T	T	T	T	T	F	F	F	T	T	T
T	T	F	F	T	T	T	F	T	F	F
T	F	T	T	T	F	F	T	F	T	T
T	F	F	T	T	F	F	T	F	T	F
F	T	T	T	F	F	F	F	T	T	T
F	T	F	T	F	F	T	F	T	F	F
F	F	T	T	F	F	F	T	F	T	T
F	F	F	T	F	F	F	T	F	T	F
			8	1	7	6	4	2	5	3

On the left, we systematically display the eight different possible assignments of truth-values to the three atoms. On the right we record the calculation of the truth-value of the wff using (as before) the convention that we write the value of each subformula under the connective whose effect is being calculated. The supplementary numbers at the foot of the columns indicate the order in which we've worked out the columns. We've first copied across the assignments of values to atoms. Then at subsequent stages we evaluate subformulae of increasing complexity, until – at the final stage, yielding the underlined values – we get to evaluate the whole wff.

Here's another example, this time involving four atoms, and so requiring sixteen lines of working:

[¬{(R ∨ ¬Q) ∧ ¬S} ∨ {¬(¬P ∧ Q) ∧ S}]

Note again how the little dodge introduced in §8.5 of using different styles of brackets can promote readability: the main connective is evidently the second disjunction. And this time, as in the last examples in §9.7, let's not bother repeating the values of the atoms on the right of the table (which saves time even if it perhaps slightly increases the risk of error):

Formula: $[\neg\{(R \lor \neg Q) \land \neg S\} \lor \{\neg(\neg P \land Q) \land S\}]$

P	Q	R	S	¬	R	¬Q	∧	¬S	∨	¬	¬P	∧	S
T	T	T	T	T	T	F	F	F	T	T	F	F	T
T	T	T	F	F	T	F	T	T	F	T	F	F	F
T	T	F	T	T	F	F	F	F	T	T	F	F	T
T	T	F	F	T	F	F	F	T	T	T	F	F	F
T	F	T	T	T	T	T	F	F	T	T	F	F	T
T	F	T	F	F	T	T	T	T	F	T	F	F	F
T	F	F	T	T	T	T	F	F	T	T	F	F	T
T	F	F	F	F	T	T	T	T	F	T	F	F	F
F	T	T	T	T	T	F	F	F	T	F	F	T	F
F	T	T	F	F	T	F	T	T	F	F	F	T	F
F	T	F	T	T	F	F	F	F	T	F	F	T	F
F	T	F	F	T	F	F	F	T	T	F	F	T	F
F	F	T	T	T	T	T	F	F	T	T	T	F	T
F	F	T	F	F	T	T	T	T	F	T	T	F	F
F	F	F	T	T	T	T	F	F	T	T	T	F	T
F	F	F	F	F	T	T	T	T	F	T	T	F	F
				9	4	1	7	2	10	6	3	5	8

Take careful note of the system used in listing the possible assignments of values to a sequence of atoms. The last atom in the sequence gets alternating Ts and Fs; the previous one gets alternating blocks of two Ts and two Fs; moving leftwards, we get next alternating blocks of four Ts and four Fs; and so it goes. This system is guaranteed to cover all the possible valuations. Any other systematic layout will do, of course. But ours is pretty much the standard convention.

Suppose there are n different atoms in the wff A. There will be $2 \times 2 \times \cdots \times 2$ (n times) $= 2^n$ different possible valuations of these n atoms. We can then always construct a *full truth-table* for A, working out the truth-values of all the increasingly large subformulae of A until we reach the overall value of A for each possible valuation of the atoms. As n increases, however, the amount of labour required will explode exponentially.

11.4 'Possible valuations'

Before continuing, an important word of clarification here. In the last section, we set out a couple of complete truth-tables without any attention to the *interpretations* of the atoms in our example wffs. Truth-functionality means precisely that the truth-values of complexes can be settled by attending only to the *truth-values* of their atoms. Yet shouldn't interpretations come back into the story after all? Suppose that (according to the code-book currently in play) 'P' has the content *All brothers are male*, and therefore expresses a necessary truth. In this case there is no valuation which respects the meaning of 'P' yet assigns it the value *false*. Or again, suppose that 'P' has the content *Jack is married*, and 'Q' the content *Jack is single*. In this case there is no valuation which respects the meanings of 'P' and 'Q' yet gives them the same truth-value.

Return then to our sample wff '¬(P ∧ ¬(¬Q ∨ R))' which contains three atoms. Maybe not all eight different assignments of values to these atoms are genuinely possible ones: won't that depend on the interpretations of the atoms? So shouldn't we take that into account when setting out the truth-table for an interpreted wff?

No. Or at least, not given the wider setting of our interest in **PL**. For our target in developing this language is to give an account of the logic of arguments *in so far as that depends on the presence of the connectives 'and', 'or' and 'not'*. Given this target, we can and should temporarily ignore the facts about the meaning of individual atoms, and concentrate only on those constraints on the overall assignments of truth-values to wffs which are due to the truth-functional meanings of the connectives '∧', '∨' and '¬'. Hence, given our particular target, we should put no restrictions on the valuation of atoms. That means that we will continue to treat any assignment of T or F to the atoms 'P', 'Q', etc., as equally legitimate *for our special purposes*, whatever the meanings of the atoms. In short, we will continue to allow *any* valuation of atoms.

11.5   Short cuts

With a bit of practice, we can learn to skip much of the laborious detail involved in setting out full truth tables. For a start, it quickly becomes easy to do some of the working 'in our heads'. Suppose the formula to be evaluated is

(¬(P ∧ ¬Q) ∨ (R ∨ (Q ∧ P))).

This wff is the disjunction of the subformulae '¬(P ∧ ¬Q)' and '(R ∨ (Q ∧ P))', and it is relatively easy to work out the truth values of these disjuncts on a given valuation of the atoms in your head. That gives us the entries under the subformulae in the following table, leaving just the value of the whole disjunction to be worked out at the final step:

P	Q	R	(¬(P ∧ ¬Q) ∨ (R ∨ (Q ∧ P)))		
T	T	T	T	T	T
T	T	F	T	T	T
T	F	T	F	T	T
T	F	F	F	F	F
F	T	T	T	T	T
F	T	F	T	T	F
F	F	T	T	T	T
F	F	F	T	T	F
			1	3	2

We can, however, speed things up even more by leaving out any redundant working. In particular, note these two obvious truisms:

- Once a disjunct has been shown to be true, the whole disjunction must be true (so it is redundant to evaluate the other disjunct).
- Once a conjunct has been shown to be false, the whole conjunction must be false (so it is redundant to evaluate the other conjunct).

Hence, in our last example, as soon as we have evaluated the first disjunct, we can get to the following:

P	Q	R	(¬(P ∧ ¬Q) ∨ (R ∨ (Q ∧ P)))	
T	T	T	T	T
T	T	F	T	T
T	F	T	F	?
T	F	F	F	?
F	T	T	T	T
F	T	F	T	T
F	F	T	T	T
F	F	F	T	T
			1	

So we only need to evaluate the second disjunct on just *two* lines in order to complete the table.

Here's another example. Suppose we are faced with the wff

(¬{Q ∨ [(S ∧ R) ∨ ¬(R ∨ ¬Q)]} ∧ {P ∧ ¬ S})

Bracket-counting confirms that the main connective is indeed the second '∧'. So this is a conjunction whose second conjunct '{P ∧ ¬ S}' is simple to evaluate:

P	Q	R	S	(¬{Q ∨ [(S ∧ R) ∨ ¬(R ∨ ¬Q)]} ∧ {P ∧ ¬ S})	
T	T	T	T	F	F
T	T	T	F	?	T
T	T	F	T	F	F
T	T	F	F	?	T
T	F	T	T	F	F
T	F	T	F	?	T
T	F	F	T	F	F
T	F	F	F	?	T
F	T	T	T	F	F
F	T	T	F	F	F
F	T	F	T	F	F
F	T	F	F	F	F
F	F	T	T	F	F
F	F	T	F	F	F
F	F	F	T	F	F
F	F	F	F	F	F
				1	

Since the big conjunction must be false whenever its second conjunct is false, that has immediately settled the value of the wff on twelve of the sixteen lines. Which leaves just four lines to go. The second and fourth lines are relatively easy too. '{Q ∨ ((S ∧ R) ∨ ¬(R ∨ ¬Q))}' is a disjunction whose first disjunct is simply 'Q'. Hence whenever 'Q' is true, that curly-bracketed subformula will be true. So its negation is false. Thus we can quickly fill in two more lines of working:

P	Q	R	S	(¬{Q	∨	[(S ∧ R) ∨ ¬(R ∨ ¬Q)]}	∧	{P ∧ ¬S})
T	T	T	T				F	F
T	T	T	F	F T	T		F	T
T	T	F	T				F	F
T	T	F	F	F T	T		F	T
T	F	T	T				F	F
T	F	T	F		F	?	?	T
T	F	F	T				F	F
T	F	F	F		F	?	?	T
...	...	...	...	4	2	3		1

Very quickly we are left with just *two* final lines of working to fill out. And then the table is speedily completed (do it!).

To be sure, even with these short cuts (which you soon get the hang of using), working out complete truth-tables for complex formulae is a tedious business. Fortunately, when it comes to using truth-tables in earnest, we rarely need to consider other than pretty simple formulae. There's no need for panic!

11.6 Truth-functional equivalence

Suppose we evaluate '((P ∧ ¬Q) ∨ (¬P ∧ Q))' and '((P ∨ Q) ∧ ¬(P ∧ Q))' for each possible assignment of values to 'P' and 'Q'. Suppressing the detailed working, we get

P	Q	((P ∧ ¬Q) ∨ (¬P ∧ Q))	((P ∨ Q) ∧ ¬(P ∧ Q))
T	T	F	F
T	F	T	T
F	T	T	T
F	F	F	F

The two wffs thus share the same values on each valuation: the full tables with the working laid out in all its detail will of course be different, but the resulting table of final values for the whole wffs are exactly the same. We will say that wffs like this are *(truth-functionally) equivalent*. A little more carefully:

> The **PL** wffs *A* and *B* are truth-functionally equivalent just if, on each valuation of all the atoms occurring in them, *A* and *B* take the same value.

Equivalent wffs *A* and *B* are true in exactly the same circumstances. That means, for example, that if *A* can be inferred to be true on the basis of certain premisses, then so can *B*. Likewise, if *A* can be used as a premiss in drawing a certain logical conclusion, then *B* could equally well be used for the same purposes. In short, equivalent wffs have the same logical powers.

So when translating ordinary claims into **PL** for logical purposes, it cannot matter greatly which of two truth-functionally equivalent translations we choose. Suppose then we want to translate 'Either Jack drank beer or he drank cider' where the 'or' is in this case intended to be *exclusive*. Using 'P' for *Jack drank beer* and 'Q' for *Jack drank cider*, we want a translation which comes out true when exactly one of 'P' and 'Q' is true: we have just seen that '((P ∧ ¬Q) ∨ (¬P ∧ Q))' or '((P ∨ Q) ∧ ¬(P ∧ Q))' will do equally well.

That last point fulfils a much earlier promise to show how to express exclusive disjunction in **PL** (§7.4). We could have introduced a new symbol '⊕' into the language with the stipulation that a wff of the form $(A \oplus B)$ is true when exactly one of *A* and *B* is true. But we can now see that this would be strictly speaking redundant; we can capture exclusive disjunction using the three connectives already to hand. More precisely: for any wff we could construct using '⊕' in an extension of **PL**, there is a truth-functionally equivalent wff already available in original, unaugmented **PL**.

11.7 Expressive adequacy

That last point generalizes. Take *any* possible truth-functional combination of atoms (which we might perhaps express using a new connective added to **PL**): then we can construct an equivalent wff *using just the original three connectives of* **PL**. This result is a very important one – for it shows that **PL**, despite its meagre resources, can express a lot more than you might guess. Its resources are enough to capture *any* truth-functional way of combining propositions, and so the language is adequate for the whole of the logic of truth-functions.

The proof of this result is pretty straightforward, but is interesting enough to be worth setting out in a little detail. (This is our first significant 'metalogical' result: i.e., it is a general result *about* our formal logical apparatus.)

We start with a trivial notational point and then a definition. First, we are going to be writing down some expressions which look like the following

$(A \wedge B \wedge C)$
$\{A \vee B \vee C \vee D \vee E\}$

when strictly speaking – given our rule that each occurrence of '∧' and '∨' comes with its own set of brackets – we ought to write something like

$((A \wedge B) \wedge C)$
$((((A \vee B) \vee C) \vee D) \vee E)$

Still, the particular choice of internal bracketing for long unmixed conjunctions or for long unmixed disjunctions evidently doesn't affect the truth-value of the resulting wffs (see Exercise 11D). So leaving out the bracketing in such cases will for once do no real harm, and will make for readability.

Now for our definition. Take any set of atomic wffs, as it might be {P, Q, S, P'}; a *basic conjunction* is formed by taking each of the atoms in turn, either naked or negated but not both, and then (if we start with more than one atom) conjoining the results, yielding e.g.

$$(P \land \neg Q \land \neg S \land P')$$
$$(\neg P \land \neg Q \land S \land \neg P')$$
$$(\neg P \land \neg Q \land \neg S \land \neg P')$$

A basic conjunction is true on one and only one assignment of values to its constituent atoms. In our examples, the first wff is true just when $P \Rightarrow T$, $Q \Rightarrow F$, $S \Rightarrow F$, $P' \Rightarrow T$; the second true just when $P \Rightarrow F$, $Q \Rightarrow F$, $S \Rightarrow T$, $P' \Rightarrow F$; and the third when $P \Rightarrow F$, $Q \Rightarrow F$, $S \Rightarrow F$, $P' \Rightarrow F$. In an obvious sense, then, a basic conjunction *corresponds* to a particular valuation of its atoms. (Incidentally, note that our definition allows a basic conjunction to consist in a single atom, if that's all we start off from!)

With those preliminaries, here's the proof we want.

- A way of constructing a sentence out of **PL** atoms is truth-functional if fixing the truth-values of the constituent atoms is enough to determine the value of the complex sentence. Hence, we can write down a truth-table displaying *how* the value of the truth-functional construction depends on the values of its components. So our task is effectively this: to show that for any given truth-table, we can write down a **PL** wff with exactly that truth-table. Now, there are an unlimited number of different modes of truth-functional composition of atoms. But however many atoms are in question, one of three things must happen in the resulting truth-table for the combination: either it has 'F's all the way down; or it has 'T' on exactly one line; or it has 'T' on more than one line. Take these cases in turn.

- *First case* When the truth-function is invariably false, it is easy to write down a **PL** wff using the same atoms which is also invariably false. Just conjoin or disjoin some contradictions. If '!(P, Q, R, S)' is defined to be a function of those four atoms which is false on every valuation, then e.g.

  $$((P \land \neg P) \lor (Q \land \neg Q) \lor (R \land \neg R) \lor (S \land \neg S))$$

  is, careless bracketing apart, a corresponding wff of **PL** which involves the same atoms and is also always false.

- *Second case* When the truth-function in question takes the value *true* on exactly one valuation of atoms, simply write down the **PL** basic conjunction that corresponds to that particular valuation, and again we have a wff with the desired truth table. For example, if '%(P, Q, R, S)' is true just when $P \Rightarrow F$, $Q \Rightarrow T$, $R \Rightarrow F$, $S \Rightarrow T$, then

  $$(\neg P \land Q \land \neg R \land S)$$

  is a wff with the right truth-table.

- *Third case* When the truth-function in question takes the value *true* on a number of different valuations of atoms, then (a) write down the **PL** basic conjunction corresponding to each valuation of atoms which makes the function come out true, and then (b) disjoin the results – i.e. write down a long disjunction, where each disjunct is one of those basic conjunctions. This wff will come out true when and only when some disjunct is true, i.e. when one of those basic conjunctions is true, i.e. on the lines where the target truth-function is true. So our **PL** wff has the desired truth-table.

For example, the following table defines a perfectly good truth-function of the three atoms 'P', 'Q' and 'R':

P	Q	R	$(P, Q, R)
T	T	T	F
T	T	F	T
T	F	T	F
T	F	F	F
F	T	T	T
F	T	F	T
F	F	T	T
F	F	F	T

This dollar truth-function might not be a terrifically *interesting* one: still, we might want to express it, and we *can* express it in **PL**, without having to expand the language. Write down a basic conjunction corresponding to each assignment of values where (P, Q, R) is true, and then disjoin them. That yields

$$\{(P \wedge Q \wedge \neg R) \vee (\neg P \wedge Q \wedge R) \vee (\neg P \wedge Q \wedge \neg R) \vee (\neg P \wedge \neg Q \wedge R) \vee (\neg P \wedge \neg Q \wedge \neg R)\}$$

This is custom built to have the right truth-table, for it comes out true on each line where one of the basic conjunctions is true. By construction, then, it is truth-functionally equivalent to (P, Q, R). It may not be the *simplest* way of capturing the truth-function $ in **PL** – '$(\neg P \vee (Q \wedge \neg R))$' does the job, as you can easily verify – but it is certainly *one* way.

We can sum all that up as follows in another definition and a result:

> A set of connectives is *expressively adequate* if a language containing just those connectives is rich enough to express all truth-functions of the atomic sentences of the language (i.e. rich enough to enable us to construct a formula with any given truth-table). We have proved that *the standard set of **PL** connectives* $\{\wedge, \vee, \neg\}$ *is expressively adequate.*

11.8 'Disjunctive normal form'

Just for the record, and because you might come across the idea elsewhere, here's a corollary of our result about expressive adequacy. (But you can skip this!)

Start with some more definitions. Let's say that a wff is a *primitive* wff if it is either an atom or a negated atom (sometimes the alternative term 'a *literal*' is used). And let's say that a wff is a *basic** conjunction if it consists of a conjunction of one or more primitives. The big difference between a basic and a basic* conjunction is that the latter allows the same atom to appear both negated and unnegated, i.e. it allows contradictions. So basic conjunctions are basic*, but not always vice versa. (Note again we allow basic* conjunctions with only one 'conjunct'!) Further, we say that a wff is in *disjunctive normal form* (DNF) if it consists of one or more disjuncts each of which is a basic* conjunction. (Note that we also allow the limiting case of a single 'disjunct'; in other words, a basic* conjunction by itself counts as being in DNF.)

In these terms, then, what we have just shown is that *any truth-functional combination of atoms can be expressed by an equivalent wff in DNF*. As we saw, in the first type of case in the proof in the last section, where the combination is always false, we can use a disjunction of basic* conjunctions which are contradictions. In the second type of case, where the combination is true on just one assignment of values to atoms, we use a single basic (and hence basic*) conjunction. And in the third type of case, where the combination is true on more than one assignment of values to atoms, we use a disjunction of two or more basic (and hence again basic*) conjunctions.

Since any PL wff expresses a truth-functional combination of atoms, that shows that any wff is truth-functionally equivalent to one in DNF.

11.9 Other adequate sets of connectives.

The key adequacy result is the one proved in the §11.7. But it is worth noting that we can do even better: the set of connectives $\{\wedge, \neg\}$ is adequate by itself. For a wff of the form $(A \vee B)$, which rules out A and B being both false, is equivalent to the corresponding wff of the form $\neg(\neg A \wedge \neg B)$. So, given a wff involving occurrences of '$\wedge$', '$\vee$' and '$\neg$', we can always construct an equivalent wff involving just '$\wedge$' and '$\neg$': replace the disjunctions with equivalent negation-conjunction combinations.

We can likewise show that the set of connectives $\{\vee, \neg\}$ is adequate: simply note the result that $(A \wedge B)$ is equivalent to $\neg(\neg A \vee \neg B)$, and then the argument goes similarly.

However, it is easily seen that $\{\wedge, \vee\}$ is *not* expressively adequate. Looking at the truth-tables for those connectives, we see that any wff built using just conjunction and disjunction must be true when its atoms are all true. So using conjunction and disjunction we can never get anything equivalent to a wff like '¬P' or '(¬P ∨ ¬Q)' which happens to be *false* when its atoms are true. Hence there are truth-functions which can't be expressed using conjunction and disjunction alone.

The result that *any* truth-function can be expressed using just a pair of connectives is surely enough to satisfy any sensible taste for economy. But it is mildly diverting to show in conclusion there are in fact *single* connectives which, just by themselves, are expressively adequate.

Given that $\{\wedge, \neg\}$ is expressively adequate, it is enough to find a single connective in terms of which both those connectives can in turn be defined. Clearly it must be at least a two-place connective, and one candidate is the following:

A	B	$(A \downarrow B)$
T	T	F
T	F	F
F	T	F
F	F	T

An expression of the form $(A \downarrow B)$ says *neither A nor B*, i.e. that the connected expressions are both false: so we have

  $\neg A$ is equivalent to $(A \downarrow A)$, since that says that A and itself are both false.

  $(A \wedge B)$, which says that neither $\neg A$ nor $\neg B$, is equivalent to $(\neg A \downarrow \neg B)$ and so can be expressed by $((A \downarrow A) \downarrow (B \downarrow B))$.

Thus any **PL** wff built up by conjunction and negation can in principle be laboriously rewritten using just the down-arrow. Hence, putting everything together, any truth-function can be expressed using $\{\wedge, \vee, \neg\}$; and wffs using only these three connectives can be recast into formulae using just $\{\wedge, \neg\}$; and these latter can in turn be recast using just '$\downarrow$'. So that single connective is expressively adequate. QED.

  We can similarly introduce the up-arrow which stands to the down-arrow as '$\vee$' stands to '$\wedge$': i.e. while $(A \downarrow B)$ signals that both $\neg A$ and $\neg B$, $(A \uparrow B)$ will say that either $\neg A$ or $\neg B$, i.e. it is false just when A and B are both true. It is immediate that $\neg A$ equivalent to $(A \uparrow A)$, so that $(A \vee B)$ is equivalent to $((A \uparrow A) \uparrow (B \uparrow B))$. That means that any wff built up using only $\{\vee, \neg\}$, and hence any wff at all, is re-expressible using just the up-arrow. (Incidentally, the up-arrow was first noted by H. Sheffer in 1913, and was written by him as '|' – hence the common label, 'the Sheffer stroke'.)

11.10 Summary

- A way of forming a complex sentence out of one or more constituent sentences is *truth-functional* if fixing the truth-values of the constituent sentences is *always* enough to determine the truth-value of the complex sentence.

- All wffs of **PL** are truth-functions of their atoms. The value of a wff is fully determined by fixing the values of the atoms – and we can calculate and set out in a truth-table the values that the wff takes on each possible assignment of values to the atoms. This is tedious mechanical labour (though there are short-cuts that, if judiciously chosen, can reduce the work)

- Conversely, *any* truth-function of atoms can be expressed using the resources of **PL**.

- Indeed, *any* truth-function of atoms can be expressed using just '$\wedge$' and '$\neg$', or just '$\vee$' and '$\neg$', or even solely the up-arrow or down-arrow.

Exercises 11

A   Give truth-tables for the following wffs (i.e. calculate the value of the wff for every assignment of values to the atoms: you can, however, use short-cuts).

1. $\neg(P \wedge \neg P)$
2. $(P \wedge \neg(P \wedge Q))$
3. $((R \vee Q) \vee \neg P)$
4. $(\neg(P \wedge \neg Q) \wedge \neg\neg R)$
5. $((P \wedge Q) \vee (\neg P \vee \neg Q))$
6. $\neg((P \wedge \neg Q) \vee (\neg R \vee \neg(P \vee Q)))$
7. $(((P \vee \neg Q) \wedge (Q \vee R)) \vee \neg\neg(Q \vee \neg R))$
8. $(\neg(\neg P \vee \neg(Q \wedge \neg R)) \vee \neg\neg(Q \vee \neg P))$
9. $(\neg((R \vee \neg Q) \wedge \neg S) \wedge (\neg(\neg P \wedge Q) \wedge S))$

B   Give equivalent wffs in DNF for each of **A.1** to 7. (Read off the truth-table and use the construction in §11.7.)

C   Show that '↓' and '↑' are the only two-place connectives which are expressively adequate taken by themselves.

D   Show that '$((P \wedge Q) \wedge R)$' and '$(P \wedge (Q \wedge R))$' have the same truth-table. Show, more generally, that pairs of wffs of the forms $((A \wedge B) \wedge C)$ and $(A \wedge (B \wedge C))$ will have the same truth-table. Generalize to show that how you bracket an unmixed conjunction, $A \wedge B \wedge C \wedge D \wedge \ldots$, doesn't affect truth-values. Show similarly that it doesn't matter how you bracket an unmixed disjunction.

E   Show that for any A, B, $\neg(A \wedge B)$ is equivalent to $(\neg A \vee \neg B)$, and $\neg(A \vee B)$ is equivalent to $(\neg A \wedge \neg B)$. Along with the equivalences of $\neg(\neg A \wedge \neg B)$ with $(A \vee B)$, and $\neg(\neg A \vee \neg B)$ with $(A \wedge B)$, these are called *De Morgan's Laws*. Use De Morgan's Laws to show that every **PL** wff is equivalent to one where the only negation signs attach to atoms.

F   Instead of denoting the truth-value 'T' and 'F', use '1' and '0'; and instead of writing the connectives '∧', '∨', write '×' and '+'. What rule of binary arithmetic corresponds to our rule for distributing the 'T' and 'F's to a sequence of atoms when constructing a truth-table? Explore the parallels between the truth-functional logic of '∧' and '∨' and binary arithmetic. What is the 'arithmetic' correlate of negation?

Tautologies

In this chapter, we note a particularly important class of truth-functional proposition, namely the tautologies. Their status illuminates the strong notion of necessity that is involved in logical theory.

12.1 Tautologies and contradictions

For any wff of **PL**, we can mechanically calculate its truth-value for each and every valuation of its atoms. Usually, the wff will be true on some valuations and false on others. However, we occasionally get a uniform result:

> A wff of **PL** is a *tautology* if it takes the value *true* on every valuation of its atoms.
>
> A wff is a *contradiction* if it takes the value *false* on every such valuation.

The following, for example, are three very simple tautologies:

(1) ¬(P ∧ ¬P)
(2) (P ∨ ¬P)
(3) ((P ∧ Q) ∨ (¬P ∨ ¬Q))

As is easily checked, it doesn't matter what value we assign to 'P' and 'Q', these wffs are invariably true. And even without doing the calculations, we can see that these wffs *should* always be true:

(1) must hold because a proposition and its negation cannot both be true (since negation flips values): this is Aristotle's 'Law of Non-contradiction';

(2) reflects our stipulation that interpreted wffs of **PL** take one truth-value or the other; there is no middle option: this is the 'Law of Excluded Middle';

(3) obtains because either 'P' and 'Q' are both true or at least one of them is false.

But tautologies need not be quite as obvious as this! Here, for example, is another, as we can confirm by doing a truth-table:

(4) (¬{¬(P ∧ Q) ∧ ¬(P ∧ R)} ∨ ¬{P ∧ (Q ∨ R)})

As the bracketing reveals, this is a disjunction, each of whose disjuncts is of the form '¬{...}'. Evaluating the simpler disjunct first and then completing the working as necessary, we get

P	Q	R	(¬{¬(P ∧ Q) ∧ ¬(P ∧ R)} ∨ ¬{P ∧ (Q ∨ R)})
T	T	T	T F T F T F T T
T	T	F	T F T F T F T T
T	F	T	T T F F F F T T F T T
T	F	F	T T F F
F	T	T	T T F
F	T	F	T T F
F	F	T	T T F
F	F	F	T T F
			9 5 4 8 7 6 10 3 2 1

So the wff is indeed a tautology. And there can, of course, be tautologies involving arbitrarily many connectives.

If *A* always takes the value *true*, then ¬*A* will always take the value *false*. Hence we have the following simple result: *the negations of tautologies are contradictions*. Conversely, *the negations of contradictions are tautologies*.

Thus '¬((P ∧ Q) ∨ (¬P ∨ ¬Q))', the negation of the tautology (3), counts as a contradiction. Note therefore that, in our customary but slightly stretched usage of the term, not all contradictions have the form (*A* ∧ ¬*A*), where a wff is directly conjoined with its negation (compare §5.4).

12.2 Generalizing about tautologies

There is, of course, nothing special here about the Ps and Qs in the examples above. For example, '¬(R ∧ ¬R)' and '¬(P″ ∧ ¬P″)' are equally good instances of the law of non-contradiction. Indeed, *any* wff of the logical form

(1′)   ¬(*A* ∧ ¬*A*)

is a tautology, whether *A* stands in for an atom or a complex wff.

Why so? Well, whatever wff we replace *A* by, and whatever valuation we give its atoms, the wff will be either true or false. And the same calculation which shows that '¬(P ∧ ¬P)' is true whatever the value of 'P' will equally show that ¬(*A* ∧ ¬*A*) is true whatever the value of *A*.

The point generalizes. Take any tautology. Systematically replace its atoms with schematic variables like '*A*', '*B*', etc. (i.e. do it *uniformly* – same atom, same replacement variable). That gives us a template for wffs, that displays the basic form of the tautology, and encodes all the information required to show that the wffs of this form are indeed tautologies. So any instance of this template, got by uniformly substituting wffs for the schematic variables, will be another tautology.

So, the claim that e.g. '¬(P ∧ ¬P)' or '((P ∧ Q) ∨ (¬P ∨ ¬Q))' is a tautology is implicitly general in two ways. First, it remains true *however we interpret the atoms*. That's because the status of a wff as a tautology depends only on truth-table calculations which don't involve any assumptions about what its atoms actually mean. And second the claim remains true *however we (uniformly) substitute wffs for atoms.*

12.3 A point about 'form'

We just used the phrase 'any wff of the logical form $\neg(A \wedge \neg A)$'. Much earlier we also talked about logical form (§3.5), and at that point we left the idea vague and inchoate. Now we can do much better. For when we say e.g. that a **PL** wff has the form $\neg(A \wedge \neg A)$, we just mean that the wff can be constructed from the negation sign, followed by a left-hand bracket, some wff *A*, the conjunction sign, the negation sign, *A* again, and a closing right-hand bracket. Or equivalently, we mean the wff can be produced from the template '$\neg(A \wedge \neg A)$' by replacing each occurrence of the schematic variable by the same wff.

In short, the notion of form here is a *syntactic* one, a matter of the surface shape of a wff. By contrast, when we previously talked informally about the form of some vernacular claims, we found ourselves tempted to make the notion in part a semantic one (recall, we found it tempting to say that 'All dogs have four legs' and 'Any dog has four legs' in some sense have the same logical form, despite the surface differences, because they mean much the same). The semantic character of the informal notion made it slippery: just how close in meaning do sentences have to be to share logical form? We left such questions on hold. But we can now explain why we don't need to answer them for our purposes in the rest of this book. *Henceforth, when we talk about 'form', we will be making use of a notion that can be characterized purely syntactically,* i.e. it is a notion having to do with the surface pattern in a wff or collection of wffs.

Of course, this syntactic notion is of *interest* to us because, in **PL**, surface form shadows semantic structure in the intuitive sense (§§9.5, 9.6). But the point is that we can *define* form in a quite unproblematic syntactic way.

12.4 Tautologies and necessity

Recall again our stipulation in §9.3. We are taking it that atoms of **PL** are such as to be either determinately true or determinately false. So if we run through all the different assignments of values to the relevant atoms and construct a full truth-table for a wff *W*, then we exhaust the possible ways things might turn out. And if each such possibility makes *W* true, then *W* is necessarily true. In sum, truth-functional tautologies are necessary truths.

But not all necessary truths encoded by wffs of **PL** are truth-functional tautologies. Suppose we read the atom 'Q'''' as expressing the proposition that *all sisters are female.* Then 'Q'''' expresses a necessary truth. Yet it obviously isn't a tautology; no mere atom can be a tautology, for it won't be true on a valuation which assigns the value F to the atom. To be sure, a valuation which assigns F to 'Q'''' won't respect the intended content of this atom. But for reasons we sketched before (§11.4), we are *not* requiring that valuations of atoms respect their meanings: i.e. for our purposes we can treat the assignment of T or F to any atom as equally allowed, whatever its content – that's because we want to isolate those logical facts that *don't* depend on the meanings of atoms.

The tautologies, then, are the necessary truths of **PL** whose necessity can be derived purely from the meaning of the connectives. In other words, the *content* of the atoms doesn't matter in making e.g. '$((P \land Q) \lor (\neg P \lor \neg Q))$' a tautology; rather, it's because that wff has the form $((A \land B) \lor (\neg B \lor \neg A))$ that it is a tautology. *So there is a good, sharp, technical sense in which we can say that tautologies are true in virtue of their logical form* (i.e., to put it more carefully, they are true in virtue of the way they are formed using the connectives).

We can also describe claims in ordinary English as being tautologies, if they involve truth-functional connectives and their natural translation into **PL** is a tautology. And these claims too might be said rather more loosely to be true in virtue of their logical form.

12.5 A philosophical aside about necessity

Philosophers have drawn a number of distinctions within the family of notions to do with kinds of necessity and possibility. (To keep things simple, we'll consider the distinctions as they apply within the class of truths; there are similar distinctions to be made among the falsehoods.)

- *Necessary vs. contingent*   There does seem to be an intuitive distinction between those claims that just contingently happen to be true and which could have been false, and those claims which are necessarily true (in the strong sense of being true in all conceivable possible worlds: see §2.1). So it happens to be the case that *Jack has a sister*; that's a contingent truth about the actual world. But it is necessarily true that *if Jack has a sister, then he isn't an only child* (that is true of any possible situation). Again, it just happens to be the case that *there are nine planets* (if we still count Pluto) but it is surely necessary that *eight plus one is nine*.

- *A priori vs. a posteriori*   The previous distinction is a metaphysical distinction, concerning how things *are*; our second distinction is an epistemological one, a distinction among the sorts of grounds on which we can *know* propositions to be true. And again there does seem to be an intuitive distinction between those claims that require relevant experiences for a knower to establish them as true, and those claims that can be shown to be true without relying on any particular course of experience. It needs empirical investigation to establish that *Jack has a sister*; that's only knowable 'a posteriori' (on the basis of relevant experiences – maybe seeing her being born to Jack's mother, or being reliably informed of the fact). But by pure thought alone, we can spot that *if Jack has a sister, then he isn't an only child*: we can know that 'a priori'. Again, it is a posteriori that *there are nine planets* (experience of the world is needed to establish that) but it is knowable a priori, it is a 'truth of reason', that *eight plus one is nine*. Of course, you *can* get to know arithmetical truths by empirical means, as when we justifiably rely on a pocket calculator. But the claim is that you don't *need* to rely on empirical enquiry to justify a belief that eight plus one is nine, because rational reflection is enough by itself.

Or so the traditional story goes.

Of course, the story raises a number of questions. How is that it some propositions are true however the world is (true, as they say, with respect to any possible world)? Why shouldn't there be, so to speak, unco-operative worlds? How can we know something to be true just by sitting in our armchairs and thinking, without checking the truth against the world? Why do the same examples so often invite treatment both as necessary truths and as knowable a priori? – do the notions really pick out the same class of truths?

There's a once-popular theory (though much less popular today) that aims to answer such questions. Supposedly, we can make a third distinction:

* *Analytic vs. synthetic*   Some claims are 'merely verbal', i.e. they are true in virtue of the meanings of the words involved. These are said to be 'analytic' truths. Other truths, the non-analytic ones which aren't merely verbal, are said to be 'synthetic'. It is analytic, true merely in virtue of the meanings of the words involved, that *if Jack has a sister, then he isn't an only child.* Again, once the meanings of 'eight', 'nine' etc. are fixed, that fixes that *eight plus one is nine* is true.

And the theory is that the necessary truths are exactly the a priori truths which are exactly the analytic truths. That is to say, some propositions remain true however the world is because they say nothing substantive about the world, but are merely verbal truths: and that's how we can know them a priori, merely by reflection – because they are empty of substantive content. Which is a nicely demystifying theory about necessity and apriority.

Note, by the way, the difference in status among the distinctions here. The necessary/contingent and the a priori/a posteriori distinctions are supposed to be graspable pre-theoretically, and have a fairly intuitive appeal. We think: some things happen to be so, others couldn't have been otherwise; some truths can only be established empirically, others can be justified by pure thought. The analytic/synthetic distinction, by contrast, belongs to a philosophical *theory* about what grounds these other, intuitively appealing, distinctions.

It's a neat theory. And whatever the *general* difficulties with identifying the necessary with the analytic, the **PL** tautologies do seem to provide us with a plausible case, where (rightly or wrongly) the idea of analyticity looks particularly attractive. We stipulated the meanings of the three connectives precisely in order to make **PL** wffs truth-functional. And it is a direct result of our stipulations that a wff will take the value it does on a given valuation of its atoms. Hence (in particular) it is a direct result of our stipulations that certain wffs come out true on every valuation. In this case, at least, it might seem that there is therefore nothing very puzzling about the resulting necessary truths; they are just a spin-off from our linguistic stipulations. There is nothing mysterious either about our knowledge of the necessities here; we don't need to inspect any worlds at all to calculate that a certain wff comes out true, however things are.

It is a very moot question whether all other candidates for necessary truth can also usefully be regarded as analytic, as somehow resulting from linguistic conventions in the same sort of way. For example, are the deep theorems of number theory really, in the last analysis, 'merely verbal', simply distant spin-offs from the basic linguistic conventions governing number talk?

Luckily, we don't need to face up to such questions here. All we need now is the thought (no doubt itself still in need of further elucidation but surely pointing in the right direction) that the necessary truth of tautologous wffs looks to be relatively unmysterious.

12.6 Summary

- Wffs of **PL** which are true on every valuation of their atoms are *tautologies*. Wffs that are false on every valuation are *contradictions*: contradictions do not have to be of the form $(A \land \neg A)$.

- A **PL** tautology remains a tautology whatever the meaning of the atoms; and uniformly substituting wffs for the atoms will yield further tautologies.

- **PL** tautologies are necessary truths: but not all necessary truths expressible in **PL** are tautologies.

- The necessity of tautologies is arguably unpuzzling, a result of the semantic rules governing the connectives.

Exercises 12

Which of the following wffs are tautologies, which contradictions, which neither?

1. $(P \land ((\neg P \lor Q) \land \neg Q))$
2. $((Q \land (\neg\neg P \lor R)) \lor \neg(P \land Q))$
3. $(\{P \lor \neg(Q \land R)\} \lor \{(\neg P \land Q) \land R\})$
4. $(\{P \lor (Q \land \neg R)\} \lor \neg\{(\neg P \lor R) \lor Q\})$
5. $(\{P \land (\neg Q \lor \neg R)\} \lor \neg\{(P \lor \neg R) \lor \neg Q\})$
6. $\neg(\{\neg(P \land \neg R) \land \neg(Q \land \neg S)\} \land \neg\{\neg(P \lor Q) \lor (R \lor S)\})$

Tautological entailment

In this pivotal chapter, all our work on the language **PL** – at long last! – gets put to real use in the assessment of arguments. We introduce the crucial 'truth-table test' for demonstrating the validity of some inferences whose essential logical materials are conjunction, disjunction and negation.

13.1 Two introductory examples

Consider the argument

    A   Jack is logical. It isn't the case that Jack is logical but Jill isn't. Hence Jill is logical.

Rendered into **PL** via a suitable code-book, the argument can be re-expressed with premisses 'P' and '¬(P ∧ ¬Q)', and conclusion 'Q'. Now, for each possible assignment of values to the atoms, let's evaluate the premisses and the conclusion, setting out all the results in a table as follows:

P	Q	P	¬(P ∧ ¬Q)	Q
T	T	T	T	T
T	F	T	F	F
F	T	F	T	T
F	F	F	T	F

(Here and henceforth, we will use a double line to separate off the list of possible valuations of atoms from the part of the table which records the values of the premisses on each valuation, and then use another double line before giving the corresponding value of the conclusion.) Look along the rows. We find that *there is no assignment that makes the premisses true and the conclusion false.* Hence there is no way the world could be with respect to 'P' and 'Q' that would make the premisses true and conclusion false. So argument **A** must be valid.

That was intuitively obvious at the outset: but note that we have now *proved* that the argument is valid, by examining all the different ways things might turn out as far as the truth-values of the premisses and conclusion are concerned.

Take another example. Consider the argument

B   Either Jack or Jill went up the hill. It isn't the case that Jack went up
       the hill and Jo didn't. Hence it isn't the case that Jo went up the hill
       and Jill didn't.

This argument can be rendered into **PL** with premisses '$(P \lor Q)$' and '$\neg(P \land \neg R)$',
and conclusion '$\neg(R \land \neg Q)$'. Again, we will evaluate each premiss and the
conclusion for each assignment of values to the three propositional atoms, yielding the next table (we just give the final results of the calculations).

P	Q	R	$(P \lor Q)$	$\neg(P \land \neg R)$	$\neg(R \land \neg Q)$
T	T	T	T	T	T
T	T	F	T	F	T
T	F	T	T	T	F
T	F	F	T	F	T
F	T	T	T	T	T
F	T	F	T	T	T
F	F	T	F	T	F
F	F	F	F	T	T

This time, inspection reveals that there *is* an assignment which makes the prem-
isses of **B** true and the conclusion false, namely $P \Rightarrow T$, $Q \Rightarrow F$, $R \Rightarrow T$. This
assignment evidently represents a way the world could be (i.e. Jack and Jo's going
up the hill without Jill) which makes the premisses true and conclusion false. So
the argument is invalid. That was also probably obvious at the outset. But we
have now proved the point by a mechanical search through the possibilities.

The rest of this chapter sharpens and generalizes the simple but powerful ideas
that we've just illustrated.

13.2 Tautological entailment in PL

First, let's give a formal definition:

> The **PL** wffs A_1, A_2, ..., A_n *tautologically entail* the wff C if and only if
> there is no valuation of the atoms involved in the A_i and C which makes
> A_1, A_2, ..., A_n simultaneously all true and yet C false.

If a bunch of premisses tautologically entails some conclusion, we'll also say that
the inference in question is *tautologically valid*.

The terminology here should seem quite natural. Compare: C is a tautology if
it is true on any valuation; C is tautologically entailed by some bunch of wffs A_i
if it is true on any valuation which makes all the A_i true.

And what is the relation between this technical notion of tautological entailment and our informal idea of logical entailment (§2.1)? The key result is:

> If the **PL** wffs A_1, A_2, ..., A_n tautologically entail C, then A_1, A_2, ..., A_n
> logically entail C.

In other words, if the inference from a bunch of premisses to conclusion C is tautologically valid, then it is deductively valid. Here's a proof (you can skip it on a first reading: and to avoid clutter, we'll just run through the case where there is a single premiss *A*, as the generalization to the many-premiss case is easy).

> Suppose that the inference from *A* to *C* is *invalid*. Then there must be some possible situation which would make *A* true and *C* false. But when *A* and *C* are wffs of **PL**, a situation can only make *A* true and *C* false by imposing certain values on their atoms (remember, we are assuming that atoms are determinately true or false; and the values of the atoms are all that matter for fixing the values of truth-functional wffs). Hence if the inference from *A* to *C* is invalid, there must be a valuation of their atoms which makes *A* true and *C* false – and that means that *A* doesn't tautologically entail *C*. So, putting that the other way about, if *A does* tautologically entail *C*, then the inference from *A does* logically entail *C* in the classical sense. QED.

The converse result, however, does *not* always hold. Even if an argument is not tautologically valid, it might yet be valid for some other reason. For example, take the reading of **PL** where 'R' expresses the proposition that Jo is a sister and 'S' expresses the proposition that Jo is female. 'R' doesn't tautologically entail 'S' – witness the valuation 'R' ⇒ T, 'S' ⇒ F. Yet the inference from premiss 'R' to conclusion 'S' is still deductively valid: there is no possible world in which someone is a female sibling without being female. Showing that an inference is not tautologically valid only shows it is plain invalid if one of the truth-value assignments that makes the premises true and conclusion false corresponds to a real possibility, as it did in case **B**.

Of course, the valuation 'R' ⇒ T, 'S' ⇒ F does not respect the intended meaning of the atoms on the described interpretation: but no matter (see §10.4). When we do a truth-table, we always run through a complete set of valuations for the relevant atoms, because the aim is to see which inferences are valid *purely in virtue of the way the three connectives work*, without paying any regard to the meaning of the atoms.

So much, then, for basic matters of principle: now for practicalities. How do we establish whether a bunch of wffs tautologically entails some conclusion?

There are various ways to do this, and we'll meet a much neater way of tackling the task in Chapter 16. But the most *direct* way to settle the matter is by a brute force calculation, exactly as in the last section. We simply set out every valuation of the relevant atoms in a table, and then laboriously track through calculating the values of the various wffs on the successive valuations until either we hit a 'bad' valuation which does make the premises true and conclusion false, or else we run out of possibilities. In short, we run a *truth-table test*.

It has to be admitted that the brute-force truth-table method is often a highly inefficient technique, sometimes quite ludicrously so. Take the fifty premises '¬P, ¬P′, ¬P″, ¬P‴, …,¬P″···″': it is obvious that these don't tautologically entail the conclusion 'Q'. But suppose we set out a truth table (with all the 'P's listed on the left before the 'Q'). Then we have to wait until the very *last* assignment in the truth-table, which assigns F to all the 'P's and to 'Q', before we get a line making all the '¬P' premises true and the conclusion 'Q' false. So if we had set out to do an entirely mechanical truth-table test, even if we could write down a complete line per second, it would take *half a million years* before we got to that last line of the full 2^{51} line truth-table! That's why we will soon move on to look at a much more elegant (and *usually* much faster) way of testing for tautological entailment. For the present chapter, however, we stick to using brute force.

13.3 Expressing inferences in PL

A brief aside. We've just described the argument **A**, when rendered in **PL**, as having premises 'P' and '¬(P ∧ ¬Q)' and conclusion 'Q'. But how would we present such a one-step inference *in PL itself* (remember, this is supposed to be a language in which we can phrase real, contentful arguments: cf. §7.4)?

An argument in English might consist of a list of premises, and then a conclusion signalled by an inference marker like 'so', or 'hence', or 'therefore'. Now, an inference marker does not *describe* what follows as being the conclusion of an inference: rather, it conventionally *signals* that an inference is being drawn. **PL** users should presumably have some similar way of signalling when an inference is being drawn. That's why, in §8.1, we put into our list of the language's symbols the familiar sign '∴' (to be read 'therefore') to use as an inference marker in **PL**. Further, though it isn't strictly necessary, we'll use the comma in **PL** in order to punctuate lists of premises. We quietly put these symbols to use in §9.1.

So, with these additional devices in place, we can neatly present the **PL** renditions of the two arguments above as follows:

> **A**   P, ¬(P ∧ ¬Q) ∴ Q
> **B**   (P ∨ Q), ¬(P ∧ ¬R) ∴ ¬(R ∧ ¬Q)

Rather oddly, counting '∴' as part of **PL** is not standard; but we are going to be very mildly deviant. However, our policy fits with our general view of the two-stage strategy for testing inferences. We want to be able to translate vernacular arguments into a language in which we can still express arguments; and that surely means not only that the individual wffs of **PL** should be potentially contentful, but also that we should have a way of stringing wffs together which signals that an inference is being made.

13.4 Truth-table testing in PL

Next example: is this inference tautologically valid?

> **C**   (P ∨ Q), (R ∨ ¬P), (¬¬R ∨ ¬Q) ∴ R

Tautological validity doesn't depend on the interpretation of the atoms: so imagine whatever reading of the atoms you like, though they need *some* interpretation if this is to be a real argument. There are three different atoms and so eight different possible assignments of values to the atoms to consider. Here's the complete truth-table (again omitting the working for the premises and the conclusion, and just giving the overall values):

P	Q	R	(P ∨ Q)	(R ∨ ¬P)	(¬¬R ∨ ¬Q)	R
T	T	T	T	T	T	T
T	T	F	T	F	F	F
T	F	T	T	T	T	T
T	F	F	T	F	T	F
F	T	T	T	T	T	T
F	T	F	T	T	F	F
F	F	T	F	T	T	T
F	F	F	F	T	T	F

A quick check shows that there is again no 'bad line', i.e. no valuation which makes the premisses true and the conclusion false. So the premisses do indeed tautologically entail the conclusion.

Note, however, that as we searched through all the possible valuations, we could have speeded up the working considerably by using a couple of simple tactics:

• A line where the *conclusion* is *true* cannot be a bad one – i.e. one with true premisses and a false conclusion. So an obvious tactic is to evaluate the conclusion first (at least, when this is not much more complex than other wffs yet to be evaluated), and we can then avoid doing any further work evaluating the premisses on the lines where the conclusion comes out true.

• Similarly, a line where a *premiss* is *false* cannot be a bad one: so another obvious tactic is to evaluate any simple premiss, and we can then avoid doing any further work evaluating the other premisses or the conclusion on the lines where the evaluated premiss comes out false.

Adopting these tactics, suppose we had (1) evaluated the conclusion of argument C, and then (2) we had evaluated the simple first premiss on the four lines which are still potentially 'bad' (i.e. where the conclusion is false). That would have given us the following partial table, where we now only have to fill in at most six slots to complete the task of checking for 'bad' lines:

P	Q	R	(P ∨ Q)	(R ∨ ¬P)	(¬¬R ∨ ¬Q)	R
T	T	T				T
T	T	F	T	?	?	F
T	F	T				T
T	F	F	T	?	?	F
F	T	T				T
F	T	F	T	?	?	F
F	F	T				T
F	F	F	F			F
			2			1

Proceeding to evaluate (stage 3) the next simplest premiss, we find that it is false on two of the relevant valuations, ruling those out as candidate bad lines:

P	Q	R	(P ∨ Q)	(R ∨ ¬P)	(¬¬R ∨ ¬Q)	R
T	T	T				T
T	T	F	T	F		F
T	F	T				T
T	F	F	T	F		F
F	T	T				T
F	T	F	T	T	F	F
F	F	T				T
F	F	F	F			F
			2	3	4	1

So at stage (4) we are left with just one candidate bad line, and the final premiss evaluates as false on that line, to reveal that there is *no* bad line and hence that the argument is valid.

Here's another example of our short cuts in operation. Consider the argument

D   (¬P ∨ R), (P ∨ Q), ¬(Q ∧ ¬S) ∴ (R ∨ S)

There are four atoms, and hence this time sixteen possible valuations to consider. Are there any bad lines where the premisses are true and conclusion false? Start the table again by (1) evaluating the conclusion, and we find that it is true on no fewer than twelve lines; so that immediately leaves only four potential bad lines. Next, evaluate the simplest premiss on those four lines – stage (2); and we reach the following:

P	Q	R	S	(¬P ∨ R)	(P ∨ Q)	¬(Q ∧ ¬S)	(R ∨ S)
T	T	T	T				T
T	T	T	F				T
T	T	F	T				T
T	T	F	F	?	T	?	F
T	F	T	T				T
T	F	T	F				T
T	F	F	T				T
T	F	F	F	?	T	?	F
F	T	T	T				T
F	T	T	F				T
F	T	F	T				T
F	T	F	F	?	T	?	F
F	F	T	T				T
F	F	T	F				T
F	F	F	T				T
F	F	F	F		F		F
					2		1

We are down to just three candidate bad lines. Evaluating the next simplest premiss (stage 3) eliminates two more, and finally (stage 4) we knock out the last candidate bad line: see the completed table. Hence **D** is another valid argument.

P	Q	R	S	(¬P ∨ R)	(P ∨ Q)	¬(Q ∧ ¬S)	(R ∨ S)
T	T	T	T				T
T	T	T	F				T
T	T	F	T				T
T	T	F	F	F	T		F
T	F	T	T				T
T	F	T	F				T
T	F	F	T				T
T	F	F	F	F	T		F
F	T	T	T				T
F	T	T	F				T
F	T	F	T				T
F	T	F	F	T	T	F	F
F	F	T	T				T
F	F	T	F				T
F	F	F	T				T
F	F	F	F		F		F
				3	2	4	1

For a further example, what of the inference

E   ¬((¬S ∧ Q) ∧ (R ∨ Q)), (P ∧ R), ¬(Q ∧ S) ∴ ((P ∨ Q) ∧ ¬(R ∨ S)) ?

This time, the conclusion is a relatively complex wff, so let's first evaluate the simplest premiss. That yields:

P	Q	R	S	¬((¬S ∧ Q) ∧ (R ∨ Q))	(P ∧ R)	¬(Q ∧ S)	((P ∨ Q) ∧ ¬(R ∨ S))
T	T	T	T		T		
T	T	T	F		T		
T	T	F	T		F		
T	T	F	F		F		
T	F	T	T		T		
T	F	T	F		T		
T	F	F	T		F		
T	F	F	F		F		
F	T	T	T		F		
F	T	T	F		F		
F	T	F	T		F		
F	T	F	F		F		
F	F	T	T		F		
F	F	T	F		F		
F	F	F	T		F		
F	F	F	F		F		
					1		

We can ignore all twelve lines where the premiss turns out to be false, for *they* cannot be bad lines (where premisses are true and conclusion false). So we now tackle the remaining wffs in order of increasing complexity:

P	Q	R	S	¬((¬S ∧ Q) ∧ (R ∨ Q))	(P ∧ R)	¬(Q ∧ S)	((P ∨ Q) ∧ ¬(R ∨ S))
T	T	T	T		T	F	
T	T	T	F	F	T	T	
T	T	F	T		F		
T	T	F	F		F		
T	F	T	T	T	T	T	F
T	F	T	F	T	T	T	
T	F	F	T		F		
T	F	F	F		F		
F	T	T	T		F		
F	T	T	F		F		
F	T	F	T		F		
F	T	F	F		F		
F	F	T	T		F		
F	F	T	F		F		
F	F	F	T		F		
F	F	F	F		F		
				3	1	2	4

To explain: when we evaluate the next simplest premiss (stage 2), that rules out the first line from being a bad one. Then (3) we evaluate the first premiss on the remaining three lines, which leaves just two potential bad lines with all true premisses. So we at last must start evaluating the conclusion (4). But we find that the first potential bad line indeed has a false conclusion. *And one bad line is bad enough*: we can stop there: there is no need to carry on further. The argument is not tautologically valid.

For more worked examples, tackle the Exercises! But for the moment, let's just summarize the idea of the truth-table test, together with our recommended devices for cutting down working.

> To determine tautological validity by a truth-table test, we do a brute-force search, looking for a valuation which makes all the premisses true and conclusion false. If there is such a 'bad line' in the table, the argument is invalid; if there is no such line the argument is valid.
>
> We can suspend work on any line of the truth-table as soon as a premiss turns out to be false on that valuation, or as soon as the conclusion turns out to be true (such a line cannot be a bad one).
>
> If we find a bad line, we needn't proceed any further. One bad line in the table is enough to show that the inference is not tautologically valid.

13.5 Vernacular arguments again

We motivated the introduction of formalized languages such as **PL** by recommending a two stage strategy for evaluating arguments presented in ordinary English: first reformulate the vernacular argument in some appropriate formalized language, then assess the validity of the formalized argument. We started putting the strategy into practice with a couple of very simple examples in §13.1 above. Here, to repeat, are the principles we used.

- Suppose the vernacular English inference I_E can be recast into an inference I_{PL} in **PL** while remaining adequately faithful to the original, at least as far as anything relevant to truth and validity is concerned. Then, if I_{PL} is a tautological entailment, I_{PL} is deductively valid, and hence the original inference I_E is valid too.

- But if I_{PL} isn't a tautological entailment, things get more complex, for it is then left open whether I_{PL} (and so I_E) is valid or not. The cases where I_{PL} is tautologically invalid yet is still informally valid will be the cases where the 'bad' valuations don't correspond to genuinely possible situations. If an inference I_{PL} is tautologically invalid, it is a purely mechanical exercise to find the bad valuations which make the premisses true and conclusion false. But it is not in general a mechanical matter to decide whether any such bad valuation corresponds to a genuinely possible situation which reveals the inference to be deductively invalid in the informal sense. In some cases the decision may be easy; but it needn't be.

So much for the general principles again. Back to some examples. So consider:

F   Either Jack is a philosopher or he is a physicist. Jack isn't both a philosopher and a fool. He is not a physicist. So Jack is not a fool.

Translating this into **PL**, using 'P' for 'Jack is a philosopher', 'Q' for 'Jack is a physicist', 'R' for 'Jack is a fool', we get an argument which can be symbolized

F′ $(P \lor Q), \neg(P \land R), \neg Q \therefore \neg R$

Is this tautologically valid? If we first evaluate the conclusion (1), and then (2) the simplest premiss on the potentially bad lines (i.e. where the conclusion is false) we are left with just two remaining potential bad lines. Finishing the job (3) gives us

P	Q	R	$(P \lor Q)$	$\neg(P \land R)$	$\neg Q$	$\neg R$
T	T	T			F	F
T	T	F				T
T	F	T	T	F	T	F
T	F	F				T
F	T	T			F	F
F	T	F				T
F	F	T	F	T	T	F
F	F	F				T
			3	3	2	1

There is no line where the premisses are true and the conclusion false: so the argument is tautologically valid. Hence the argument (as encoded in **PL**) is plain valid. But is the translation accurate? Perhaps the first premiss is intended as an exclusive disjunction. However, this wouldn't in fact make a difference to the argument's tautological validity (check that claim). So on either reading, the original argument is valid too.

What about this argument:

G   It is not the case that either Jack or Jill will be at the party. At least one of Jack and James will be at the party. Hence it isn't the case that James and not Jill will be at the party.

Using the obvious translation, and setting out a truth-table (with the same 'conclusion-first' short cut) we get:

P	Q	R	¬(P ∨ Q)	(P ∨ R)	¬(R ∧ ¬Q)
T	T	T			T
T	T	F			T
T	F	T	F	T	F
T	F	F			T
F	T	T			T
F	T	F			T
F	F	T	T	T	F
F	F	F			T
			3	2	1

In this case, there *is* a 'bad line'. So the argument is not tautologically valid. And the bad assignment where P ⇒ F (Jack won't be at the party) and Q ⇒ F (Jill won't be there) and R ⇒ T (James *is* at the party) is evidently a genuine possibility – so we've found a way the premisses could be true and the conclusion false: the **PL** argument is invalid. Assuming that there are no worries about the translation, the original argument is invalid too.

Consider thirdly this example:

H   Jill's either in the library or she's in the coffee bar. And she's either not in the library, or she'll be working on her essay. It sure isn't the case that she's there in the coffee bar but not talking to her friends. So she's either working on her essay or talking to friends.

Is the inference here valid? Read the **PL** atoms to encode the following messages:

P: Jill is in the library.
Q: Jill is in the coffee bar.
R: Jill is working on her essay.
S: Jill is talking to her friends.

Then a possible translation for the first premiss is simply '(P ∨ Q)' – for after all, the premisses don't stop the coffee bar being in the library! If we take the 'And' at the beginning of the second premiss as just stylistic glue without logical function, then that premiss translates as '(¬P ∨ R)'. The third premiss translates as '¬(Q ∧ ¬S)', though we lose in translation the rhetorical flourish of 'It *sure* isn't the case that' when we replace it by plain negation, and we also lose any fine differences of tone between 'but' and plain conjunction. Finally, we could translate the conclusion as '(R ∨ S)'. So, *one* rendition of the original argument into **PL** is:

H′   (P ∨ Q), (¬P ∨ R), ¬(Q ∧ ¬S) ∴ (R ∨ S)

But compare §13.4 D: the premises here do tautologically entail the conclusion (for recall, the order in which premises are presented can't make any difference to questions of validity). So the **PL** inference is a valid one. Hence assuming the adequacy of the translation, the original argument is valid too.

But there are complications. It is probably more natural to suppose that the initial disjunctive premiss is meant exclusively; so it might be said that our first shot at a translation into **PL** is inadequate, and ((P ∨ Q) ∧ ¬(P ∧ Q)) would render the first premiss better. However, if that's the only change we make, the argument will remain tautologically valid. Adding a wff as a conjunct to a premiss has the same effect as adding the wff as a separate premiss (why?): and adding an extra premiss cannot destroy validity (why? see §13.8 below).

What, however, if the disjunctive conclusion is also intended exclusively? Then the argument translates as

H″   ((P ∨ Q) ∧ ¬(P ∧ Q)), (¬P ∨ R), ¬(Q ∧ ¬S) ∴ ((R ∨ S) ∧ ¬(R ∧ S))

and this is quickly shown by a truth-table test *not* to be tautologically valid. The valuation P ⇒ T, Q ⇒ F, R ⇒ T, S ⇒ T makes the premises true and the conclusion false. And this remains the case even if the second premiss is also read as an exclusive disjunction. So H″ isn't a tautological entailment. Moreover, the bad valuation which shows this corresponds to a situation which is evidently fully possible – namely the situation where Jill is in the library, not the coffee bar, working on her essay but talking to her friends at the same time. So *this* inference is not only non-tautological but plain invalid. In sum, then, our sample vernacular inference is valid if the conclusion is meant inclusively, invalid otherwise.

13.6 'Validity in virtue of form'

Let's gather together a number of observations about tautological entailment:

- To emphasize again: there are **PL** inferences which are valid entailments, but not tautological entailments. But the tautologically valid inferences are those that hold irrespective of the particular interpretations of the atoms, just in virtue of the distribution of the logical connectives in the premises and conclusion.

- Note that an inference like **A** or **C** remains tautologically valid however we re-interpret the atomic letters. And if we systematically replace the atoms in **A** or **C** with other wffs, again the result is a tautologically valid argument. So there is an implicit double generality in saying that e.g. **A** and **C** are valid (in short, what we said about tautologies in §12.2 evidently carries over to the idea of tautological validity).

- We suggested in the last chapter that the necessity of tautological wffs is relatively unmysterious, in some sense resulting from the stipulations governing the meanings of the connectives. Similarly, the validity of tautologically valid arguments is equally unmysterious (it isn't some metaphysical superglue that ties premisses to conclusions; it's a matter of the meanings of the connectives).

- Since the *content* of the atoms is irrelevant, we might say that the tautological entailments hold in virtue of their logical *form* (see §12.4 where we made a similar point about tautologies). The notion of form here is entirely transparent; for in **PL** where syntactic structure exactly reflects semantic structure, we can go by surface form. Start with a tautological entailment. Systematically replace its atoms with schematic variables, and this gives the form of the inference in the sense of encoding all the information that is needed to show that the inference is tautologically valid. And instances of this schematic form got by systematically substituting wffs for the variables will all be tautologically valid.

Note, that last point helps us see exactly how the truth-table test relates to the informal technique for the evaluation of inferences that we described in Chapter 4. Looking for a 'bad line' in a table is a stripped-down version of looking for a counterexample. To explain: recall that a counterexample is an inference of the same form as the target inference up for assessment, which has true premisses and a false conclusion. In the case of a **PL** argument, that would mean abstracting away from the particular atoms and then finding a substitution instance of the resulting schematic form with true premisses and a false conclusion. But in **PL**, all that matters for the truth of premisses and falsity of conclusion is the values of the atoms. So what ultimately matters about a counterexample is the values the atoms take in it. We can therefore streamline things by skipping the step of actually writing down the full counterexample and just latch on to the corresponding *valuation* of the atoms which makes the premisses true and conclusion false. *So in PL terms, looking for a counterexample comes down to looking for a 'bad line' in the truth-table.*

Now, as we noted in Chapter 4, failing to find a counterexample doesn't normally demonstrate validity. But in the special case of **PL** arguments, we can do a systematic search through all the relevantly different types of counterexample – i.e. all the different valuations of atoms that they produce. So in this special case, failure to find a counterexample, i.e. failure to find a bad valuation, *does* demonstrate validity.

To return to the point about 'form':

- We complained before (§3.6) about over-generalizing the thought that valid arguments are valid because of their form. But, just as the concept of a tautology gives sharp content to the idea of truth-in-virtue-of-form (§12.4), *the concept of tautological entailment gives a sharp content to the idea of validity-in-virtue-of-form for one class of arguments.*

As applied to arguments couched in **PL**, then, the idea of validity-in-virtue of form is clear-cut. What of arguments framed in English? Well, in so far as a vernacular argument can be well represented by a tautologically valid **PL** translation, it too might readily be counted as tautologically valid and so, in the explained sense, valid in virtue of form. (But questions of translation can be rather indeterminate, so the question whether a **PL** tautological entailment adequately captures the original vernacular argument can in turn be rather indeterminate.)

13.7 '⊨' and '∴'

That's the main business of this chapter done. However, there's another bit of standard shorthand symbolism we really should introduce and comment on. Then in the next section we'll use the symbolism in stating some general results about tautological validity. (But by all means skip this and the next section on a first reading).

It is standard to abbreviate the claim that the wffs $A_1, A_2, ..., A_n$ tautologically entail C by writing

(1) $A_1, A_2, ..., A_n \models C$

(The symbol '⊨' here is often called the 'double turnstile' – and perhaps a useful mnemonic is to think of it as made up by superimposing the two 't's from 'tautological'.)

Note, this symbol is being added to *English*. It is a metalinguistic shorthand for talking about the relationship of some wffs in **PL**, and it doesn't belong to **PL** itself. This point can get obscured because it is both convenient and absolutely conventional to write, e.g.

(2) $(P \lor Q), \neg P \models Q$

rather than write (as we should do, since we are talking metalinguistically)

(3)   '$(P \lor Q)$', '$\neg P$' $\models$ 'Q'

In other words, it is conventional to take '⊨' as – so to speak – generating its own quotation marks.

There is a very important contrast, then, between (2) and

(4) $(P \lor Q), \neg P \therefore Q$

(4) is an *argument couched in some dialect of* **PL**. So it makes *three* assertions, with the inference marker '∴' signalling that the third assertion is being inferred from the other two. By contrast, (2) is not an argument at all, it is a *single statement made in shorthand English*. It's short for

(2′)   The **PL** inference whose premisses are '$(P \lor Q)$' and '$\neg P$' and
        whose conclusion is 'Q' is tautologically valid.

So don't confuse '⊨' with '∴'!

Finally, note again that (1) says that C is always true, on the condition that the A_i are true: so

(5) $\models C$

is naturally read as saying that C is true on no condition at all, i.e. that C is a tautology.

13.8 Some simple metalogical results

In §§13.1–5, we've demonstrated a number of logical results, showing that various arguments are either tautologically valid or invalid. Finally in this chapter, we go up a level, so to speak, and give some *metalogical* results – i.e. results *about* the idea of tautological validity.

First, some further terminology that will come into its own later:

> A set of **PL** wffs A_1, A_2, ..., A_n is *satisfiable* or *tautologically consistent* if there is at least one possible valuation of the relevant atoms which makes the wffs all true together. The set is otherwise *unsatisfiable* or *tautologically inconsistent*.

To start, note that

(i) $A_1, A_2, ..., A_n \vDash C$ if and only if $A_1, A_2, ..., A_n, \neg C$ are tautologically inconsistent.

This is immediate. $A_1, A_2, ..., A_n \vDash C$ if and only if there is no valuation which makes the A_i true and C false, i.e. just when there is no valuation which makes the A_i and $\neg C$ all true, i.e. just when that set of wffs is unsatisfiable.

Next, note three more results:

(ii)   For any B, if $A_1, A_2, ..., A_n \vDash C$ then $A_1, A_2, ..., A_n, B \vDash C$.

(iii)   If C is a tautology, then $A_1, A_2, ..., A_n \vDash C$ for any A_i.

(iv)   If the wffs $A_1, A_2, ..., A_n$ are tautologically inconsistent, then $A_1, A_2, ..., A_n \vDash C$ for any C.

If a valuation of the relevant atoms which makes the A_i all true makes C true as well, then a valuation which makes the A_i all true and B as well will still make C true (note, if any of the atoms in B are new ones, their values cannot affect the values of the other wffs – see §9.6). In other words, (ii) adding further premisses cannot destroy a tautological entailment. Again, if C is true on any valuation of its atoms, then it is true on any valuation of a wider collection of atoms, and so true on any valuation which makes the A_i true. Hence (iii) a tautology is tautologically entailed by any bunch of premisses. Checking (iv) is left as an exercise. (These three results reflect the fact that tautological validity is a 'classical' notion of validity in the sense we explored in §6.1. Tautologically valid inference doesn't require any relation of relevance between the premisses and the conclusion.)

Next, recall the notion of truth-functional equivalence introduced in §11.6. A and B are equivalents just if they take the same value on all valuations of their atoms. So whenever A is true, B is, and vice versa. Hence we have

(v) A and B are equivalents just when $A \vDash B$ and $B \vDash A$.

And finally let's note a result that justifies the use of reductio ad absurdum arguments to establish tautological entailments (for the general idea of such arguments, see again §5.4):

(vi)   If C is a contradiction, and $A_1, A_2, ..., A_n, B \vDash C$, then $A_1, A_2, ..., A_n \vDash \neg B$

Suppose $A_1, A_2, ..., A_n, B \vDash C$. Then whenever $A_1, A_2, ..., A_n$ and B are all true, C is true too. But if C is a contradiction, C is never true on any valuation. So given C *is* a contradiction, then $A_1, A_2, ..., A_n$ and B can never all be true together on any valuation. But if $A_1, A_2, ..., A_n$ and B can never be true together, then whenever $A_1, A_2, ..., A_n$ *are* true, B is false. Which shows, as we wanted, that $A_1, A_2, ..., A_n \vDash \neg B$.

13.9 Summary

- Given an argument involving just 'and', 'or', 'not' as its relevant logical materials, we can evaluate it as follows:

  1. Represent the argument in **PL** using '$\wedge$', '$\vee$' and '$\neg$' as the connectives.
  2. Consider all the possible ways of assigning truth-values to the atoms in the premisses and conclusion.
  3. Work out in each case (for each possible assignment) the truth-values of the premisses and conclusion of the argument.
  4. Ask: Is there a *bad assignment*, i.e. one that makes the premisses all true and yet the conclusion false?
  5. If there is no bad assignment, then the argument is *tautologically valid*. If there is a bad assignment, then the argument is *tautologically invalid*.

- Steps (2) to (5) involve an entirely mechanical procedure; a computer can test whether an inference presented in **PL** is tautologically valid.

- As before, we are assuming that we are interpreting atoms of **PL** to express the sort of propositions which are determinately true or false. Tautological validity implies plain validity. Tautological invalidity does not always imply plain invalidity, but means that the original argument, if valid at all, must depend for its validity on something other than the distribution of the propositional connectives in the premisses and conclusion.

- The test procedure, however, can be speeded up if we note that a line on the truth-table where any premiss is false or the conclusion is true cannot be a bad line. Such lines can be ignored in the search for bad assignments.

- We introduced the symbol '$\vDash$' as an augmentation of English, to express the relation of tautological entailment. It is important to distinguish the metalinguistic '$\vDash$' from the object-language inference marker '$\therefore$'.

Exercises 13

A   Use the truth-table test to determine which of the following arguments are tautologically valid:

  1. $\neg(P \vee Q) \therefore \neg P$
  2. $(P \wedge Q) \therefore (P \vee \neg Q)$
  3. $(P \vee \neg Q), \neg P \therefore \neg \neg Q$

4.   (P ∧ S), ¬(S ∧ ¬R) ∴ (R ∨ ¬P)
5.   P, ¬(P ∧ ¬Q), (¬Q ∨ R) ∴ R
6.   (P ∨ Q), ¬(P ∧ ¬R), (R ∨ ¬Q) ∴ R
7.   (¬P ∨ ¬(Q ∨ R)), (Q ∨ (P ∧ R)) ∴ (¬P ∨ Q)
8.   (P ∨ Q), ¬(Q ∧ ¬¬R) ∴ ¬(R ∨ P)
9.   (P ∧ (Q ∨ R)) ∴ ((P ∧ Q) ∨ (P ∧ R))
10.   (P ∨ Q), (¬P ∨ R), ¬(Q ∧ S) ∴ ¬(¬R ∧ S)

B   Evaluate the following arguments:

1.   Either Jack went up the hill or Jill did. Either Jack didn't go up the hill or the water got spilt. Hence, either Jill went up the hill or the water got spilt.

2.   It isn't the case the Jack went up the hill and Jill didn't. It isn't the case that Jill went up the hill and the water got spilt. So either Jack didn't go up the hill or the water got spilt.

3.   Either Jack is lazy or stupid. It isn't true that he is both lazy and a good student. Either Jack isn't a good student or he isn't stupid. Hence he isn't a good student.

4.   Either Jill hasn't trained hard or she will win the race. It isn't true that she'll win the race and not be praised. Either Jill has trained hard or she deserves to lose. Hence either Jill will be praised or she deserves to lose.

C   (Now read §§13.7–8 if you took up the invitation to skip them on a first reading.) Which of the following are true; which false; and why?

1.   Any two tautologies are truth-functionally equivalent.
2.   If *A* tautologically entails *C*, then *A* entails any wff truth-functionally equivalent to *C*.
3.   If *A* and *B* tautologically entail *C* and *B* is a tautology, then *A* by itself tautologically entails *C*.
4.   If $A \vDash B$, then $\neg A \vDash \neg B$.
5.   If $A \vDash B$, then $\neg B \vDash \neg A$.
6.   If *A* is a contradiction, then $A \vDash B$, for any *B*.
7.   If the set of wffs $A_1, A_2, ..., A_n$ is unsatisfiable, the set $\neg A_1, \neg A_2, ..., \neg A_n$ is satisfiable.
8.   If the set of wffs $A_1, A_2, ..., A_n$ is unsatisfiable, then $A_1, A_2, ..., A_n \vDash B$, for any *B*.
9.   If $A \vDash C$, then $\vDash (\neg A \vee C)$.
10.   If *A*, *B*, *C* are tautologically inconsistent, then $A \vDash (\neg B \vee \neg C)$.
11.   If *A*, *B* are tautologically inconsistent, so are ¬*A*, ¬*B*.
12.   If the wffs *A*, *B* together make a satisfiable set, and *B*, *C* do too, then the wffs *A*, *C* together make a satisfiable set.

Propositional logic

Let's review the state of play again:

- We have explained why it can be advantageous to adopt a two-stage strategy in logical theory – first translating inferences into a well-designed artificial language (eliminating obscurity and ambiguity), and then evaluating the neatly regimented results.

- We have seen how to execute the first stage of this strategy in a very plausible and attractive way for arguments whose essential logical materials are the propositional connectives 'and', 'or' and 'not', by deploying the artificial truth-functional language **PL**.

- We then developed a tight definition of 'tautological validity' for arguments in **PL**; this is validity in virtue of the distribution of the connectives. And we have seen how we can mechanically test for tautological validity, by the truth-table test.

- We are assuming we are interpreting atoms of **PL** to express the sort of propositions which are determinately true or false. Then, if an argument is tautologically valid, it is valid in the informal sense – though not always vice versa.

Along the way, we have started the work of tidying some of the loose ends we noted in the last Interlude.

- We acknowledged in the previous Interlude that we hadn't given a very sharp characterization of the general notion of deductive validity. We now, at any rate, have a very clear notion of tautological validity that captures the intuitive notion as applied to one class of arguments.

- We also acknowledged earlier that we hadn't given any clear account of the idea of a proposition, nor of what counts as the 'logical form' of a proposition or the 'pattern' of an inference. But we can at least now talk crisply and accurately about forms of **PL** wffs or schematic patterns of **PL** inferences: we can display such patterns by replacing atoms with metalinguistic variables. The surface forms of **PL** wffs encode all the semantic facts (about the distribution of connectives) that matter for the tautological validity of **PL** inferences, transparently and without any ambiguity. So we don't need to dig any deeper. We don't need here any murkier notion of 'underlying' logical form. Nor indeed need we fuss about distinguishing (interpreted) wffs from the propositions they express.

Since **PL** is expressively adequate – i.e. has the resources to express any possible truth-functional way of combining atoms – there's a good sense in which our discussion so far covers the whole of the logic of truth-functions. Something that we haven't talked about yet is how we might express *proofs* (chains of tautologically valid inference steps) in **PL**. But in a way, we don't need to. The whole point of using proofs, recall, is to enable us to warrant complex inferences by breaking down the big inference step from premisses to conclusion into a number of smaller steps, each one of which is indubitably a valid move. However, we've just found a nicely mechanical and direct way of demonstrating tautological validity. Run a truth-table test and that settles the matter! So why bother with proofs?

Still, despite the fact that we have – in a sense – already dealt with the whole logic of truth-functions, we are going to proceed over the next few chapters to discuss one more truth-function in some detail. We'll be examining the so-called 'material conditional' which is a truth-functional connective that purports to capture (something of) the ordinary language conditional construction 'if *A*, then *B*'. Conditional propositions of that kind are so important in reasoning that it would be odd indeed to have no way of treating them in our logic. But as we'll see, they raise a whole raft of contentious problems (which is why we've kept their treatment sharply separate from the relatively uncontentious truth-functional logic of 'and', 'or' and 'not').

And despite the fact that it is – in a sense – otiose to lay down formal procedures for chaining together **PL** inferences into proofs, we'll also spend some time developing what can be regarded as one proof system for propositional logic (though we'll introduce it as a way of doing truth-tables faster). That's because later, when investigating 'quantificational logic' we'll *need* to consider proof systems (there is no equivalent of a mechanical truth-table test for establishing quantificational validity).

Overall, our menu

the basic syntax/semantics of **PL**
truth-table testing
discussion of the material conditional
a proof system

is, in one form or another, in one ordering or another, pretty much the standard offering in the first part of any elementary text on formal logic. Together, its elements constitute *classical propositional logic*.

PLC and the material conditional

We have seen how to evaluate arguments whose logical materials are the three connectives 'and', 'or' and 'not'. Can our truth-table techniques be extended to deal with arguments involving the conditional construction 'if ..., then ...'? Easily, *if* the core meaning of the conditional can be captured by a truth-function. The present chapter explores the only candidate truth-function, the so-called material conditional.

14.1 Why look for a truth-functional conditional?

Consider the following simple arguments:

A   If Jack bet on Eclipse, then Jack lost his money. Jack *did* bet on Eclipse. So Jack lost his money.

B   If Jack bet on Eclipse, then Jack lost his money. Jack did *not* lose his money. So Jack did not bet on Eclipse.

C   If Jack bet on Eclipse, then Jack lost his money. So if Jack didn't lose his money, he didn't bet on Eclipse.

D   If Jack bet on Eclipse, then Jack lost his money. If Jack lost his money, then Jack had to walk home. So if Jack bet on Eclipse, Jack had to walk home.

E   Either Jack bet on Eclipse or on Pegasus. If Jack bet on Eclipse, he lost his money. If Jack bet on Pegasus, he lost his money. So, Jack lost his money.

These are all intuitively valid inferences, and are presumably valid in virtue of what 'if ..., then ...' means (given the meanings of 'not' and 'either ... or ...').

Contrast the following arguments:

F   If Jack bet on Eclipse, then Jack lost his money. Jack lost his money. So Jack bet on Eclipse.

G   If Jack bet on Eclipse, then Jack lost his money. Jack did not bet on Eclipse. So Jack did not lose his money.

These are plainly fallacious arguments. The premisses of **F** can be true and conclusion false; foolish Jack may have lost his money by betting on Pegasus, another hopeless horse. For the same reason, the premisses of **G** can be true and conclusion false.

It would be good to have a systematic way of evaluating such arguments involving conditionals. Now, on the face of it, 'if ..., then ...' seems to be a connective like '... and ...' and '... or ...', at least in respect of combining two propositions to form a new one. So the obvious question is: can we directly extend our techniques for evaluating arguments involving those other two-place connectives to deal with the conditional as well?

Here is a quick reminder of how we dealt with arguments involving conjunction and disjunction.

- The strategy has two stages: (a) we translate the arguments into **PL**, and then (b) we evaluate the translated arguments, by running the truth-table test.

- Step (a) presupposes that the ordinary language connectives 'and' and 'or' are close in their core meanings to the corresponding connectives of **PL**. (If the translation were too loose, then a verdict on the **PL** rendition couldn't be carried back into a verdict on the original argument.)

- Step (b) depends on the **PL** connectives being truth-functional. The truth-table test requires that, given any assignment of values to the relevant propositional atoms, we can always work out whether or not that assignment makes the premisses and conclusion false.

So, if the approach of the previous chapters is to be extended to cover arguments involving the conditional, we need to find some way of rendering conditional statements with tolerable accuracy using a dyadic truth-function. Can we do the trick?

14.2   Introducing the material conditional

First, let's introduce some entirely standard terminology:

> Given a conditional *if A then C*, we refer to the 'if' clause *A* as the *antecedent* of the conditional, and to the other clause *C* as the *consequent*.
>
> An inference step of the form *if A then C, A; hence C* is standardly referred to by the label *modus ponens*. An inference step of the type *if A then C, not-C; hence not-A* is called *modus tollens*.

Those two medieval Latin labels for the valid inference-moves in our examples **A** and **B** above are still in such common use that you just need to learn them.

Now, it is quite easy to see that there is only one truth-function which is a serious candidate for capturing conditionals: this much is agreed on all sides. We'll first give a longer argument which identifies this truth-function by exhaustion of cases, and then a second, shorter, argument for the same conclusion.

(1) Let's for the moment informally use the symbol '⊃' to represent a truth-function which is intended to be conditional-like. The issue is how to fill in the '?'s in its truth-table. To repeat, there must be no indeterminate slots left in this table: if there were, we wouldn't be able to apply the truth-table test to this conditional connective, and the whole point of the current exercise would be thwarted.

A	C	$(A \supset C)$
T	T	?
T	F	?
F	T	?
F	F	?

Step one Suppose we claim *if A, then C*. Then plainly we are ruling out the case where the antecedent *A* holds and yet the consequent *C* doesn't hold – i.e. we are ruling out the situation $A \Rightarrow$ T, $C \Rightarrow$ F. Putting it the other way about, if we *do* have $A \Rightarrow$ T, $C \Rightarrow$ F, then it would plainly be wrong to claim *if A, then C*. Hence, if '⊃' is to be conditional-like, $A \Rightarrow$ T, $C \Rightarrow$ F must likewise make $(A \supset C)$ false. Hence the entry on the second line of the table has to be F.

A	C	$(A \supset C)$
T	T	?
T	F	F
F	T	?
F	F	?

Step two A main point (maybe *the* main point) of having a conditional construction in the language is to set up modus ponens inferences, from the premisses *if A then C* and *A* to the conclusion *C*. In asserting *if A then C*, I am as it were issuing a promissory note, committing myself to assent to *C* should it turn out that *A is* true.

But if modus ponens inferences are ever to be sound, it must be possible for the two premisses *if A then C* and *A* to be true together. Hence the truth of *A* can't always rule out the truth of *if A then C*.

Similarly, if '⊃' is to be conditional-like, and so feature in useful modus ponens inferences, the truth of *A* can't always rule out the truth of $(A \supset C)$. That means that the first line of the table for '⊃' *can't* also be F. Hence, the entry on first line needs to be T.

Step three The table for $(A \supset C)$ must therefore be completed in one of the ways (a) to (d).

We can now argue simply by elimination:

A	C	$(A \supset C)$			
T	T	T			
T	F		F		
F	T	T	T	F	F
F	F	T	F	T	F
		a	b	c	d

- Column (*d*) can't be right: for then the full truth-table would be identical to the table for $(A \wedge C)$, and *if*s aren't *and*s.

- Column (*b*) can't be right because that would make $(A \supset C)$ truth-functionally equivalent to plain *C*; and conditionals aren't always equivalent to their consequents (compare '*If* Jack bet on Eclipse, he lost his money' and plain 'Jack lost his money').

- Column (*c*) won't do either: for that column would make $(A \supset C)$ always equivalent to $(C \supset A)$, and real conditionals are not in general reversible like that (compare 'If Jo is a man, Jo is a human being' and 'If Jo is a human being, Jo is a man').

• And so ... cue drums and applause ... the winner is column (*a*).

(2) Here's a shorter argument for the same result. A proposition like *If Jo is tall and dark then Jo is tall* is a necessary truth. Now suppose we translate it **PL**-style, using '⊃' for a truth-functional conditional. We get, say, '((P ∧ Q) ⊃ P)'. This too should be a necessary truth, true in virtue of the truth-functional meanings of '∧' and '⊃'; so it should in fact be a *tautology*. Consider, therefore, the following truth-table:

P	Q	(P ∧ Q)	P	((P ∧ Q) ⊃ P)
T	T	T	T	T
T	F	F	T	T
F	T	F	F	T
F	F	F	F	T

In the right-hand column we record the desired result that this conditional is a tautology. The middle columns give the values of the antecedent '(P ∧ Q)' and the consequent 'P' of that conditional. And by inspection, to get from the assignments in the *middle* columns to the values in the *final* column, a truth-functional conditional (*A* ⊃ *C*) must be true when *A* and *C* are both true; when *A* is false, and *C* true; and when *A* is false and *C* false. That fixes three lines of the table. Which leaves the remaining case when *A* is true, *C* is false: in that case (*A* ⊃ *C*) must be false (otherwise *all* conditionals will be tautologous). So, again, ...

The only candidate that even gets to the starting line as an approximately conditional-like truth-function is the *material conditional*, defined by:

A	*C*	(*A* ⊃ *C*)
T	T	T
T	F	F
F	T	T
F	F	T

Why '*material* conditional'? Let's not worry about that. It's one of those standard labels that has stuck around, long after the original connotation has been forgotten. (This truth-function is also sometimes known as the *Philonian conditional*, after the ancient Stoic logician Philo, who explicitly defined it.)

14.3 '⊃' is conditional-like

A bit later (§14.5 below), we'll officially add the horseshoe symbol '⊃' to **PL**, adjusting the syntax and semantics of that language to get an extended language we'll call **PLC**. But for the moment we'll proceed casually, pretend that '⊃' is already part of our language, and quickly check that if we translate the examples in §14.1 using '⊃', and run a truth-table test, we get the right verdicts. Thus,

A   If Jack bet on Eclipse, then Jack lost his money. Jack *did* bet on Eclipse. So Jack lost his money.

Translation: (P ⊃ Q), P ∴ Q

Truth-table test:

P	Q	(P ⊃ Q)	P	Q
T	T	T	T	T
T	F	F	T	F
F	T	T	F	T
F	F	T	F	F

There are no bad assignments, where all the premisses are true and the conclusion is false: so argument **A** is tautologically valid as well as being intuitively valid. Skipping the next two simple cases (left as exercises!), we next treat

D   If Jack bet on Eclipse, then Jack lost his money. If Jack lost his money, then Jack had to walk home. So if Jack bet on Eclipse, Jack had to walk home.

Translation: (P ⊃ Q), (Q ⊃ R) ∴ (P ⊃ R)

Truth-table test:

P	Q	R	(P ⊃ Q)	(Q ⊃ R)	(P ⊃ R)
T	T	T			T
T	T	F	T	F	F
T	F	T			T
T	F	F	F		F
F	T	T			T
F	T	F			T
F	F	T			T
F	F	F			T
			2	3	1

Here we use the now familiar short cut of evaluating the conclusion first, and then only pursuing potential bad lines, i.e. lines where the conclusion is false. There are no bad lines, so again the argument is tautologically valid.

The next argument exemplifies a common form of inference, often called *argument by cases*:

E   Either Jack bet on Eclipse or on Pegasus. If Jack bet on Eclipse, he lost his money. If Jack bet on Pegasus, he lost his money. So, Jack lost his money.

Translation: (P ∨ Q), (P ⊃ R), (Q ⊃ R) ∴ R

Truth-table test:

P	Q	R	(P ∨ Q)	(P ⊃ R)	(Q ⊃ R)	R
T	T	T				T
T	T	F	T	F		F
T	F	T				T
T	F	F	T	F		F
F	T	T				T
F	T	F	T	T	F	F
F	F	T				T
F	F	F	F			F
			2	3	4	1

Again the argument is tautologically valid. So turn next to

F   If Jack bet on Eclipse, then Jack lost his money. Jack lost his money. So Jack bet on Eclipse.

We noted before that the inference here is a gross fallacy (traditionally called *affirming the consequent*). And the translation of **F** is tautologically *invalid*.

Translation: (P ⊃ Q), Q ∴ P

Truth-table test:

P	Q	(P ⊃ Q)	Q	P
T	T	T	T	T
T	F	F	F	T
F	T	T	T	F
F	F	T	F	F

The bad line here evidently corresponds to a possible state of affairs – Jack's not betting on Eclipse yet losing his money all the same – so that confirms that the original argument is plain invalid.

Likewise for

G   If Jack bet on Eclipse, then Jack lost his money. Jack did not bet on Eclipse. So Jack did not lose his money.

That's another horrible fallacy (traditionally called *denying the antecedent*). Translating and running a truth-table test confirms this.

In sum, we see from examples **A** to **G** that the representation of some ordinary conditionals using the truth-functional conditional '⊃' yields the right verdicts about validity and invalidity. So at least in some respects, '⊃' approximates to the vernacular 'if ..., then ...'. In the next chapter, we consider at length just how close this approximation is.

14.4 'If', 'only if', and 'if and only if'

So far, we have only discussed 'if' conditionals, i.e. conditionals of the forms

(i)   If *A* then *C*.

We can also write 'if' conditionals in the form

(i′)   *C*, if *A*.

We take these two versions, without further ado, to be mere stylistic variants. However, there are other kinds of conditional which need to be discussed, in particular 'only if' conditionals. Consider the pair

(ii)   *A* only if *C*.
(ii′) Only if *C*, then *A*.

Again, we'll take this second pair to be trivial variants of each other. The interesting question concerns the relation between (i) *if A then C* and (ii) *A only if C*.

Consider the following line of thought. Given *if A then C*, this means that *A*'s truth implies *C*, so we will only have *A* if *C* obtains as well – i.e. *A only if C* holds. Conversely, given *A only if C*, that means that if *A* is true, then, willy-nilly, we get *C* as well. So plausibly, (i) and (ii) imply each other. To take an example, consider

Hi   If Einstein's theory is right, space-time is curved.

Hii   Einstein's theory is right only if space-time is curved.

These two do indeed seem equivalent.

However, we can't always equate *if A then C* with the corresponding *A only if C* in this simple way. Compare:

Ii   If I drink malt whisky, then I get stupid.
Iii   I drink malt whisky only if I get stupid.

The latter, but not the former, surely suggests that I have to be stupid first before I make things worse by turning to the whisky. And note another complexity: these conditionals are naturally read to be about general habits, not about particular situations. But we'll set aside such quirks of the vernacular. Take the cases where *A only if C* and *if A then C* do come to the same; and then both will be rendered into our extension of **PL** using the material conditional, by something of the shape $(A \supset C)$, with the antecedent and consequent translated appropriately.

Be careful: translating 'only if' conditionals does get many beginners into a tangle! The basic rule is: first put 'only if' conditionals into the form *A only if C* and then replace the 'only if' with the horseshoe '⊃', to get $(A \supset C)$. The conditional *A if C* is of course translated the other way about, since that is trivially equivalent to *if C then A* and so is to be rendered by $(C \supset A)$.

What then about the two-way conditional or *biconditional*, i.e. *A if and only if C* (quite often, in philosophical English, this is written *A iff C*)? This is equivalent to the conjunction of *A only if C* and *A if C*. The first conjunct we've just suggested should be translated into our formal language as $(A \supset C)$; the second conjunct is to be rendered by $(C \supset A)$. Hence the whole biconditional can be rendered by $((A \supset C) \wedge (C \supset A))$

However that is a bit cumbersome. And so, although it isn't necessary, it is common to introduce another dyadic truth-functional connective, '$\equiv$', to replace that construction. This, then, is the so-called *material biconditional*; and a simple calculation shows that its truth-table will be as displayed.

Finally, we should pause to connect the use of conditionals and biconditionals with talk of *necessary* and *sufficient* conditions. A condition C is necessary for A to hold just when we have A *only if* C. For example, it is a necessary condition of Jo's being a sister that Jo is female – for Jo is a sister only if she is female. But being female isn't a sufficient condition for being a sister: some

A	C	$(A \equiv C)$
T	T	T
T	F	F
F	T	F
F	F	T

females are only children. A condition C is sufficient for A to hold just when we have *if C then A*. For example, it is a sufficient condition for Jo's being a sister that she is a female twin – for if Jo is a female twin, then she is a sister. But being a female twin isn't a necessary condition for being a sister: some sisters aren't twins. Finally, C is a necessary *and* sufficient condition for A just when A *if and only if* C.

14.5 The official syntax and semantics of PLC

The material conditional – like any truth-function – can be expressed using the existing resources of **PL**. For example it is easily checked that $(\neg A \vee C)$ has the right truth-table to do the job. Likewise $\neg(A \wedge \neg C)$ has the same truth-table.

Similarly the material biconditional can be expressed using '$\wedge$', '$\vee$' and '$\neg$'. For example, we could directly conjoin two translations of *if A then C* and *if C then A* to get something of the form $((\neg A \vee C) \wedge (\neg C \vee A))$. Or looking at the truth-table for the biconditional, we see that $((A \wedge C) \vee (\neg A \wedge \neg C))$ also has the right table. So we certainly don't *need* to add special symbols for the conditional or biconditional.

A	C	$(\neg A \vee C)$
T	T	T
T	F	F
F	T	T
F	F	T

But there are trade-offs between various kinds of simplicity here. On the one hand, there is the pleasing economy of having just three connectives in our language: the cost is that the translation of conditionals and biconditionals into standard **PL** isn't particularly pretty. On the other hand, if we add new symbols for the conditionals, we thereby get neater translations: the cost will be that when we develop formal proof devices for dealing with arguments things get a bit more complicated (e.g. we need more inference rules). On balance, however, it is worth taking the second trade: so we'll now officially expand **PL** by adding two new connectives for the material conditional and biconditional. We'll call the new language **PLC**. Its additional syntactic and semantic rules should be obvious, but here they are, for the record.

Syntax   The full alphabet of **PLC** is the same as that for **PL** (see §8.1), with the addition of the two symbols '⊃' and '≡'.

The rules for what counts as an *atomic* wff in **PLC** stay the same. But we need to add two new rules for building complex molecular wffs with '⊃' and '≡'. Syntactically, however these behave like the other dyadic connectives. So the rules (in short form) are W1 to W4 as before (§8.1), plus the two new rules:

> (W1)   Any atomic wff is a wff.
> (W2)   If A is a wff, so is $\neg A$.
> (W3)   If A and B are wffs, so is $(A \wedge B)$.
> (W4)   If A and B are wffs, so is $(A \vee B)$.
> (W5)   If A and B are wffs, so is $(A \supset B)$.
> (W6)   If A and B are wffs, so is $(A \equiv B)$.
> (W7)   Nothing else is a wff.

Semantics   We'll say quite a bit more about the pros and cons of the translation of '⊃' as 'if' in the next chapter. Here we'll concentrate on valuations (see §9.3). The rules for evaluating wffs with the old three connectives stay the same, of course. So we just need to add:

(P4)   For any wffs, A, B: if $A \Rightarrow$ T and $B \Rightarrow$ F then $(A \supset B) \Rightarrow$ F; otherwise $(A \supset B) \Rightarrow$ T.

(P5)   For any wffs, A, B: if both $A \Rightarrow$ T and $B \Rightarrow$ T, or if both $A \Rightarrow$ F and $B \Rightarrow$ F, then $(A \equiv B) \Rightarrow$ T; otherwise $(A \equiv B) \Rightarrow$ F.

In Chapter 9, we gave rules in this style first and then summed them up in truth-tables. This time, we've already given the truth-tables for '⊃' and '≡', and these new semantic rules evidently reflect those tables.

Finally, just for the record, the notions of a tautology and of tautological entailment are defined for **PLC** exactly as they were for **PL**. A **PLC** wff is a tautology if it is true on all valuations of its atoms; a **PLC** inference is tautologically valid if there is no valuation of the atoms in the premisses and conclusion which makes the premisses true and conclusion false.

14.6   '⊢', '∴', '⊃', '≡', and '→'

Recall that in §13.7 we drew a very sharp distinction between the object language inference marker '∴' and the metalinguistic sign '⊢' for the entailment relation. We now have another sign we must also distinguish very clearly from *both* of these, namely the object language conditional '⊃'. Let's spell things out carefully.

As we stressed before

(1) $(P \wedge Q) \therefore Q$

is a mini-*argument* expressed in **PL** (or in **PLC**, since this extended language includes the original one). Someone who asserts (1), assuming that some content has been given to the atoms, is asserting the premiss, asserting the conclusion, and indicating that the second assertion is based on the first. Compare the English *Jack is a physicist and Jill is a logician. Hence Jill is a logician.*

By contrast

(2) $((P \wedge Q) \supset Q)$

is a *single* proposition in **PLC**; someone who asserts (2) asserts neither '$(P \wedge Q)$' nor 'Q'. Compare the English *If Jack is a physicist and Jill is a logician, then Jill is a logician.*

(3) $(P \wedge Q) \vDash Q$

is another single proposition, but this time in our extended English metalanguage, and the **PLC** wffs there are not used but mentioned. For (3) just abbreviates the English

(3')   '$(P \wedge Q)$' tautologically entails 'Q'.

Note too the difference between the grammatical category of the **PLC** connective '$\supset$' and the English-augmenting sign '$\vDash$'. The horseshoe sign is a *connective*, which needs to be combined with two sentences (wffs) to make a compound sentence. The double turnstile expresses a *relation* between a premiss (or a set of premisses) and a conclusion, and needs to be combined with two denoting expressions referring to wffs (or sets of wffs) to make a sentence.

There's a bad practice of blurring these distinctions by talking, not of the 'material conditional', but of 'material implication', and then reading $(A \supset B)$ as '*A materially implies B*'. If we also read $A \vDash B$ as '*A logically implies B*', this makes it sound as if $(A \supset B)$ is just the same sort of claim as $A \vDash B$, only weaker. But not so. Compare (2) and (3) again: one involves an object-language connective, the other a metalanguage relation.

The following important result, however, does link 'logical implication' in the sense of entailment to 'material implication':

E/C $A \vDash C$ if and only if $\vDash (A \supset C)$.

Or in plain words: the one premiss argument $A \therefore C$ is tautologically valid iff (i.e. if and only if) the corresponding material conditional $(A \supset C)$ is a tautology.

Proof: $A \vDash C$, i.e. $A \therefore C$ is tautologically valid,
   iff there is no assignment of values to the atoms which appear in A and C which makes A true and C false (by definition),
   i.e. iff there is no assignment of values to the atoms which appear in A and C which makes $(A \supset C)$ false (by the truth-table for '$\supset$'),
   i.e. iff every assignment of values to the atoms which appear in A and C makes $(A \supset C)$ true,
   i.e. iff $(A \supset C)$ is a tautology (by definition).

By similar reasoning, we can show that the argument $A_1, A_2, ..., A_n \therefore C$ is valid just when $\{(A_1 \wedge A_2 \wedge ... \wedge A_n) \supset C\}$, properly bracketed, is a tautology.

Here's a related result, relating equivalence to the biconditional:

E/B A and C are truth-functionally equivalent if and only if $\vDash (A \equiv C)$.

Or in plain words: two wffs *A* and *C* have the same truth-table if and only the corresponding single wff (*A* ≡ *C*) is a tautology.

Proof: *A* and *C* are truth-functionally equivalent
    iff every assignment of values to the atoms which appear in *A* and *C* gives the same values to *A* and to *C* (by definition),
    i.e. iff every assignment of values to the atoms which appear in *A* and *C* makes (*A* ≡ *C*) true (by the truth-table for '≡'),
    i.e. iff (*A* ≡ *C*) is a tautology (by definition).

Finally, we should note that '→' is very often used as an alternative symbol for the conditional. But I recommend the convention of using '⊃' when the truth-functional material conditional is definitely meant, keeping '→' for more general use.

14.7 Summary

- The only truth-functional connective that is a candidate for translating the conditional of ordinary discourse is the so-called *material conditional*. If we represent this connective by '⊃', then (*A* ⊃ *C*) is true just when *A* is false or *C* is true.

- A number of ordinary-language arguments using conditionals can be rendered using '⊃' while preserving their intuitive validity/invalidity.

- In many cases, corresponding pairs of the form *if A then C* and *A only if C* are equivalent. (But of course *if A then C* and *only if A then C* are not equivalent and their conjunction yields a biconditional.)

- The only truth-functional connective that is a candidate for translating the biconditional *A if and only if C* is the material biconditional. Representing this by '≡', (*A* ≡ *C*) is true just when *A* and *C* are both true or both false.

- Since **PL** is expressively complete, we don't *need* to add new symbols to the language to express the material conditional and material biconditional. But it is convenient to do so. The extended language, with the symbols '⊃' and '≡' added in the obvious ways, will be termed **PLC**.

- It is important to distinguish the metalinguistic '⊢' from the object-language conditional '⊃' and the object-language inference marker '∴'.

Exercises 14

A   Given

        'P' expresses *Plato is a philosopher*
        'Q' expresses *Quine is a philosopher*
        'R' expresses *Russell is a philosopher*
        'S' expresses *Socrates is a philosopher*

translate the following into **PLC** as best you can:

1. Quine is a philosopher only if Russell is.
2. If either Quine or Plato is a philosopher, so is Russell.
3. Only if Plato is a philosopher is Russell one too.
4. Quine and Russell are both philosophers only if Socrates is.
5. Russell's being a philosopher is a necessary condition for Quine's being one.
6. Plato is a philosopher if and only if Quine isn't.
7. If Plato is a philosopher if and only if Socrates is, then Russell is one too.
8. Only if either Plato or Russell is a philosopher are both Quine and Socrates philosophers.
9. That Socrates is a philosopher is a sufficient condition for it to be a necessary condition of Quine's being a philosopher that Russell is one.
10. Quine is a philosopher unless Russell is one.
11. Provided that Quine is a philosopher, Russell is one too.

B   Use truth-tables to determine which of the following arguments are tautologically valid:

1.   P, (P ⊃ Q), (Q ⊃ R) ∴ R
2.   P, (P ≡ Q), (Q ≡ R) ∴ R
3.   (P ⊃ R) ∴ ((P ∧ Q) ⊃ R)
4.   ¬R, (¬P ⊃ R) ∴ P
5.   (P ≡ R) ∴ (¬R ≡ ¬P)
6.   (P ⊃ R) ∴ (¬R ⊃ ¬P)
7.   ¬R, (P ⊃ R), (¬P ⊃ Q) ∴ Q
8.   (P ∨ Q), (P ⊃ R), ¬(Q ∧ ¬R) ∴ R
9.   (R ≡ (¬P ∨ Q)), ¬(P ∧ ¬R) ∴ ¬(¬R ∨ Q)
10. (¬P ∨ Q), ¬(Q ∧ ¬R) ∴ (P ⊃ R)
11. (P ∧ ¬R), (Q ⊃ R) ∴ ¬(P ⊃ Q)
12   ¬(P ≡ (Q ∧ R)), (S ∨ ¬Q) ∴ ¬(S ⊃ P)

C   Which of the arguments that you have just shown to be tautologically valid correspond to intuitively valid arguments when '⊃' is replaced by 'if ..., then ...', and '≡' is replaced by '... if and only if ...'?

D   Which of the following are true and why?

1.   If *A* doesn't tautologically entail *B*, then ⊨ ¬(*A* ≡ *B*).
2.   If *A* and *B* tautologically entail each other, then ⊨ (*A* ≡ *B*).
3.   If ¬*A* ⊨ ¬*B*, then ⊨ (*B* ⊃ *A*).
4.   If *A*, *B* ⊨ *C* then *A* ⊨ (*B* ⊃ *C*)
5.   If ⊨ (*A* ⊃ *B*) and ⊨ (*B* ⊃ *C*), then ⊨ (*A* ⊃ *C*).
6.   If ⊨ (*A* ⊃ *B*) and ⊨ (*A* ⊃ ¬*B*), then ⊨ ¬*A*.
7.   If ⊨ (*A* ≡ *B*), then *A*, *B* are tautologically consistent.
8.   If ⊨ (*A* ≡ ¬*B*), then *A*, *B* are tautologically inconsistent.

More on the material conditional

The last chapter introduced the material conditional. This chapter discusses more carefully its relation with conditionals in ordinary language. We begin by drawing a key distinction between types of conditional.

15.1 Types of conditional

Consider the following memorable example (due to David Lewis):

(1)   If kangaroos had no tails, they would topple over.

How do we decide whether this is true? By imagining a possible world very like our actual world, with the same physical laws, and where kangaroos are built much the same except for the lack of tails. We then work out whether the poor beasts in such a possible world would be unbalanced and fall on their noses. In short, we have to consider not the actual situation (where of course kangaroos do have tails), but other possible ways things might have been. Let's call a conditional that invites this kind of evaluation (i.e. evaluation by thinking about other ways things might have been) a *possible-world conditional*.

To introduce another memorable example, compare the following:

(2)   If Oswald didn't shoot Kennedy in Dallas, someone else did.
(3)   If Oswald hadn't shot Kennedy in Dallas, someone else would have.

To evaluate (2), note that in the actual world, Kennedy was shot. Someone must have done it. Hence, if not Oswald, someone else. So (2) is true. Now in fact (let's suppose), Oswald *did* shoot Kennedy, acting alone. Then, to decide whether (3) is true, we have to consider a non-actual possible world, a world like this one except that Oswald missed. Keeping things as similar as we can to the actual world, Oswald would still be acting alone. There would still be no conspiracy, no back-up marksmen. In such a possible situation, Kennedy would have left Dallas unscathed. Hence (3) is false.

Given that they take different truth-values, the possible-world conditional (3) therefore must diverge in content from the simple conditional (2).

The logic of possible-world conditionals is beyond the scope of this book, but one thing is clear: *possible-world conditionals are not truth-functional*. A truth-functional conditional has its actual truth-value fixed by the *actual* truth-values of its constituent sentences. By contrast, the truth-value of a possible-world conditional depends on what happens in *other* possible scenarios; so its truth-value can't be fixed just by the this-worldly values of the antecedent and consequent.

Hence, quite uncontroversially, the truth-functional rendition can at most be used for conditionals like (2).

What *is* very controversial, though, is the range of 'conditionals like (2)'. Cases like (3) are conventionally called *subjunctive* conditionals, for supposedly they are couched (according to some traditional grammars) in the subjunctive mood. (2), and all the conditionals in arguments **A** to **G** in §14.1 above, are called *indicative* conditionals (being framed in the indicative mood). The trouble is that the intended indicative/subjunctive division seems to carve up conditionals in an unhelpful way. For compare (3) with

(4)   If Oswald doesn't shoot Kennedy in Dallas, someone else will.

Someone who asserts (4) before the event – believing, perhaps, that a well-planned conspiracy is afoot – will after the event assent to (3). Likewise, someone who denies (4) before the event – knowing that Oswald is a loner – will deny (3) after the event. Either way, it seems that exactly the same type of possible world assessment is relevant for (3) *and* (4): we need to consider what occurs in a world as like this one as possible given that Oswald doesn't shoot Kennedy there. Plausibly, then, we might well want to treat (3) and (4) as belonging to the *same* general class of possible-world conditionals. But, in traditional grammatical terms, (4) is *not* subjunctive. So the supposed distinction between subjunctive and indicative conditionals divides the past interpretation of (3) from the future interpretation of (4) in a pretty unnatural manner.

We can make a related point in passing. Someone who asserts (2) leaves it open whether Oswald shot Kennedy; but someone who asserts (3) thereby implies that the antecedent is actually false, and that Oswald didn't miss. For that reason, conditionals like (3) are often called *counterfactual* conditionals. But (4) is *not* a counterfactual in this sense: someone who asserts (4) again leaves it open whether Oswald will miss. So the counterfactual/non-counterfactual distinction again unnaturally divides (3) from (4).

We can't pursue these matters further here. The point, to repeat, is that it isn't obvious exactly where and how to draw the line between those possible-world conditionals like (3) which have to be set aside as *definitely* not apt for a truth-functional treatment, and those 'indicative' conditionals like (2) which are still in the running.

15.2   In support of the material conditional

We noted that the material conditional (like any truth-function) can be expressed with the unaugmented resources of **PL**; the two simplest ways of capturing the truth function using the original connectives are $\neg(A \wedge \neg C)$ and $(\neg A \vee C)$. Do these equivalent ways of expressing the truth-function make it look more or less like a conditional? Well, consider the following two lines of thought:

- *If A then C* rules out having *A* true while *C* is false; hence, borrowing the **PL** symbols for brevity, *if A then C* implies the corresponding $\neg(A \wedge \neg C)$. Suppose conversely that we believe that $\neg(A \wedge \neg C)$, i.e. that we think that we don't have *A* true while *C* is false. That seems to imply that if *A is* true, then *C* won't be false, or in other words, *if A then C*. So, apparently, *if A then C* implies and is implied by $\neg(A \wedge \neg C)$. Which is to say that the ordinary language conditional is equivalent to the material conditional.

- Suppose we hold that *if A then C,* and put that together with the law of excluded middle ($\neg A \vee A$). Then we either have $\neg A$, or we have *A* and so *C*. In other words, *if A then C* implies ($\neg A \vee C$). Conversely, suppose we hold ($\neg A \vee C$); if *A* is in fact true, then – since that rules out the first disjunct – we can conclude *C* must be true as well. So, apparently, *if A then C* implies and is implied by ($\neg A \vee C$). Which is again to say that the ordinary language conditional is equivalent to the material conditional.

These arguments are enough to lead many logicians to hold that at least *some* ordinary 'if ..., then ...' propositions *are* fundamentally truth-functional.

15.3 Against identifying vernacular and material conditionals

Do the arguments of the previous section really establish that the core meaning of indicative conditionals is truth-functional (i.e. that these ordinary language conditionals are material conditionals)?

Note that a material conditional, a proposition with the same truth-table as ($\neg A \vee C$), can be true irrespective of any connection between the matters mentioned by *A* and *C*. All that's required is *A* is false and/or *C* is true. You might hold that this simple observation is already enough to sabotage the claim that ordinary 'if's can be treated as equivalent to material conditionals. For don't 'if's signal some kind of real *connectedness* between the topics of the antecedent and consequent?

No, not always. Suppose I remember that one of Jack and Jill has a birthday in April (I know that they threw a party last Easter, but I just can't recall whose birthday it was celebrating). You say, a bit hesitantly, that you think that Jack has an October birthday. So I respond

(1)   If Jack *was* born in October, then Jill was born in April.

Surely in the context that's a perfectly acceptable conditional, though I don't for a moment think that there is any causal or other connection between the facts about their birth months. All I'm saying is that either you are wrong about Jack, or else Jill was born in April: i.e., borrowing the **PL** symbols again, I'm just committing myself to

(2)   ($\neg$Jack was born in October $\vee$ Jill was born in April).

So it seems that perfectly acceptable conditionals may be no-more-than-material.

Of course, when we assert conditionals, we often *do* think that there is (for example) a causal connection between the matters mentioned in the antecedent and the consequent ('If you press that button, it will turn on the computer'). But equally, when we assert disjunctions, it is often because we think there is some mechanism ensuring that one disjunct or the other is true ('Either the e-mail will be sent straight away or you'll get a warning message that the e-mail is queued'). However, even if our *ground* for asserting that disjunction is our belief in an appropriate mechanism relating the disjuncts, what we actually *say* is true so long as one or other disjunct holds (exclusively, in our example). Likewise, our *reason* for asserting a conditional may be a belief in some mechanism that ensures that if the antecedent holds the consequent does too. It could still be the case that what we *say* is true just so long as the material conditional holds.

So far, the material conditional theory survives. But now compare the following (we'll borrow '⊃' to signal the material conditional in English):

(3)   Wordsworth didn't write *Pride and Prejudice.* So, (Wordsworth wrote *Pride and Prejudice* ⊃ Wordsworth wasn't a novelist).

(4)   Wordsworth didn't write *Pride and Prejudice.* So, if Wordsworth wrote *Pride and Prejudice* then he wasn't a novelist.

Grant (3)'s premiss that Wordsworth didn't write *Pride and Prejudice.* The conclusion is of the form $(A \supset C)$ with a false antecedent, so the conclusion must be true too. Hence the inference in (3) is valid. That is no more puzzling than the validity of the inference from $\neg A$ to $(\neg A \vee C)$; indeed, (3) in effect just *is* that trivial inference. By contrast, the inference in (4) looks decidedly unhappy. Many will say: its conclusion looks unacceptable (being the author of *Pride and Prejudice* would surely be enough to make you a novelist!): so how can it validly follow from the true premiss that Wordsworth didn't write *Pride and Prejudice?*

Generalizing, the inference pattern

(M)   *not-A; so $(A \supset C)$*

is unproblematically and trivially reliable. On the other hand, inferences of the type

(V)   *not-A; so if A then C*

strike us as very often unacceptable (for indicative conditionals as well as subjunctive ones). Equating the vernacular conditional in (V) with the material conditional in (M) has the highly implausible upshot that the two inference patterns should after all be exactly on a par.

Now we seem to be in a mess. On the one hand, we presented in the last section what seemed a good argument for saying that *if A then C* implies and is implied by $\neg(A \wedge \neg C)$, and then another good argument for saying that *if A then C* implies and is implied by $(\neg A \vee C)$, where $\neg(A \wedge \neg C)$ and $(\neg A \vee C)$ are just two ways of expressing the material conditional. Yet here, on the other hand, we've emphasized what looks to be a *very* unwelcome upshot of treating ordinary 'if's as equivalent to material '⊃'s. What to do?

15.4 Robustness

Go back to a remark made in passing in §14.2. I said: when we claim *if A then C*, we are normally issuing a promissory note, committing ourselves to assent to C should it turn out that *A* is true.

Now imagine that I currently hold *not-A*, and on that ground alone hold ($\neg A \vee C$). Would I in *this* case be prepared to go on to issue a promissory note committing myself to assent to C should it turn out that in fact *A* after all? No! If it turns out that *A*, I'd *retract* my current endorsement of *not-A*, and so *retract* the disjunction ($\neg A \vee C$). Hence, if I accept ($\neg A \vee C$) – or equivalently ($A \supset C$) – on the grounds that *not-A*, then I won't be prepared to endorse the promissory conditional *if A then C*.

Let's say, following Frank Jackson, that my assent to ($\neg A \vee C$), or equivalently ($A \supset C$), is *robust* with respect to A if the discovery that A is true won't lead me to withdraw my assent. Then the point we've just made comes to this: if my assent to ($\neg A \vee C$), or equivalently ($A \supset C$), isn't robust with respect to A, I won't accept the force of the §15.2 argument from ($\neg A \vee C$) to *if A then C*.

Exactly similarly, suppose I endorse something of the form $\neg(A \wedge \neg C)$ because I believe $\neg A$. Again, I'll take it back if I discover than in fact A holds. So again I won't accept the force of the §15.2 argument from $\neg(A \wedge \neg C)$ to *if A then C*.

This observation suggests a new account of ordinary, indicative conditionals, an account which both recognizes the appeal of the arguments in §15.2 yet also recognizes the intuitive distinction between the inferences (M) and (V) that we stressed in §15.3. The suggestion is that ordinary conditionals are not just material conditionals, rather they are *material conditionals presented as being robust* (i.e. robust with respect to their antecedents).

Let's explain that more carefully. Compare:

(1) Jack came to the party.
(2) Even Jack came to the party.

These plainly differ in overall meaning. And yet it seems that the worldly state of affairs they report is one and the same, namely Jack's presence at the party. The difference between (1) and (2) doesn't seem to reside in what they tell us about the world. Rather, the word 'even' functions rather like a verbal highlighter – or a verbalized exclamation mark – typically used to signal the speaker's feeling that Jack's presence was surprising or out of the ordinary.

Similarly, compare the following:

(3) Jack came to the party and had a dreadful time.
(4) Jack came to the party but had a dreadful time.

Again, the worldly state of affairs these two report seems to be the same. The difference between them is that the use of (4) signals, very roughly, that Jack's going to the party without enjoying it was against expectation.

Likewise, it is now being suggested that the difference between these –

(5) (Jack came to the party $\supset$ Jack had a good time)
(6) If Jack came to the party, then he had a good time.

– is again *not* a difference in what they say about the world. Each is strictly speaking true so long as its antecedent is false and/or consequent true. Rather the difference arises because someone who uses 'if' signals that their claim is robust with respect to the antecedent. In other words, the speaker who uses (6) rather than (5) undertakes not to withdraw the conditional should it turn out that the antecedent is true and Jack *did* come to the party.

Go back to that example where I recall the Easter birthday party which shows that one of Jack and Jill has an April birthday. We argued that it could be natural to endorse

(7)   If Jack was born in October, then Jill was born in April.

That's nicely in keeping with the current theory, since (in the circumstances) although (7) may be no-more-than-material in propositional content, it *is* robust.

Now since, on this theory, the core content of *if A then C* is exactly $(A \supset C)$, it is no surprise that – quietly assuming robustness – we can construct plausible arguments as in §15.2, from $(A \supset C)$ to *if A then C*, as well as vice versa. But note too that, while we can accept the unproblematic inference

(M)   *not-A; so* $(A \supset C)$

we won't *robustly* endorse $(A \supset C)$ if our ground for believing the material conditional is simply the falsity of the antecedent. Hence, we *won't* want to signal that the conclusion of argument (M) is being held robustly. In other words – according to our theory which equates ordinary-language conditionals with material conditionals signalled as being robust – we won't want to say something of the form

(V)   *not-A; so if A then C*

Which accords exactly with the point of §15.3.

In summary, the 'robust material conditional' theory is *one* theory about ordinary 'if ... then ...' which neatly accommodates both the attractions of the §15.2 arguments for thinking that indicative conditionals are material conditionals, and also the §15.3 considerations that press against a pure equivalence.

15.5 'Dutchman' conditionals

The previous section's theory about indicative conditionals is neat: but is it right? Is asserting a conditional always tantamount to giving a promissory note to accept the consequent should the antecedent turn out to be true? What about the following conditionals? –

(1)   If George W. Bush is a great philosopher, then pigs can fly.
(2)   If George W. Bush is a great philosopher, then I'll eat my hat.

Or, to take a paradigm example, consider

(3)   If George W. Bush is a great philosopher, then I'm a Dutchman.

Here I assert the conditional inviting you to apply modus tollens, i.e. to infer from my assertion *if A then C*, together with the obvious truth of *not-C*, that *not-A*. And my grounds for asserting the conditional are none other than my belief in *not-A*. I commit myself to no more than the corresponding ($\neg A \lor C$). Were I to discover that the antecedent A is in fact true – discover that Bush has a secret other life as the pseudonymous author of some much admired philosophical works – then I'd have to take back my scornful conditionals.

In short, these 'Dutchman' conditionals are mere material conditionals which are not robustly asserted. What does that show?

You could take it as counting against the 'robust material conditionals' theory. But equally, you could take the opposing line. After all, 'Dutchman' conditionals do strike us pre-theoretically as exceptional cases, as being a jokey idiom. And our theory explains why: they are conspicuously non-robust, and so offend against one aspect of the usual rules governing the assertion of conditionals, which is why they seem non-standard. So, you might hold that far from being refuted by the behaviour of 'Dutchman' conditionals, our theory explains it.

But we can't pursue this tangled issue any further here. We've already followed the twists and turns of arguments about the nature and variety of conditionals far enough. Two points, however, have clearly emerged.

First, it should now be clear why, when introducing truth-functional logic, we *didn't* attempt to deal with conditionals right from the start. Rendering 'if ..., then ...' by a truth-function raises complex problems of a quite different order to the issues raised by treating 'and', 'or and 'not' as truth-functions. To clutter the initial discussion of **PL** inferences with these problems would have been very distracting; while to have introduced the material conditional straight off without much comment would have seemed at best mystifying, at worst a cheat.

Secondly, on the 'robust material conditional' theory, the rendition of indicative conditionals by '⊃' captures that aspect of the content of ordinary indicative conditionals which has to do with matters of truth and falsehood, and hence the aspect which logic (the study of truth-preserving inferences) needs to focus on. And we can now see that, on any theory, there is some considerable plausibility in at least treating indicative conditionals as having a truth-functional core, even if there are other aspects of the use of vernacular conditionals that are missed if we try to insist on a bald equation between 'if' and '⊃'.

The moral, then, is this:

> *If* you want to be able to regiment arguments involving the vernacular indicative conditional into a form that you can test using the truth-table test, *then* there is nothing for it but to use the material conditional '⊃'. *If* you want to be able to apply the truth-table test, then the price is that you have to re-express the premises and conclusion using '⊃'. The rendition preserves the validity/invalidity of arguments in at least some central cases. But arguably the translation is not always an accurate one (even once we've set aside the possible world conditionals for alternative treatment). Something – in some cases, you may think, too much – may be lost in translation.

Whatever you think about all the considerations in this chapter, the material conditional has to come with the warning *use with care*.

15.6 Summary

- 'Subjunctive' or 'counterfactual' conditionals are possible-world conditionals, and are not truth-functional.

- Even restricting ourselves to indicative conditionals, there are various valid inferences involving the truth-functional conditional whose ordinary-language counterparts look dubious or positively invalid.

- The 'robust material conditional' theory gives one possible explanation of some of these divergences. (Note, however, the theory is a contentious one!)

- The material conditional must come with a Logical Health Warning.

Exercises 15

A   (a)   The pattern of inference *C; so* $(A \supset C)$ is trivially valid. What about the inference *C; so if A then C*? (Consider: given *C is* true, then it is true whether *A* is or not, so in particular it is true if *A* is. But also compare the inference *Bush will win the election; so if there is a huge financial scandal involving the Bush family then Bush will win the election*.)

(b)   The pattern of inference $(B \supset C)$; *so* $((A \wedge B) \supset C)$ is trivially valid. What about the inference *if B then C; so if A and B then C*? (Consider, 'If you strike this match, it will light; so if you wet this match and strike it, it will light.')

B   Consider the variant language **PLF** whose *only* one- or two-place connective is '$\supset$', but which has – as well as the usual atoms, 'P', 'Q', etc. – the additional atomic wff '$\bot$' which is governed by the semantic rule that it *always* takes the value F (so there is no option about how to evaluate this atom).

(a)   Show that negation, disjunction, and conjunction can be expressed in **PLF**. (Hint: start by considering material conditionals relating two propositions, one of which is '$\bot$'.)

(b)   What does it mean to say that **PLF** is expressively adequate? Is it true?

(c)   You are familiar with the idea of a one-place truth-functional connective (like negation) and a two-place connective (like conjunction): so what does it mean to talk of an *n*-place truth-functional connective? What, then, would be a 0-place truth-functional connective? Can we think of '$\bot$' as a 0-place connective?

(The symbol '$\bot$' is often called the *falsum*; hence the label '**PLF**'.)

Introducing PL trees

In previous chapters we defined what it is for **PL** (and **PLC**) inferences to be tautologically valid. We then introduced one way of testing for tautological validity, namely the brute force truth-table test, where we search through every assignment of values to the relevant atoms to see if there is one which makes the premisses true and the conclusion false. In this chapter, we introduce another way of testing for tautological validity, using 'truth trees'. We begin with an informal exploration via examples: and for the moment we will concentrate on **PL** inferences (we'll extend the story to cover conditionals in Chapter 18).

16.1 'Working backwards'

Consider the following argument (here and henceforth you can give the atoms whatever content you like to make things meaningful):

A $\qquad$ (P ∧ ¬Q), (P′ ∧ Q′), (¬P″ ∧ ¬Q″) ∴ R

This is plainly not tautologically valid. But suppose we want to confirm this using a truth-table. There are seven different atoms involved in the argument, so using the direct truth-table test would mean writing down a 128-line table, and checking each valuation until we find a line where we have true premisses and a false conclusion. Fortunately, slogging through like this is utterly unnecessary. For consider the following chain of reasoning.

 The inference in **A** is tautologically *invalid* if and only if there is a valuation which makes the premisses true and conclusion false, i.e. a valuation where

$$
\begin{array}{lc}
(1) & (P \wedge \neg Q) \Rightarrow T \\
(2) & (P' \wedge Q') \Rightarrow T \\
(3) & (\neg P'' \wedge \neg Q'') \Rightarrow T \\
(4) & R \Rightarrow F
\end{array}
$$

(for the convention of dropping quotes round the wff preceding '⇒', see §10.2: but do remember that (1) etc. are still metalinguistic). (1) will hold just so long as we have both

$$
\begin{array}{lc}
(5) & P \Rightarrow T \\
(6) & \neg Q \Rightarrow T
\end{array}
$$

since a conjunction is only true if its conjuncts are true. Similarly, to guarantee (2) and (3) we need

(7)	$P' \Rightarrow T$
(8)	$Q' \Rightarrow T$
(9)	$\neg P'' \Rightarrow T$
(10)	$\neg Q'' \Rightarrow T$

To get these assignments (4) to (10), we just need the following valuation of the seven atoms:

$$P \Rightarrow T, P' \Rightarrow T, P'' \Rightarrow F, Q \Rightarrow F, Q' \Rightarrow T, Q'' \Rightarrow F, R \Rightarrow F$$

So we have very speedily located a valuation of the atoms in **A**, custom designed to make the premises true and conclusion false. Hence the inference is indeed tautologically invalid.

What we've done here is to 'work backwards'. Instead of starting with the whole set of possible valuations for the atoms, and then working *forward* through each possibility, calculating the truth-values of the premises and conclusion as we go, looking for a 'bad' valuation all the while, we try starting from the other end. We suppose for the sake of argument that there *is* a 'bad' valuation where the premises are true and the conclusion false. Then we try to work *backwards* towards a valuation of the atoms that has that unfortunate consequence. If we can do the trick, then obviously the inference is not valid.

A few more simple examples should help make the general strategy clear. So consider the following inference:

B $(P \wedge Q), \neg(\neg R \vee S) \therefore \neg(R \wedge Q)$

This will be invalid if and only if there is a valuation where

(1)	$(P \wedge Q) \Rightarrow T$
(2)	$\neg(\neg R \vee S) \Rightarrow T$
(3)	$\neg(R \wedge Q) \Rightarrow F$

To ensure that (1) holds, we evidently need

(4)	$P \Rightarrow T$
(5)	$Q \Rightarrow T$

To get (2), we need

(6)	$(\neg R \vee S) \Rightarrow F$

and since a disjunction is false only when both disjuncts are false, that means

(7)	$\neg R \Rightarrow F$
(8)	$S \Rightarrow F$

And, from (7), that means we need

(9)	$R \Rightarrow T$

Finally, from (3) we have

(10)	$(R \wedge Q) \Rightarrow T$

which requires, consistently with what we had before, that

(11) $R \Rightarrow T$
(12) $Q \Rightarrow T$

In sum, then, we have backtracked to the valuation

$$P \Rightarrow T, Q \Rightarrow T, R \Rightarrow T, S \Rightarrow F$$

which is expressly constructed to yield the truth-value assignments (1), (2) and (3), and thus make the premisses of **B** true and the conclusion false. Hence this second sample argument is again invalid.

Now consider this even simpler inference:

C $(P \wedge \neg Q) \; \therefore \; \neg(Q \wedge R)$

As before, this is invalid if and only if there is a valuation which makes the premiss true and the conclusion false, i.e. a valuation where

(1) $(P \wedge \neg Q) \Rightarrow T$
(2) $\neg(Q \wedge R) \Rightarrow F$

From the specification (1) we can infer

(3) $P \Rightarrow T$
(4) $\neg Q \Rightarrow T$

Next, (2) implies

(5) $(Q \wedge R) \Rightarrow T$

which in turn implies

(6) $Q \Rightarrow T$
(7) $R \Rightarrow T$

But plainly, we can't make '$\neg Q$' true (as (4) requires), and simultaneously make 'Q' true (as (6) requires). Hence the supposition that **C** is invalid puts contradictory demands on the valuation of the relevant atoms. And if that supposition leads to contradiction, **C** must be *valid* after all (that's a reductio argument: see §5.4).

Even in this last mini-example, the 'working backwards' technique saves time as against writing out an eight-line truth-table. Here's another example where the saving is even greater:

D $\neg(\neg P \vee Q), \neg(\neg R \vee S) \; \therefore \; (R \vee \neg P)$

A brute force truth-table evaluation would require examination of the sixteen possible valuations of the relevant atoms, and barring short cuts, forty-eight evaluations of wffs. Compare, however, the following simple chain of reasoning. Suppose there is a valuation such that

(1) $\neg(\neg P \vee Q) \Rightarrow T$
(2) $\neg(\neg R \vee S) \Rightarrow T$
(3) $(R \vee \neg P) \Rightarrow F$

Now, to get (1) and (2), we need

$$(4) \qquad\qquad (\neg P \vee Q) \Rightarrow F$$
$$(5) \qquad\qquad (\neg R \vee S) \Rightarrow F$$

Again, for a disjunction to be false, both its disjuncts must be false. So it follows that for (3) to hold we'd need

$$(6) \qquad\qquad R \Rightarrow F$$

while for (5) to hold we'd need

$$(7) \qquad\qquad \neg R \Rightarrow F$$

Contradiction! In sum, a valuation will only make the original premises of **D** all true if it makes 'R' true (as (7) requires), and it will only make the conclusion false if it makes 'R' false (as (6) says). That's impossible. So **D** must be tautologically valid.

16.2   Branching cases

To summarize so far. Testing an argument for tautological validity is a matter of deciding whether there are any 'bad' valuations of the relevant atoms, i.e. valuations which make all the premises true and yet the conclusion false. One approach (embodied in the truth-table test) is to plod through every possible valuation, to see if a 'bad' one turns up. It is usually quicker, however, to approach the same decision task from the other end. We assume there *is* a bad valuation that makes the premises and the conclusion false, and then work out what this 'bad' valuation must look like. If we can do the trick, that shows the entailment doesn't hold. On the other hand, if working backwards from the assumption of invalidity leads to contradictory results, then (by a reductio ad absurdum argument) that shows that there is no bad valuation after all and the argument is valid.

Unfortunately, however, when following through this new strategy in practice, things are typically rather more complicated than our examples so far reveal. Consider next this plainly bogus mini-argument:

$$\textbf{E} \qquad\qquad (P \vee Q) \therefore P$$

Working backwards by assuming that the premiss is true and the conclusion is false, we have

$$(1) \qquad\qquad (P \vee Q) \Rightarrow T$$
$$(2) \qquad\qquad P \Rightarrow F$$

This time we cannot deduce from (1) the values of 'P' and 'Q'. All that (1) tells us is that (at least) *one* of these is true, but not *which*. So here we must consider the possible alternatives in turn. We can perspicuously represent the two cases we need to examine by drawing a forking pathway for our inferences, thus:

$$(1) \qquad\qquad (P \vee Q) \Rightarrow T$$
$$(2) \qquad\qquad P \Rightarrow F$$

$$(3) \qquad\qquad P \Rightarrow T \qquad\qquad Q \Rightarrow T$$

We know that, given (1), at least one of the forking alternatives at line (3) must hold; so we now examine the options in turn. The assignment P ⇒ T on the left-hand fork gives one way of guaranteeing (1). But this alternative has in fact already been ruled out, since we earlier have P ⇒ F at step (2). Taking the left-hand fork leads to contradiction. Let's 'close off' a path like this where a wff is assigned inconsistent values by writing a marker like '✳' at its foot, to signal that we've hit an absurdity. So we get

(1)   (P ∨ Q) ⇒ T
(2)   P ⇒ F
                                  ⟋‾‾‾
(3)   P ⇒ T   Q ⇒ T
                        ✳

However, while taking the left-hand fork leads straight to absurdity, there is no contradiction on the right-hand fork. Taking that fork gives us the valuation P ⇒ F, Q ⇒ T which makes the premiss of the argument true and conclusion false. Hence argument **E** is shown to be invalid.

We will work through some more examples to show how we typically need to trace down forking alternatives when searching backwards for a 'bad' valuation. So consider next the valid 'disjunctive syllogism' inference

F   (P ∨ Q), ¬P ∴ Q

Starting by assigning truth to the premisses and falsehood to the conclusion, we have:

(1)   (P ∨ Q) ⇒ T
(2)   ¬P ⇒ T
(3)   Q ⇒ F
                                  ⟋‾‾‾
(4)   P ⇒ T   Q ⇒ T

At (4) we have begun considering the alternative ways of verifying (1). We don't know what values 'P' and 'Q' have. But we know that one of them is true. This time we see that the left-hand option is inconsistent with (2), which of course implies that 'P' is false; and the right-hand option is inconsistent with (3). So *both* branches lead to absurdity:

(1)   (P ∨ Q) ⇒ T
(2)   ¬P ⇒ T
(3)   Q ⇒ F
                                  ⟋‾‾‾
(4)   P ⇒ T   Q ⇒ T
                        ✳   ✳

Hence there is no consistent way of making the premisses and negation of the conclusion all true; so the argument is indeed valid.

For another example, consider

G   ¬(P ∧ Q), (P ∧ R) ∴ ¬(R ∨ Q)

We start the same way, supposing the premises true and conclusion false:

(1)   ¬(P ∧ Q) ⇒ T
(2)   (P ∧ R) ⇒ T
(3)   ¬(R ∨ Q) ⇒ F

To make these true, we definitely need

(4)   (P ∧ Q) ⇒ F   (from 1)
(5)   P ⇒ T   (from 2)
(6)   R ⇒ T   (from 2)
(7)   (R ∨ Q) ⇒ T   (from 3)

But now we can make no more firm deductions. (4) tells us that the conjunction of 'P' and 'Q' is false, so one of the conjuncts must be false. But we aren't told which, so we need to consider the alternative possibilities, thus

(8)   P ⇒ F   Q ⇒ F
                        *

Here the left hand fork leads immediately to contradiction, as we already have P ⇒ T at (4). So we must close off that path with the absurdity marker to indicate that no coherent valuation will emerge on this route, and turn our attention to the right-hand path. On this path, there is no explicit contradiction yet; however, we still have to deal with the implications of (7). This says that at least one, though we don't yet know which, of 'R' and 'Q' must be true. We again need to consider forking alternatives, so we'll get the tree:

(1)   ¬(P ∧ Q) ⇒ T
(2)   (P ∧ R) ⇒ T
(3)   ¬(R ∨ Q) ⇒ F
(4)   (P ∧ Q) ⇒ F
(5)   P ⇒ T
(6)   R ⇒ T
(7)   (R ∨ Q) ⇒ T

(8)   P ⇒ F   Q ⇒ F   (from 4)
                   *

(9)   R ⇒ T   Q ⇒ T   (from 7)
                                          *

The new right-hand fork leads immediately to contradiction, since taking that route would require 'Q' to be both false and true; hence that branch too can be closed off with an absurdity marker. But the remaining path down, where

P ⇒ T, Q ⇒ F, R ⇒ T

leaves us with a consistent valuation custom designed to make (1), (2) and (3) all hold (check that it does!), thus showing G to be invalid.

16.3   Signed and unsigned trees

Let's summarize again. We can test an argument for tautological validity by assuming that the inference is invalid and then 'working backwards', trying to find a consistent valuation which warrants that assumption. At various points, we may have to consider alternative valuations. For example, if we have arrived at $(A \lor B) \Rightarrow T$, then we have to consider the alternatives $A \Rightarrow T$ and $B \Rightarrow T$. If, twist and turn as we might, *all* such alternatives lead in the end to contradictory assignments of values to atoms, then we know that the original assumption that the inference is invalid is incorrect. In other words, if all branches of the 'tree' on which we set out the working can be closed off with absurdity markers, then the argument is valid. Conversely, if at least one alternative route through the tree involves no contradiction (even when all the complex wffs are fully unpacked), then the assumption that the argument is invalid is vindicated.

Now, so far, our 'working backwards' arguments we've been illustrating have been set out in the form of *signed trees* (we'll speak of 'trees' here even if there is – as in §1 – only a trunk without any branching). The trees are 'signed' in the sense that the wffs that appear on them are all explicitly assigned a truth-value, either T or F. In the interests of neatness and economy, we are now going to move to using *unsigned* trees.

As a major step towards doing this, note that by a couple of simple tricks, we can easily ensure that every wff which appears on a 'working backwards' tree in fact gets the uniform value T. Here's how:

- *Trick one*   Instead of starting a tree by assuming that the premisses of the argument to be tested are true and the conclusion false, assume that the premisses and the *negation* of the conclusion are all true.

- *Trick two*   Thereafter, avoid ever writing something of the form '$C \Rightarrow F$' by always jumping instead to something of the form '$\neg C \Rightarrow T$'.

So, for example, when we meet on a tree a claim of the kind

$$(a) \qquad\qquad \neg(A \lor B) \Rightarrow T$$

don't proceed via the obvious consequence

$$(b) \qquad\qquad (A \lor B) \Rightarrow F$$

to

$$(c) \qquad\qquad A \Rightarrow F$$
$$(d) \qquad\qquad B \Rightarrow F$$

but instead (trick two) jump straight from (a) to

$$(c^*) \qquad\qquad \neg A \Rightarrow T$$
$$(d^*) \qquad\qquad \neg B \Rightarrow T$$

And so on. The resulting 'T'-only trees are only very slightly less natural than the 'T-and-F' trees we've been constructing so far.

Let's quickly illustrate by re-doing a couple of examples in this style, with 'T'-only trees: the changes required are marginal. So return first to

D   ¬(¬P ∨ Q), ¬(¬R ∨ S) ∴ (R ∨ ¬P)

Starting this time by assigning truth to the premises and the *negation* of the conclusion, and we have:

(1)	¬(¬P ∨ Q) ⇒ T
(2)	¬(¬R ∨ S) ⇒ T
(3)	¬(R ∨ ¬P) ⇒ T

Now, let's twice apply the rule just given for negated disjunctions ('jump from ¬(A ∨ B) ⇒ T to add both ¬A ⇒ T and ¬B ⇒ T'), and we immediately get

(4)	¬¬R ⇒ T	(from 2)
(5)	¬S ⇒ T	(from 2)
(6)	¬R ⇒ T	(from 3)
(7)	¬¬P ⇒ T	(from 3)
	*	

We can close off the tree with an absurdity marker because of (6) and (4): no wff and its negation can be true together on a valuation. Since the tree starting from the supposition that the premises and negation of the conclusion are true leads to contradiction, the argument must be valid. (And why are there two negation signs at line (4)? Because our rule said that if a negated disjunction is true, then each negated disjunct is true; and if a disjunct already starts with a negation sign, as in '¬R' in (2), then it acquires another when negated.)

For a second 'T'-only tree, return to

G   ¬(P ∧ Q), (P ∧ R) ∴ ¬(R ∨ Q)

Starting with the assumption that the premises and negated conclusion are true we get:

(1)	¬(P ∧ Q) ⇒ T	
(2)	(P ∧ R) ⇒ T	
(3)	¬¬(R ∨ Q) ⇒ T	
(4)	P ⇒ T	(from 2)
(5)	R ⇒ T	(from 2)
(6)	(R ∨ Q) ⇒ T	(from 3)

(7)   ¬P ⇒ T   ¬Q ⇒ T   (from 1)
                *

(8)   R ⇒ T   Q ⇒ T   (from 6)
                                        *

The rule applied at (7) relies on the fact that, if a negated conjunction is true, then one or other of the conjuncts must be false (so we have the rule 'given ¬(A ∧ B) ⇒ T, consider alternatives ¬A ⇒ T and ¬B ⇒ T'). Not all branches of this resulting tree lead to absurdity. So there *is* a coherent way of making the premises and negated conclusion all true together, and so argument **G** must be tautologically invalid.

It is worth quickly checking that our other sample trees can be rewritten with minimal changes as 'T'-only trees. But why care that we can do this?

Because if we make assignments of *truth* standard on trees, and avoid ever assigning falsehood, then it isn't really necessary to keep on writing down '⇒ T' against each and every wff in our back-tracking reasoning. We might as well just let it be understood as ever-present. *So from now on, let's save time and ink by using 'T'-only trees and then keep the uniform assignments of truth tacit.* Suppressing the explicit assignments leads to *unsigned* trees. We'll give some more examples in the next section.

16.4 More examples

Consider next the inference

H   (P ∨ Q), ¬(P ∧ ¬R) ∴ (Q ∨ R)

To test for validity, we start from the assumption that the premisses and the negation of the conclusion are all true: so we start off our unsigned tree

(1)	(P ∨ Q)
(2)	¬(P ∧ ¬R)
(3)	¬(Q ∨ R)

with each line ending with a now invisible '⇒ T'. Since, as we said before, a negated disjunction is true just so long as both negated disjuncts are true, the assumption that (3) is true leads to the requirement that each of '¬Q' and '¬R' is true. So we add

| (4) | ¬Q |
| (5) | ¬R |

each with their invisible '⇒ T'. Now apply the principle that a disjunction is true so long as one or other disjunct is true, to get

(6)   P   Q   (from 1)
                              ✳

The alternative on the right-hand branch immediately leads to contradiction, for we have the occurrence of both 'Q' [⇒ T] and '¬Q' [⇒ T] on that fork. So we close off that fork with the absurdity marker: no consistent valuation can be built on this branch. We've now extracted the juice from (1) and (3), so that leaves (2) to deal with. The relevant principle to invoke here is again that the negation of a conjunction is true when at least one of the negated conjuncts is true. So if '¬(P ∧ ¬R)' is true, one of '¬P' or '¬¬R' must be true. Applying that to our tree, we get branching possibilities, so the full tree will look like this:

(1)	(P ∨ Q)	
(2)	¬(P ∧ ¬R)	
(3)	¬(Q ∨ R)	
(4)	¬Q	(from 3)

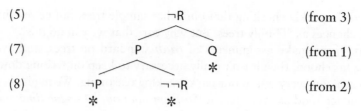

(5)   ¬R   (from 3)

(7)   P   Q   (from 1)
                                              *

(8)   ¬P   ¬¬R   (from 2)
          *   *

This time, both new forks lead to contradiction (we can't have both P ⇒ T and ¬P ⇒ T, as the left-most fork requires; nor can we have ¬R ⇒ T and ¬¬R ⇒ T as the other fork requires). So all forking paths are closed off by contradictions. That means that there is no route which discloses a consistent valuation of atoms which makes (1), (2) and (3) all take the value T. Every alternative way of trying to make good the supposition that the inference in question is invalid runs into contradiction. So the inference *is* tautologically valid after all.

In constructing this last proof of validity, note that we chose first to draw consequences from (3) before making use of (1) and (2). There was no necessity about this; and we could have chosen to proceed in a different order, e.g. as follows:

(1)   (P ∨ Q)
(2)   ¬(P ∧ ¬R)
(3)   ¬(Q ∨ R)

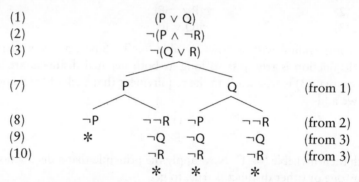

(7)   P   Q   (from 1)

(8)   ¬P   ¬¬R   ¬P   ¬¬R   (from 2)
(9)   *   ¬Q   ¬Q   ¬Q   (from 3)
(10)   ¬R   ¬R   ¬R   (from 3)
                   *   *   *

Here we have proceeded more systematically, unpacking the complex wffs at the top of the tree in turn. Starting with (1), we asked what is required for that to be true, which immediately forces us to consider branching alternatives. Then following up each of these alternatives, we considered the implications of (2) for *each* of these alternative ways of proceeding; so we had to add the forking alternatives '¬P' [⇒ T] and '¬¬R' [⇒ T] to *both* existing branches. So we now have *four* options to consider. We can close off one path because it contains contradictory wffs which can't both be true. Then we proceeded to extract the implications of (3) for the three options that remain open. For (3) to be true, both 'Q' and 'R' must be false: so we add '¬Q' [⇒ T] and '¬R' [⇒ T] to every branch that still remains open. We now end up with every route leading to absurdity, and so closed off with the absurdity marker. Hence, as we would hope, we arrive at the same result as before: **H** must be valid. (Later, in §19.3, we'll formally prove that it can never make any difference in which precise order we unpack wffs when 'working backwards'.)

Another example: what of the following inference?

I   (P ∨ (Q ∧ R)), (¬P ∨ R), ¬(Q ∨ ¬¬S) ∴ (S ∧ R)

We start off in the now familiar way, with the premises and negated conclusion:

(1)   (P ∨ (Q ∧ R))
(2)   (¬P ∨ R)
(3)   ¬(Q ∨ ¬¬S)
(4)   ¬(S ∧ R)

A negated disjunction is true only if both disjuncts are false, i.e. the negated dis-juncts are true. So (3) implies a couple of definite truth assignments

(5)   ¬Q   (from 3)
(6)   ¬¬¬S   (from 3)

Where now? Well, we know that a pair of negation signs in effect cancel each other out. So we can infer

(7)   ¬S   (from 6)

Let's now unpack the other complex wffs in turn. First we have

(8)   P   (Q ∧ R)   (from 1)

The right-hand branch doesn't close off immediately, but if we have (Q ∧ R) ⇒ T on that way of making (1) true, then we must also have Q ⇒ T and R ⇒ T. So we can add to the right hand branch thus

(9r)   Q   (from 8)
(10r)   R   (from 8)

And we've hit a contradiction between (5) and (9r) and can close off that branch. Proceeding down the left-hand branch, we've so far ensured that (1) and (3) are true. To make (2) true we need either ¬P ⇒ T or R ⇒ T. So we have

(9l)   ¬P   R   (from 2)

and the left-hand of the two branches we've added immediately closes off, since we can't have 'P' and '¬P' both true. That leaves us with only (4) to unpack. So we apply the rule for negated conjunctions again and the resulting tree from line (7) again looks like this:

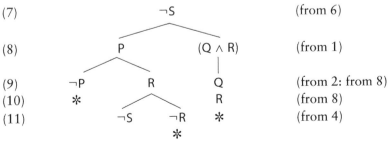

(7)   ¬S   (from 6)

(8)   P   (Q ∧ R)   (from 1)

(9)   ¬P   R   Q   (from 2: from 8)
(10)   *   R   (from 8)
(11)   ¬S   ¬R   *   (from 4)
                              *

So, all but one of the branches close. Looking at that remaining branch, however, and putting back the suppressed assignments of truth, we can read off the following valuation

$$P \Rightarrow T, \neg Q \Rightarrow T \text{ (so } Q \Rightarrow F), R \Rightarrow T, \neg S \Rightarrow T \text{ (so } S \Rightarrow F).$$

This is designed to make the premises of I true and the conclusion false (check that it does!). So the argument is tautologically invalid.

This tree technique is elegant and (usually) efficient. Our next task is to gather together the various principles we have found for tree-building, check whether we need any others, and put them together in one systematic list. Though first, let's briefly pause for breath.

16.5 Summary

- Testing an argument for tautological validity is a matter of searching for 'bad' valuations. We can do this by a brute-force inspection of all possible valuations. Or we can 'work backwards', i.e. assume there *is* a bad valuation and try to determine what it would look like. If the supposition that there is such a bad valuation leads to contradiction, then the argument in question is valid.

- The 'working backwards' method, in the general case, will necessitate inspecting branching alternative possibilities, e.g. looking at alternative ways of making a disjunction true. The resulting working is naturally laid out in the form of a downward branching tree.

- By some simple tricks, we can ensure that every wff on such a tree is assigned the value T. But then, if every wff is assigned the same value, we needn't bother to explicitly write down the value – giving us 'unsigned' trees.

- The official rules for tree-building follow in the next chapter.

Exercises 16

Although we haven't yet laid down formal rules, it is worth doing a bit of informal exploration to get more feeling for the tree method before we dive into the official story. So, see how you get on testing the following inferences for validity, using the techniques of this chapter. At this stage it might help to keep writing down explicit 'T'-only trees, appending '$\Rightarrow$ T' alongside every wff to remind yourself what you are doing and why you are doing it.

1. $(R \lor P), (R \lor Q) \therefore (P \lor Q)$
2. $\neg(P \land \neg Q), P \therefore Q$
3. $(\neg P \lor Q), (R \lor P) \therefore (R \land Q)$
4. $((P \land \neg Q) \lor R), (Q \lor S) \therefore (R \land S)$
5. $P, \neg(P \land \neg Q), \neg(Q \land \neg R), (\neg R \lor S) \therefore S$
6. $((P \land Q) \lor (R \land P)) \therefore ((Q \lor R) \lor P)$

Rules for PL trees

This chapter gives a more formal description of the tree test for tautological validity. The first section, then, is necessarily going to be rather abstract. But it will really do no more than carefully formulate the principles we've already used in the last chapter. (Those who get a bit fazed by this sort of abstraction can try skimming through the statement of the rules and the following illustrative sections a few times in quick succession, using each to throw light on the other.)

17.1 The official rules

As we have seen, our 'working backwards' method of testing for tautological validity typically requires us to explore alternatives. These explorations are most neatly set out in the form of *trees* (upside-down ones, with the 'trunk' at the top and the branches forking downwards). In fact, our trees are *binary* trees, meaning that a branch never forks more than two ways.

We'll take the idea of a downward branching binary tree to be intuitively clear: we could get mathematically precise here, but little would be gained. The thing to emphasize is that trees fan out without branches being rejoined at a lower level: as in the case of a respectable family tree, incestuous reunions are banned.

Some definitions will prove useful, though:

- Positions on a tree are conventionally called *nodes*.

- Start from the topmost node of the tree, and track down some route through the inverted tree to the bottom tip of a branch. The nodes en route form *a path* through the tree.

- In our truth-trees, the nodes are occupied by wffs. If a wff W occurs at a node on a path, we'll say more briefly that W is on the path or that the path contains W.

- If a path contains a pair of wffs of the form W, ¬W, then the path is *closed*; otherwise it is *open*.

- If a tree has an open path it is *open*; otherwise the tree is *closed*.

And before we proceed, here's a reminder of another definition:

- A *primitive* wff is a wff which is either an atom or the negation of an atom (§11.8). So a *non-primitive wff* is any wff that isn't primitive (i.e. any complex wff that features more than a solitary negation sign).

Suppose, then, that we want to determine whether the premisses A_1, A_2, ..., A_n tautologically entail the conclusion C. We begin the truth-tree in the now familiar way by placing the A_i and $\neg C$ at nodes down the initial trunk of a tree (all accompanied by an invisible '$\Rightarrow$ T'). Then we try systematically to generate a 'bad' valuation that will indeed make these initial wffs all true.

The strategy at each stage is to infer, from the assumed truth of a non-primitive wff W already on some path, *shorter* wffs which must also be true if that W is to be true (maybe we have to consider alternatives, so the path has to fork as we add them). When a path contains both a wff and its negation (both, of course, implicitly assigned the value T), then we know that this must be a contradictory dead-end and cannot lead to a consistent valuation of primitives. Such a path is *closed*, and we will mark off closed paths with the absurdity marker '$*$'. We continue extending *open* paths by unpacking the implications of non-primitive wffs on the path, with forks in the tree marking points where we have alternatives to investigate. And we keep on going until either (i) all the paths we explore contain contradictory pairs of wffs and the tree is closed, or else (ii) we have no more wffs to unpack and we are still left with an open tree with at least one open path. (We'll put that more carefully in a moment.)

What rules govern extending the tree at each stage? There are five, depending on the character of the non-primitive wff W we aim to unpack. The rules have already been illustrated informally in the previous chapter, and are summed up in the box displayed opposite. Three quick comments:

- The five rules exhaust all the cases of non-primitive wffs, as can be seen by quickly running through the possibilities. Any wff is either an atom, or a conjunction (in which case apply rule (b)) or a disjunction (in which case apply rule (d)), or else the negation of something. In the last case, the wff must either be the negation of an atom, or of a conjunction (in which case apply rule (e)) or of a disjunction (in which case apply rule (c)), or finally it's the negation of a negation (in which case apply rule (a)). So we have a rule to apply to every eventuality, until we get down to the level of primitives (atoms and negations of atoms).

- Why, in applying a rule to some wff W, must we add the appropriate (possibly alternative) wffs to the foot of *every* open path containing W? Because these different open paths containing W are supposed to indicate alternative possible valuations where W is true – and we need to consider the further implications of W's truth for each of these alternative valuations.

- Note, these rules for unpacking wffs tell us what to do with a selected non-primitive W. But they don't tell us how to make the selection. Typically, the rules can be applied in various orders to the same initial trunk of wffs. The chosen order *won't* affect the ultimate verdict of the tree test.

---

*Rules for unpacking wffs on **PL** trees*

Suppose that W is a non-primitive wff on an open branch, which is supposed true on some valuation. One of the following cases must obtain:

(a) W is of the form $\neg\neg A$. If W is true on a valuation, that means that A itself must be true: to unpack W, add A [$\Rightarrow$ T] to each open path through the tree containing W. Schematically,

$\neg\neg A$
$|$
A

(b) W is of the form $(A \wedge B)$. If a conjunction W is true on a valuation, both its conjuncts must be true: to unpack W, add both A [$\Rightarrow$ T] and B [$\Rightarrow$ T] to each open path containing W. Schematically,

$(A \wedge B)$
$|$
A
B

(c) W is of the form $\neg(A \vee B)$. If the negation of a disjunction is true on a valuation, the disjunction must be false, and hence both its disjuncts false, and thus the *negations* of both disjuncts true. So to unpack W, add $\neg A$ [$\Rightarrow$ T] and $\neg B$ [$\Rightarrow$ T] to each open path containing W. Schematically,

$\neg(A \vee B)$
$|$
$\neg A$
$\neg B$

(d) W is of the form $(A \vee B)$. If W is true on a valuation, at least one of its disjuncts must be true. We have to explore alternatives. So to unpack W, add a new fork to each open path containing W, with branches leading to the alternatives A [$\Rightarrow$ T] and B [$\Rightarrow$ T]. Schematically,

$(A \vee B)$

$A \qquad\quad B$

(e) W is of the form $\neg(A \wedge B)$. If W is true on a valuation, $(A \wedge B)$ is false, and so at least one of the conjuncts must be false, i.e. at least one of their negations must be true. We have to explore alternatives. So to unpack W, add a new fork to each open path containing W, with branches leading to the alternatives $\neg A$ [$\Rightarrow$ T] and $\neg B$ [$\Rightarrow$ T]. Schematically

$\neg(A \wedge B)$

$\neg A \qquad\quad \neg B$

Now, suppose that we have already applied the appropriate rule to some complex formula W. Then it would be pointlessly redundant to apply the rule to the same W *again*. And it would be worse than redundant to get 'stuck in a loop', repeatedly re-applying this same rule to W, time without end. Henceforth, then, we adopt the following structural rule for tree-construction:

> When the appropriate rule has been fully applied to a non-primitive wff, we 'check off' the wff (with a '√'); and a wff, once checked off, can be thought of as 'used up', i.e. as no longer available to have one of the 'unpacking' rules applied to it.

This ensures that a properly constructed tree can't go on for ever. Unpacking a wff generates *shorter* wffs: and since we are not allowed to revisit 'used' wffs, after a finite number of stages, we'll run out of new wffs to check off, and there will be no more rules to apply.

So, with that 'check off' rule in play, here more carefully are step-by-step instructions for building a *complete* tree.

(1) Start off the trunk of the tree with the premisses A_1, A_2, ..., A_n and the negation of conclusion C of the inference to be tested.

(2) Inspect any path which isn't yet closed off by an absurdity marker to see whether it involves a contradiction, i.e. for some wff W it contains both W and also the corresponding formula $\neg W$. If it does, then we *close* that path with the marker, and ignore it henceforth.

(3) If there is no unchecked non-primitive formula W left on any open path, then stop!

(4) Otherwise, we choose some unchecked non-primitive formula W that sits on an open path. We then apply the appropriate rule from (a) to (e), and add the results to the foot of *every* open path containing that W. We can then check off W.

(5) Loop back to step (2).

As just remarked, this tree-building procedure must eventually terminate. There are two possible outcomes. Either we end up with a *closed tree*, i.e. a tree such that every path through the tree closes (there is an absurdity marker at the foot of every branch). Or we end up with a *completed open* tree – i.e. one where some paths don't close, and on each open path, we have repeatedly applied the rules (a) to (e) as much as we possibly can, so that *every* non-primitive formula on the path is used up and checked off, but without always producing a contradiction.

Finally, here is the *Basic Result* about truth trees:

> If, starting a tree with the wffs A_1, A_2, ..., A_n, $\neg C$ at the top, we can construct, using rules (a) to (e), a *closed* tree, then the argument from the premisses A_1, A_2, ..., A_n to the conclusion C is tautologically valid. Conversely, if applying the rules leads to some completed tree with an *open* branch, then the argument in question is *not* tautologically valid.

We'll prove the Basic Result in Chapter 19. But it should seem intuitively correct, since it only restates the principle underlying all our worked examples in the previous chapter – namely, that if 'working backwards' from the assumption that the premisses are true and the conclusion false lands us in self-contradiction, then that assumption can't be made good and the given argument must be valid after all; while if that assumption *does* lead to a consistent assignment of truth to various primitives, then the argument in question is invalid.

17.2 Tactics for trees

A couple of pieces of tactical advice before we start working through more illustrative examples.

First, suppose that at a given point we have the option of applying either a 'straight' rule – i.e. one of (a) to (c) – or one of the 'branching' rules (d) and (e). To take a very simple example, suppose we have arrived at the following point on some branch of a tree:

$$\cdots$$
$$(P \vee Q)$$
$$(R \wedge S)$$

How should we continue? Well, if we apply the 'branching' rule (d) first, and then rule (b), we will get

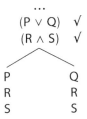

However, if we apply the rules in the reverse order, we get

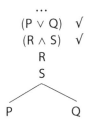

This makes absolutely no difference in principle: either way we get a (sub)tree with two paths through it, with exactly the same wffs occurring on the left path each time and exactly the same wffs occurring on the right path each time. The only difference is that, in the case where we applied the branching rule (d) first, we had to double the expenditure of ink when it came to applying rule (b), since we already had two branches to which to add 'R' and 'S'. It is quicker and neater to proceed as in the second case. Moral: it generally pays to apply non-branching rules first when one has the option.

The second bit of tactical advice is even more obvious: aim to close off branches as soon as possible. If you have the option of applying a rule which will immediately lead to a path being closed off, then it is good policy to select that option. Suppose, for example, that a tree begins

$$(P \lor Q)$$
$$(P' \lor \neg S)$$
$$(R \lor S)$$
$$\neg R$$

Then we have three options for applying the branching rule (d). If we blindly apply the rule to the first wff first, and then deal with the others in turn, we get

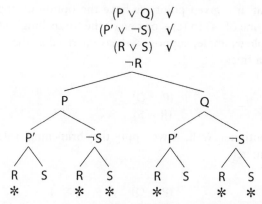

Thus, after sprawling eight branches wide, most of the tree suddenly closes off, leaving us with just two open branches. One branch says that

$$\neg R \Rightarrow T \text{ (so } R \Rightarrow F), P \Rightarrow T, P' \Rightarrow T \text{ and } S \Rightarrow T$$

The other one yields

$$\neg R \Rightarrow T \text{ (so } R \Rightarrow F), Q \Rightarrow T, P' \Rightarrow T \text{ and } S \Rightarrow T$$

And either of those valuations will make the wffs on the trunk all true.

But now suppose that we start again, and this time applying rule (d) to the premises in the opposite order. Then the completed tree looks like this:

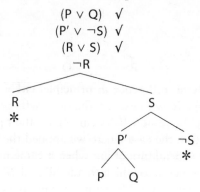

We still end up with two open paths on the tree, carrying just the same wffs as before. But by a judicious choice of the order of application for the branching rule, we have achieved a much more neatly pruned structure. Evidently, it is sensible, wherever possible, to choose similarly to apply rules in an order which closes off branches as early as possible. A bit of practice enables you to spot good choices!

('But hold on a moment! The first valuation just listed doesn't explicitly assign any value to "Q", and the second valuation doesn't tell us about "P"!' No matter: this simply means that, so long as the *rest* of the valuation holds, it doesn't matter which value the unmentioned atom has. The wffs on the trunk, in particular '(P ∨ Q)', will still come out true.)

17.3 More examples

Let's turn, then, to a couple more examples, and put the official rules of §17.1 and the tactical advice of §17.2 into effect. (For additional exercises, it is perhaps worth taking the worked examples in the previous chapter, and doing them again 'by the book'.) Consider, then, the inference

> **A**   (P ∨ Q), (R ∨ ¬Q), ¬(¬S ∧ P) ∴ (R ∨ S)

Is this tautologically valid? Well, start the 'trunk' of the tree in the usual way (for convenience, we'll continue numbering steps – though note that the numbering isn't part of the 'official' tree itself):

> (1)   (P ∨ Q)
> (2)   (R ∨ ¬Q)
> (3)   ¬(¬S ∧ P)
> (4)   ¬(R ∨ S)

We first apply the only relevant non-branching rule, i.e. (c) to line (4). That gives

> (4)   ¬(R ∨ S) √
> (5)   ¬R
> (6)   ¬S

We now have to apply branching rules to lines (1), (2) and (3). Since we already have '¬R' on the tree, it will be good policy to deal next with line (2): obviously that will create a branch that can immediately be closed off. So check off line (2) and continue

> (7)

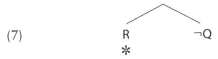

We now apply rule (d) to (1), as again one of the newly created branches will immediately close. So we can check off line (1) and add on the right the new fork

> (8)

Which leaves us again with just one open path, and the one complex wff (3) still to deal with. So finally applying rule (e) to that wff gives us the following completed tree:

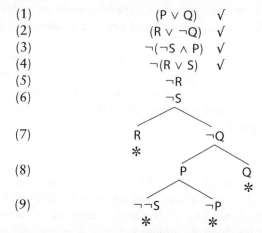

(1)	(P ∨ Q) √
(2)	(R ∨ ¬Q) √
(3)	¬(¬S ∧ P) √
(4)	¬(R ∨ S) √
(5)	¬R
(6)	¬S
(7)	R ¬Q
(8)	P Q
(9)	¬¬S ¬P

Every downwards path closes off, showing that there is no consistent way of backtracking to a valuation of atoms which makes the premisses of the given argument true and the conclusion false. Hence **A** is valid.

Note carefully the application of rule (e) at the final step. The rule is: given a negated conjunction ¬(A ∧ B) on an open path, add forks, with ¬A and ¬B on the two branches. Applied to (3), on the left fork we need to add the negation of '¬S', and that is '¬¬S', *not* 'S'. The application of (e), and likewise (c), *always* results in the prefixing of a new negation sign. We can later use (a) to strip off pairs of negation signs if we need to: but this is a further step. In more advanced work, you will encounter logical systems with a non-classical understanding of negation. In particular, you'll meet so-called intuitionistic logic (see §9.3), which lacks a rule like (a) and where '¬¬S' is not equivalent to 'S'. So it is bad to get into the habit of just ignoring double negations.

Another example. What are we to make of this inference?

B   (P ∧ (Q ∨ ¬R)), ¬((P ∧ Q) ∧ S)), (¬(P ∧ S) ∨ R) ∴ ¬S

The truth-tree for this inference starts

(1)	(P ∧ (Q ∨ ¬R))
(2)	¬((P ∧ Q) ∧ S))
(3)	(¬(P ∧ S) ∨ R)
(4)	¬¬S

NB again the doubling up of the negation signs at line (4). Let's apply the non-branching rule (b) to (1), followed by the rule (e) to (2) to add

(5)	P
(6)	(Q ∨ ¬R)
(7)	¬(P ∧ Q) ¬S

What next? Unpacking the wffs at line (7) and then line (6) will lead each time to branches which quickly close, and the tree will get to the following state:

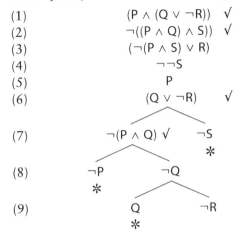

(1)   (P ∧ (Q ∨ ¬R)) √
(2)   ¬((P ∧ Q) ∧ S)) √
(3)   (¬(P ∧ S) ∨ R)
(4)   ¬¬S
(5)   P
(6)   (Q ∨ ¬R)   √
(7)   ¬(P ∧ Q) √   ¬S
                          *
(8)   ¬P   ¬Q
       *
(9)   Q   ¬R
             *

Which leaves us just with the complex wff at line (3) to deal with. So we now check that off and continue the only open branch ...

(10)   ¬(P ∧ S) √   R
                              *
(11)   ¬P   ¬S
            *   *

So the tree closes, and **B** is shown to be valid.

For a final example in this section, consider the following inference (involving five atoms, and thus requiring a thirty-two line table if one uses the brute force truth-table test for tautological validity):

C   (P ∧ (Q ∨ R)), (P ∨ ¬(P' ∨ R)), ((P' ∧ S) ∨ ¬Q) ∴ ((Q ∨ R) ∧ S)

To test this, we start off in the standard manner

(1)   (P ∧ (Q ∨ R))
(2)   (P ∨ ¬(P' ∨ R))
(3)   ((P' ∧ S) ∨ ¬Q)
(4)   ¬((Q ∨ R) ∧ S)

We will continue by applying the non-branching rule (b) to (1), and then applying rule (e) to (4): that gives us

(5)   P
(6)   (Q ∨ R)
(7)   ¬(Q ∨ R)   ¬S
        *

The path forking to the left immediately closes off, as both the wff '(Q ∨ R)' and its negation appear on that path (NB, *any* pair of contradictory wffs closes off a branch). We next deal with (3), and add to the open path accordingly:

(8)   (P′ ∧ S)   ¬Q

Here at line (8), we have on the left a complex wff to which we can apply the non-branching rule (b) again. So in accordance with our tactical policy of trying to apply such rules first, let's deal with this wff, check it off, and extend the tree as follows:

(9)   P′
(10)   S
                           *

Again the left branch closes off. How should we extend the remaining path? There are two complex unchecked wffs on that path, at lines (2) and (6). Recall the other tactic of choosing, where we can, to apply branching rules that imme- diately give us closing paths. We should evidently deal first with (6) and add to the one open path on the tree a fork with 'Q' on one branch (immediately yield- ing a contradiction) and 'R' on the other. That leaves the complex wff at line (2) to deal with. Applying rule (d) to (2) then puts the tree into the following state, now without the unofficial numbers

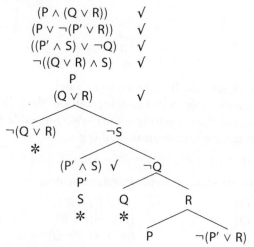

We are not quite done, since we have a new complex wff appearing on the right- hand branch at the last line; so to finish the tree, we need to apply the relevant rule (c), and check off that complex wff as follows:

                              ¬(P′ ∨ R) √
                                 ¬P′
                                 ¬R
                                  *

Which at last *does* complete our tree: we now have 'unpacked' every complex wff featuring on open paths, and we are still left at the end with an open pathway. From this branch, reading from the bottom up, we can read off the valuation P ⇒ T, R ⇒ T, ¬Q ⇒ T (so Q ⇒ F) and ¬S ⇒ T (so S ⇒ F). The value of 'P'' is left undetermined, which means that the other values should suffice to fix the initial wffs on the trunk as true, no matter what value 'P'' has. And our valuation indeed verifies (1) to (4), as is easily checked. So **C** is invalid.

17.4 Testing for tautologies

The tree test determines whether C has to be true, given the premisses A_i. As a special case, we can use the test to determine whether C is a tautology, i.e. has to be true, period. We just suppose ¬C is true, and then see if we can work backwards to extract a 'bad' valuation which will have this consequence. If we can, then C isn't a tautology; if we can't, then it is.

To take a fairly simple example, is

    D   ((P ∧ (Q ∨ R)) ∨ (¬P ∧ (¬Q ∧ ¬R)))

a tautology? Well, suppose it isn't. Suppose that there is a valuation which makes

    (1)   ¬((P ∧ (Q ∨ R)) ∨ (¬P ∧ (¬Q ∧ ¬R)))

true. Unpacking (1) and making the next few tree-building moves we get

  (2)   ¬(P ∧ (Q ∨ R))   √
  (3)   ¬(¬P ∧ (¬Q ∧ ¬R))
                          ―――――――――――――
  (4)   ¬P   ¬(Q ∨ R)   √
  (5)   ¬Q
  (6)   ¬R

We now need to apply rule (e) to (3), and place the result (remember!) at the foot of each of the two open paths it heads. So we get

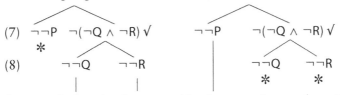

We have three marked paths that are evidently not going to close. But strictly speaking we are not quite finished yet. There are still non-primitive wffs on each of those paths, and we need to apply rule (a) in each case to strip off the double negations. So continuing we get …

                | | |
  (9)   Q   R   P

And now we *are* finished. We have a completed open tree, and so **D** is not a tautology. (Exercise: read off the open branches the 'bad' valuations which make **D** false.)

For a final worked example, consider the wff

E $(\neg(\neg(P \wedge Q) \wedge \neg(P \wedge R)) \vee \neg(P \wedge (Q \vee R)))$

Is this a tautology? Well, start the tree:

(1) $\neg(\neg(\neg(P \wedge Q) \wedge \neg(P \wedge R)) \vee \neg(P \wedge (Q \vee R)))$

What now? How do we read this slightly monstrous wff? Working from the inside, we can use different styles of brackets to make the groupings clearer, thus

(1) $\neg[\neg\{\neg(P \wedge Q) \wedge \neg(P \wedge R)\} \vee \neg\{P \wedge (Q \vee R)\}]$

which reveals (1) to be of the form $\neg[A \vee B]$, with $A = \neg\{\neg(P \wedge Q) \wedge \neg(P \wedge R)\}$, and $B = \neg\{P \wedge (Q \vee R)\}$. So we need to apply rule (c) to get

(2) $\neg\neg\{\neg(P \wedge Q) \wedge \neg(P \wedge R)\}$
(3) $\neg\neg\{P \wedge (Q \vee R)\}$

Next we obviously apply rule (a) to get rid of those paired negation signs

(4) $\{\neg(P \wedge Q) \wedge \neg(P \wedge R)\}$
(5) $\{P \wedge (Q \vee R)\}$

Both of these wffs are conjunctions so we can apply rule (b) twice to get

(6) $\neg(P \wedge Q)$
(7) $\neg(P \wedge R)$
(8)   P
(9) $(Q \vee R)$

The wffs at lines (1) to (5) have now all been checked off and 'unpacked', so we need now only worry about the wffs (6) to (9). Dealing with (6) and (7) first, we get

(10)   ¬P   ¬Q
                   *
(11)   ¬P   ¬R
                         *

Which will leave just one unchecked non-primitive wff to deal with, at line (9). So to finish the tree ...

(12)   Q   R
                         *   *

Thus all branches close off, and the wff E is a tautology (you are hereby invited to check this by running a direct truth-table test...!).

17.5 Comparative efficiency

Operating the tree test for tautological validity often involves choices: in which order do we apply the unpacking rules? But we could make it fully mechanical by laying down strict rules for choosing which wff to unpack next. That would give us two mechanical tests for validity, the original truth table test *'search for-wards' through all valuations, looking for a bad one,* and the tree test *'search backwards', starting from the assumption that there is a 'bad' valuation.* As we've seen, running the tree test on an argument is often much faster than running the truth table test. Sometimes the gain is staggering. Return to the example we gave in §13.2: take the inference from the 50 premises '¬P, ¬P′, ¬P″, ¬P‴, ...,¬P″⋯″' to the conclusion 'Q'. Even if we could write down a line a second, just setting out the 2^{51} line truth-table (before looking for a bad line) would take half a million years. Writing down a truth tree at the same rate would take less than a minute (just write down the premises, the negation of the conclusion and – hey presto! – we've got a complete open tree and the test is done).

Is the tree-test *always* more efficient? No. In the worst case scenario, trees can sprawl horribly. Take the inference with 50 premises '(P ∨ P′), (P′ ∨ P″), (P″ ∨ P‴), ...' and conclusion 'Q'. Obviously it is invalid. But running a tree-test, at least as we've so far described it, leads to an exponentially sprawling tree. When we unpack the first premise, the tree splits into two branches; when we unpack the second premise, each branch splits into two again; unpack the third premise and we get eight branches. So by the end of the tree, it has exploded into 2^{50} branches; and even if we could construct one complete downward path per second, that would be another ludicrous task.

Now, you might think that we just need to add some more bells and whistles to the tree test. After all, *we* can instantly spot that the last inference is invalid: we immediately see that the valuation which makes all the 'P's true and 'Q' false will make the disjunctive premises all true and conclusion false. And if we can instantly spot that, can't we build our way of doing the trick (whatever that is) into an improved tree test to speed things up? To which the answer is, yes, we can add a few more particular tricks that can deal with some special cases: but *there is no general-purpose trick that is guaranteed to avoid exponential sprawl.* Put it this way: given an inference with fifty-something atoms, then with truth-tables and truth trees alike, even augmented with short-cut tricks, in the worst case applying either test for tautological validity is a million-year task.

Is there some *other* mechanical test for tautological validity which does better? No. (Or at least, so it is widely believed: but the issue relates to a still unsolved problem in the foundations of the theory of computation.)

17.6 Summary

- We have set out the formal rules for building **PL** truth-trees to test inferences for tautological validity. The rules consist of step-by-step instructions for building a tree, 'unpacking' more complex wffs into simpler ones as the tree grows, checking off wffs as we use them up, and closing off any paths which contain contradictory pairs of wffs.

- The instructions for tree-building don't determine a unique tree, since we are allowed to choose which non-primitive wff to 'unpack' when there is more than one candidate.

- But it doesn't matter which choices you make: the test will always give the same verdict. If the tree closes, then the argument under test is tautologi-cally valid; if it doesn't, it isn't. (A proof of that comes later, in Chapter 19.)

- We can make life easier for ourselves by a couple of obvious tactics for tree building: apply non-branching rules before branching ones, and when pos-sible choose to apply rules that lead to a branch quickly closing.

- The tree test can be used in an obvious extension to determine whether a given wff is a tautology.

- Running a tree test is very often quicker than a truth-table test. But not always.

Exercises 17

A   Revisit Exercises 13A, and now use the tree test to determine which of the arguments listed there are tautologically valid.

B   Revisit Exercises 12 and now use the tree test to determine which of (2) to (6) are tautologies.

C   And since practice makes perfect, use the tree test to determine which of the following are true.
   1. $((P \lor Q) \land (P \lor R)) \vDash ((Q \land R) \lor P)$
   2. $((P \land \neg(P' \land R)) \lor S), (P' \lor S) \vDash (R \land S)$
   3. $P, \neg(Q \land (P' \lor R)), \neg(\neg Q \land P), ((P' \land S) \lor (R \land S)) \vDash S$
   4. $\vDash (\neg(\neg(P \land Q) \lor \neg(P \land R)) \lor \neg(P \lor (Q \land R)))$
   5. $((P \land Q) \lor (R \land P)), \neg(\neg S \land P), \neg(Q \land R) \vDash (Q \land S)$

D   How would the rules and principles of tree-building need to be re-written if we had stuck to using *signed* trees, where wffs are explicitly assigned T or F?

PLC trees

We now extend the use of the tree test to cover inferences involving the material conditional.

18.1 Rules for PLC trees

Suppose we want to extend the tree test for tautological validity to cover arguments expressed in **PLC**. We need rules that tell us how to disassemble wffs whose main connectives are '⊃' and '≡'.

A wff of the form $(A \supset B)$, which has '⊃' as its main connective, is equivalent to the corresponding wff $(\neg A \vee B)$. So if $(A \supset B)$ appears on a tree (silently assumed to be true, of course), then we should unpack it as we would unpack $(\neg A \vee B)$. So we'll need to add new forks, leading to the alternatives $\neg A$ and B.

Query: looking at the truth-table, aren't there *three* ways of making $(A \supset B)$ true – namely the way which makes A, B true, the way which makes $\neg A$, B true and the way which makes $\neg A$, $\neg B$ true. So why don't we split the tree *three* ways, adding those pairs of wffs to the respective branches? Reply: no reason other than aesthetics! We could construct trees like that; it would in the end come to exactly the same (see Exercise 18B). It is just prettier to stick to two-way splits.

To continue: a wff of the form $(A \supset B)$ is only false when A is true and B false. So if $\neg(A \supset B)$ appears on a path – i.e. $(A \supset B)$ is assumed to be false – then that requires A, $\neg B$ to be true on the same valuation.

A wff of the form $(A \equiv B)$ is true in two cases, when A and B are both true, and when A and B are both false. So we should unpack $(A \equiv B)$ disjunctively, adding new forks with A, B on one branch and $\neg A$, $\neg B$ on the other.

Finally, a wff of the form $\neg(A \equiv B)$ is also true in two cases, when A is true and B false, and when A is false and B true. So we should also unpack $\neg(A \equiv B)$ disjunctively, this time adding new forks with A, $\neg B$ on one branch and $\neg A$, B on the other.

These rules are summed up in the following box in the style of §17.1. The general principles for building trees and then reading off the results of the test remain (of course) exactly the same as before. And there is an extended Basic Result: a **PLC** argument is tautologically valid if and only if its tree closes.

Further rules for unpacking wffs on PLC trees

Suppose that *W* is a non-primitive wff on an open branch, which is supposed true on some valuation. Then, as well as the five cases (a) to (e) as listed before, there are now four further possibilities to cover:

(f) *W* is of the form (*A* ⊃ *B*). To unpack *W*, add a new fork to each open path containing *W*, with branches leading to the alternatives ¬*A* [⇒ T] and *B* [⇒ T]. Schematically,

(*A* ⊃ *B*)

¬*A*   *B*

(g) *W* is of the form ¬(*A* ⊃ *B*). To unpack *W*, add both *A* [⇒ T] and ¬*B* [⇒ T] to each open path containing *W*. Schematically,

¬(*A* ⊃ *B*)
|
A
¬*B*

(h) *W* is of the form (*A* ≡ *B*). To unpack *W*, add a new fork to each open path containing *W*, with *A* [⇒ T] and *B* [⇒ T], on one branch, and ¬*A* [⇒ T] and ¬*B* [⇒ T] on the other. Schematically,

(*A* ≡ *B*)

A   ¬*A*
B   ¬*B*

(i) *W* is of the form ¬(*A* ≡ *B*). To unpack *W*, add a new fork to each open path containing *W*, with *A* [⇒ T] and ¬*B* [⇒ T], on one branch, and ¬*A* [⇒ T] and *B* [⇒ T] on the other. Schematically,

¬(*A* ≡ *B*)

A   ¬*A*
¬*B*   *B*

18.2   Examples

We'll immediately turn to a few examples. First, let's test the following argument which we've met before (interpret the atoms as you will):

    A   (P ⊃ Q), (Q ⊃ R) ∴ (P ⊃ R)

Here's the corresponding completed tree, where we apply the non-branching rule (g) first, and omit the unnecessary line numbers.

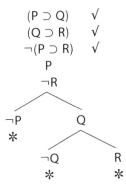

The tree closes so the argument is valid. (By this stage, you certainly shouldn't need persuading that, even in simple cases like this, trees can be a lot neater than truth tables; but compare §14.3D.)

Next, consider

   B $((P \land Q) \supset R) \therefore ((P \supset R) \lor (Q \supset R))$

Starting the tree and applying the non-branching rules for negated disjunctions and then negated conditionals first, we get

$$((P \land Q) \supset R)$$
$$\neg((P \supset R) \lor (Q \supset R)) \quad \checkmark$$
$$\neg(P \supset R) \quad \checkmark$$
$$\neg(Q \supset R) \quad \checkmark$$
$$P$$
$$\neg R$$
$$Q$$
$$\neg R$$

That just leaves the initial premiss to be unpacked. So, we check that off and continue

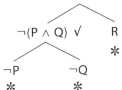

and we are done; another valid argument (perhaps rather surprisingly, see §18.4).

For another example, consider

   C $(P \equiv Q), (Q \equiv R) \therefore (P \equiv R)$

This time the tree starts

$$(P \equiv Q)$$
$$(Q \equiv R)$$
$$\neg(P \equiv R)$$

and there are no non-branching rules to apply. So let's just proceed by unpacking wffs from the top down, remembering to add the necessary new forks at each stage to *every* open branch:

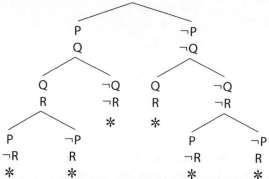

Again the tree closes and the argument is confirmed to be valid (make sure you follow exactly how this tree has been constructed).

Next, we'll test whether the following is a tautology:

D   ((P ⊃ Q) ∨ (Q ⊃ P))

As before (§17.4), to use a tree test we start with the negation of this wff and aim to find contradictions. The negated disjunction unpacks into the negated disjuncts, and then we apply rule (g) a couple of times to get

$$
\begin{array}{ll}
\neg((P \supset Q) \vee (Q \supset P)) & \checkmark \\
\neg(P \supset Q) & \checkmark \\
\neg(Q \supset P) & \checkmark \\
P & \\
\neg Q & \\
Q & \\
\neg P & \\
* &
\end{array}
$$

so D comes out as a tautology. Perhaps that's another surprising result, and we'll again comment on it in §18.4.

What about the following?

E   ({P ⊃ Q} ⊃ {(Q ≡ R) ⊃ (P ⊃ R)})

A moment's reflection suggests that this *ought* to be a tautology (why?); and indeed it is. Starting with the negation of E and applying non-branching rules first we get

$$
\begin{array}{ll}
\neg(\{P \supset Q\} \supset \{(Q \equiv R) \supset (P \supset R)\}) & \checkmark \\
\{P \supset Q\} & \\
\neg\{(Q \equiv R) \supset (P \supset R)\} & \checkmark \\
(Q \equiv R) & \\
\neg(P \supset R) & \checkmark \\
P & \\
\neg R &
\end{array}
$$

and now we apply the branching rules to the unchecked wffs to finish the tree

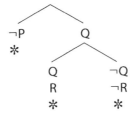

18.3 An invitation to be lazy

Now for a slightly more complex example, which incidentally also illustrates a general point about trees (whether **PL** or **PLC** trees). Consider

F   (P ⊃ (Q ⊃ R)), (S ⊃ (Q ∨ P)) ∴ (S ⊃ R)

The tree starts

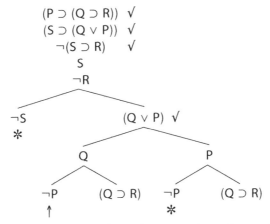

So far, we've applied the non-branching rule to the negated conclusion first, and then the two branching rules to the premisses (choosing first the application that immediately leads to a closed branch). And now let's pause here, though the tree isn't completed yet, for neither of those occurrences of '(Q ⊃ R)' has been checked off.

Note, however, that the path up from the bottom left node *is* completed: i.e. there are no unchecked wffs left on that path through the tree. Further work on the tree therefore can't close off *that* path. *Hence argument F is already shown to be tautologically invalid.* For an argument to be valid according to the tree test, *all* paths through the tree must close; that means that just *one* completed open path is enough to show invalidity. In other words, an unfinished tree with a *completed* open path demonstrates invalidity. (Compare: for an argument to be valid according to the truth-table test, *all* lines of the truth table must be 'good'. But *one* bad line is enough to show invalidity.)

Reading up that open branch, we find the primitives ¬P, Q, ¬R, S all silently assigned the value *true*. So one valuation which makes the premisses true and conclusion false is P ⇒ F, Q ⇒ T, R ⇒ F, S ⇒ T. Completing the other two currently open branches will in fact reveal another valuation which has the same effect. But if you are feeling lazy, you can leave the tree unfinished.

For a final worked example, consider the argument

G   ((¬Q ⊃ R) ⊃ (Q ≡ S)), (P ∧ R), ¬(Q ∧ S) ∴ ¬((P ∧ Q) ⊃ (R ∧ S))

We can begin a tree thus:

$$((¬Q ⊃ R) ⊃ (Q ≡ S))$$
$$(P ∧ R) \qquad √$$
$$¬(Q ∧ S)$$
$$¬¬((P ∧ Q) ⊃ (R ∧ S))√$$
$$((P ∧ Q) ⊃ (R ∧ S)) \quad √$$
$$P$$
$$R$$

```
              ¬(P ∧ Q) √                    (R ∧ S) √
                                                R
         ¬P          ¬Q                         S
          *
```

So far, so obvious (let's hope!). Now unpack the shorter of the two remaining unchecked wffs, and we have to add the two forks

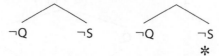

There are still three open branches. So now let's take the remaining unchecked wff, apply rule (e), and continue the left-most branch.

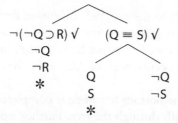

The other two branches will get similar continuations, but again we'll be lazy, as we've already found a *completed* open branch. As before, there is little point in filling out the rest of the tree, as further growth on the *other* branches won't affect this open branch. The tree is not going to close, and that settles that G too is tautologically invalid.

18.4 More translational issues

Given the meaning of '⊃', the results of the last two sections are entirely unproblematic. But a couple look a bit puzzling when we translate them back into

English, using the vernacular 'if ..., then ...'. We aren't going to reopen all the debates of Chapter 15, but we shouldn't just pass by without comment.

(1) An example due to Richard Jeffrey. Consider again the inference

B $((P \land Q) \supset R) \therefore ((P \supset R) \lor (Q \supset R))$

We've shown this to be tautologically valid. Now consider the following, got by reading 'P' as *the captain turns his key*, 'Q' as *the gunnery officer turns his key*, and 'R' as *the missile fires*, and rendering the horseshoe connective by 'if'.

> If the captain turns his key and the gunnery officer turns *his* key, then the missile fires. Hence either the missile fires if the captain turns his key, or else it fires if the gunnery officer turns *his* key.

That surely can't be a valid inference, or there would be no dual-key fail-safe systems! What's gone wrong? (Question for consideration: are the conditionals here, despite being 'indicative' in surface grammatical form, naturally read as possible-world conditionals? That is to say, we treat the conditionals as telling us what *would* happen if a key were turned, even if no key is turned in the actual world – which is why they are not truth-functional.)

(2) The conditional 'If it is boiling hot, then it is freezing cold' looks simply absurd; so does the conditional 'If it is freezing cold, then it is boiling hot'. So consider the disjunction

> Either it is freezing cold if it is boiling hot or it is boiling hot if it is freezing cold.

This looks to be a disjunction of silly falsehoods; so it should be *false*. Put 'P' for *it is freezing cold*, 'Q' for *it is boiling hot*, render 'if' by '$\supset$', and we get

D $((P \supset Q) \lor (Q \supset P))$

which, far from being false, we've just proved is a *tautology*. Doesn't this show that the 'if' here can't be rendered by '$\supset$'? (Question for consideration: when we take e.g. 'If it is boiling hot, then it is freezing cold' as plainly false are we again reading it as a possible-world conditional?)

(3) I can't resist adding another example, also due to Richard Jeffrey. Note that the inference

H $((P \supset Q) \supset R), \neg P \therefore R$

is also tautologically valid (check that with a quick tree test!). But interpret 'P' as *we advance*, 'Q' as *the enemy retreats* and 'R' as *the enemy lose*. Then this corresponds to the argument

> If the enemy retreats if we advance, then they lose. But we won't advance. So the enemy lose!

which is surely *not* valid. (You can't zap the enemy with inaction plus mere logic. Question for consideration: are the conditionals here really possible world conditionals yet again?)

And so it goes! There are numerous cases where, on the face of it, **PLC** arguments involving '⊃' which are tautologically valid (and hence plain valid) become, at first blush, problematic and dubious when translated back into English using 'if ..., then ...'. Maybe, with greater or lesser difficulty, the problematic appearances can often be explained away. But the moral needs repeating: *Treat the material conditional with great care.*

18.5   Summary

- The tree-building rules we met in Chapter 16 can be simply extended to tell us how to unpack wffs whose main connectives are '⊃' and '≡'.

- We have also noted that we can stop work on a tree once a complete open branch is found (i.e. an open branch with no unchecked non-primitives remaining). One completed open branch is enough to show than the argument being tested is invalid.

Exercises 18

A   Redo Exercises 14B using the tree method to determine which of the arguments are tautologically valid.

B   (a) Suppose we added to our rules for tree-building the following: you can at any point extend an open branch with a fork of the form

Show informally that this change would make no difference (in the sense that a tree which closes using the original rules will still close, and a tree which stays open using the original rules will remain open).

(b) Show that we could therefore replace our rule for unpacking conditionals with this more complex one:

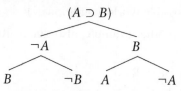

(c) Conclude that we could therefore replace our rule for unpacking conditionals with the following three-ply rule (§18.1):

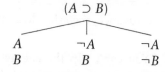

PL trees vindicated

The tree method appears to be a very elegant way of determining tautological validity. But now we ought to confirm what we've been taking for granted, namely that the tree test gets things right (we called that the 'Basic Result' about trees). So the business of this short chapter is to vindicate the tree test. It can be skipped by those who have no taste for such 'metalogical' niceties!

19.1 The tree method is sound

The Basic Result (§17.1) has two parts. The first half says:

(S) If, starting a tree with the wffs A_1, A_2, A_3, ..., A_n, $\neg C$ at the top, we can construct a *closed* tree by following the official rules for tree building, then the inference from the premises A_i to conclusion C is indeed tautologically *valid*.

Equivalently: if an inference is *invalid*, then a corresponding tree *can't* close.

Let's (re)establish a bit of terminology, before turning to the proof of (S). A set of wffs is *satisfiable* if there is a valuation of the relevant atoms which makes all the wffs in the set true. Derivatively, we will call a *path* through a tree *satisfiable* if there is a valuation (of the relevant atoms that appear in wffs on the path) which makes all the wffs on the path true.

It is immediate that *a satisfiable path through a tree cannot close*. For a closed path by definition contains a contradictory pair of wffs, W and $\neg W$; and no valuation can make such a pair true together. And that observation underlies the following ...

Proof of (S) Suppose that there is some valuation which ensures that

$$A_1 \Rightarrow T, A_2 \Rightarrow T, A_3 \Rightarrow T, \ldots, A_n \Rightarrow T, \neg C \Rightarrow T$$

and hence that the initial 'trunk' of the corresponding truth-tree is satisfiable (so the inference in question is *invalid*). Now consider what happens as we extend our tree, step by step, applying the official rules. *At each stage, a satisfiable path will get extended to a longer satisfiable path.*

That's obvious with the *non-branching* rules: for example, if there is a valuation which makes everything on a path true, including a wff $(A \wedge B)$, then when we apply rule (b) to extend that path by adding A and B, the same valuation will obviously make both these added new wffs true as well – since if the conjunction is true on a valuation, so are the conjuncts. So the extended path will remain satisfiable. You can easily check that the same is true for extensions by means of rules (a) and (c) – and for (g) as well, on those occasions when '⊃' is also in play (§18.1).

And in a case where we extend a satisfiable path by using a *branching* rule, one branch (at least) of the extended tree must still remain satisfiable. For instance, if there is a valuation which makes everything on a path true, including a wff (A ∨ B), then when we apply rule (d) to extend this path with A on one fork and B on the other, the same valuation must make at least one of these new wffs true. For if a disjunction is true on a valuation, so is one of its disjuncts. Hence, at a fork we can always choose at least one extension of a satisfiable path which remains satisfiable. And you can easily check that the same is true for extensions by means of rule (e), and likewise for the branching rules (f), (h) and (i) governing (bi)conditionals.

Thus, if we start with a satisfiable trunk, then however far down we extend the tree according to our rules, there will always remain at least one satisfiable path through it. But as we noted, a satisfiable path must stay open. Hence a tree with a satisfiable trunk can never fully close.

So, if a tree *does* close, then its trunk *isn't* satisfiable. There is no valuation which makes true all the premisses and the negation of the conclusion of the inference under test, and hence that inference is *valid*. QED.

In summary: the tree method is indeed a reliable or *sound* one (a **PL** or **PLC** argument which the appropriate tree test says is tautologically valid really is tautologically valid).

19.2 The tree method is complete

The other half of the Basic Result says:

(C) If, starting a tree with the wffs $A_1, A_2, A_3, ..., A_n, \neg C$ at the top, we can construct a complete open tree by following the official rules for tree building, then the inference from the premisses A_i to conclusion C is tautologically invalid.

To prove this result, it is enough to prove two subsidiary results, as follows.

(C_1) For any completed open path, there is a valuation which makes the *primitives* true (for the idea of a 'primitive' see §17.1).

(C_2) If there's a valuation for the atoms that appear in the wffs on a completed open path which makes the *primitives* true, then it will make *all* the wffs on the path true.

For suppose (C_1) and (C_2) are correct. By definition, if a tree headed by the A_i and $\neg C$ is open, then it has at least one completed open path through it. By (C_1), there's a valuation which satisfies the primitives on that path; hence by (C_2)

there's a valuation which satisfies *all* the wffs on that path, including the wffs on the initial trunk. That valuation therefore makes all of A_i and $\neg C$ true. So, as (C) claims, the inference from the premisses A_i to conclusion C must be tautologically invalid.

Hence we now just need to prove (C_1) and (C_2).

Proof of (C_1) Take a completed open path. Climb up it from the bottom, and construct the following valuation for the atoms which occur in the wffs on the branch:

> When you encounter a totally bare atom on the path, assign it the value T; any other atoms, that appear only inside some non-atomic wff on the path, get the value F.

Call this the *chosen* valuation.

An example: suppose that as you climb up the chosen path you encounter in sequence the wffs

$$Q, \neg P, (\neg P \wedge Q), R, (R \vee S), ((R \vee S) \vee P), ((\neg P \wedge Q) \vee S)$$

then the chosen valuation constructed as we march up the branch is

$$Q \Rightarrow T, P \Rightarrow F, R \Rightarrow T, S \Rightarrow F$$

By construction, the chosen valuation makes all the primitives on the given path true. For a primitive is either (i) a bare atom, or (ii) the negation of an atom. Case (i): the valuation trivially makes the bare atoms true, because that's how it was chosen! Case (ii): suppose the negation of an atom, say '$\neg P$', appears on the path. Since by hypothesis the path is open and contains no contradictions, the corresponding bare atom 'P' *can't* appear on the path. So the chosen valuation will make 'P' *false*: and hence the primitive wff '$\neg P$' that does appear on the branch will, as required, be *true*. QED

Next, we will give two proofs of (C_2). First, a rather brisk official proof, then a more relaxed, intuitive one.

Official proof of (C_2) Take a completed open path, and suppose the *primitives* on this path are all true on some valuation, but also that this valuation makes one or more *other* wffs on the path false. Choose one of the *shortest* of these false wffs. Since the primitives are all true by hypothesis, this shortest false wff is non-primitive. And since the path is a completed one, this wff must have been checked off, and had the corresponding rule applied to it. Now consider cases. Suppose the shortest false wff is of the form $(A \wedge B)$; then both A and B must appear on the path, and being shorter must by assumption be true. But that's impossible: we can't have $(A \wedge B)$ false yet A and B each true. Again, suppose the shortest false wff is of the form $(A \vee B)$; then either A or B must appear on the chosen path, and being shorter must by hypothesis be true. But that's impossible: we can't have $(A \vee B)$ false yet one of its disjuncts true. The other cases are similar. Hence the idea that there are false wffs on the path, and hence shortest false wffs, leads to contradiction. Thus the chosen valuation which makes the primitives true makes *every* wff on the chosen path true. QED.

That argument is brisk, but perhaps it doesn't really bring out *why* (C$_2$) holds. So here is ...

A more intuitive proof of (C$_2$)? We'll say that the wffs S_1, S_2, ... are together *truth-makers* for W if the joint truth of S_1, S_2 ... would ensure that W is true. For example, the wffs A, B are truth-makers for (A ∧ B); the wff ¬A by itself is a truth-maker for ¬(A ∧ B). And so on.

Take another look at the tree-building rules. They say, for example, that if W is of the form (A ∧ B) we add its truth-makers A and B to its path; while if W is of the form ¬(A ∧ B) we split the path, adding the truth-maker ¬A to one fork, and the alternative truth-maker ¬B to the other. Likewise for the other rules. Each rule enjoins us to extend a path (maybe forking) by adding appropriate truth-makers for non-primitive wffs.

Now, by the definition of a completed path, *every* non-primitive wff has been checked off and 'unpacked' by the appropriate rule. So that means a completed open path set must contain truth-makers for every non-primitive wff (and if those truth-makers are non-primitive, there will be further truth-makers for *them*, until we get down to the shortest possible wffs).

Suppose, then, that we can find a valuation which does indeed make true the shortest wffs on the path, i.e. it makes the primitives true. Then, truth will spread upwards. As we work back up the path, the truth of the shortest wffs (the truth-makers for the next shortest wffs) will make those next shortest wffs true, and then the next shortest, and so on up. So *everything* will be made true so long as we make the primitives true. QED.

So that establishes (C$_1$) and (C$_2$) and hence (C).

(C) says, in effect, that we have *enough* rules for tree-building. Suppose we were allowed to use only three of the rules instead of all five/nine. If a tree built using just these three rules closed, that would still indicate a valid argument. But the fact that a tree didn't close would *not* mean that the argument is invalid: it might stay open because we didn't have enough rules for dealing with (say) disjunctions. But (C) shows we have a *complete* set of tree-building rules; there are enough rules to establish that any tautologically valid **PL/PLC** inference is indeed valid.

19.3　A corollary and a further result

First, we'll quickly note a corollary of the Basic Result: as announced in §16.4, the order in which the tree-building rules are applied (when we have a choice) can't matter. Why?

Well, note that the Basic Result entails that if an argument is tautologically valid, then *any* properly built tree will close, irrespective of the order in which we apply the relevant tree-building rules. Likewise, if an argument is invalid, *any* appropriate tree will stay open, again irrespective of the order in which tree-building rules are applied.

Secondly, let's note a result which slightly extends our argument for (C). Let's say that a set Σ of **PL** wffs is *saturated* if the following conditions obtain:

(a) if a wff of the form $\neg\neg A$ is in Σ, then so is A;
(b) if a wff of the form $(A \wedge B)$ is in Σ, then so are both A and B;
(c) if a wff of the form $\neg(A \vee B)$ is in Σ, then so are both $\neg A$ and $\neg B$;
(d) if a wff of the form $(A \vee B)$ is in Σ, then so is at least one of A and B;
(e) if a wff of the form $\neg(A \wedge B)$ is in Σ, then so is at least one of $\neg A$ and $\neg B$.

Likewise a set of **PLC** wffs Σ is saturated if it satisfies (a) to (e) plus the conditions

(f) if a wff of the form $(A \supset B)$ is in Σ, then so is at least one of $\neg A$ and B;
(g) if a wff of the form $\neg(A \supset B)$ is in Σ, then so are both A and $\neg B$;
(h) if a wff of the form $(A \equiv B)$ is in Σ, then so are either both A and B or both $\neg A$ and $\neg B$;
(i) if a wff of the form $\neg(A \equiv B)$ is in Σ, then so are either both A and $\neg B$ or both $\neg A$ and B.

(Why 'saturated'? Because such a set is saturated with truth-makers, containing a truth-maker for every non-primitive wff in it.)

We'll also say that Σ is *syntactically consistent* if it contains *no* pairs of wffs of the form A, $\neg A$. (Compare the semantic definition of logical consistency in §2.1.)

The conditions (a) to (i) will have an entirely familiar look. They correspond exactly to the rules for building up a path on a **PL** (**PLC**) tree; so it is immediate that the wffs on a completed open path form a syntactically consistent saturated set. The set will be consistent, since an open path contains no contradictions, and saturated because checking off every non-primitive wff on the path in order to complete the tree ensures that (a) to (i) are fulfilled on the path. So our earlier result that a completed open path is satisfiable is just a version of the rather more general result that *every syntactically consistent saturated set Σ of PL (PLC) wffs is satisfiable* – i.e. there is a valuation which makes all the wffs in Σ true.

The proof of this result goes exactly as before. First, we define the *chosen* valuation for Σ: if the atom A appears naked in Σ, assign it the value T; otherwise assign that atom the value F. It is simple to show – as in (C_1) – that this chosen valuation makes all the primitives in Σ true, and then – as in (C_2) – that a valuation which makes all the primitives in Σ true will make every wff in Σ true.

The advantage of talking about 'consistent saturated sets of wffs' as opposed to 'sets of wffs on a completed open branch of a tree' is that it emphasizes that the arrangement of wffs onto a tree isn't of the essence here. It's the structure described by (a) to (i) above that ensures that the set, if consistent, is satisfiable.

And, if we abstract from the case where the wffs are to be written out on a common-or-garden finite tree, then we can even allow the case where the consistent saturated set Σ contains infinitely many wffs. The proof that consistent saturated sets are satisfiable doesn't depend on any assumption of finitude. Of course, in the case of propositional logic, the consideration of infinite sets of wffs is mostly of 'technical' interest. But matters are different, as we will see in Chapter 30, when we turn to thinking about more complex logical systems.

19.4   Summary

- We have shown that the tree test is *sound*: i.e. if an argument is judged valid by the test – i.e. the appropriate tree closes – then it is indeed tautologically valid.

- We have shown that the tree test is *complete*: i.e. if an argument is tautologically valid, then the tree test will reveal it as such by the appropriate tree closing. Equivalently, if an appropriate tree stays open, the argument under test is invalid.

- The proof of these results shows that it doesn't matter in which order we apply the tree-building rules (when we have a choice).

Exercises 19

A   Show how to extend the following sets of wffs into consistent saturated sets (*hint*: think of the given sets as 'trunks' of trees; you want to extend the trunks to some complete open path).

   1.   ¬Q, (P ⊃ Q), (¬¬R ⊃ (P ∧ R))

   2.   ¬(Q ∧ (P′ ∨ R)), ¬(¬Q ∧ P), ((P′ ∧ S) ∨ (R ∧ S))

   3.   (Q ⊃ (P ∨ R)), ¬(P ⊃ S), ((P ∨ S) ∧ (¬Q ⊃ S)), ((S ∨ R) ⊃ ¬P)

In each case, what is the 'chosen valuation' for the set? Check that it does indeed make true all the wffs in the set.

B   Rewrite the arguments (C₁) and (C₂) to turn them into explicit arguments for the thesis that every syntactically consistent saturated set Σ of **PL** (**PLC**) wffs is satisfiable. Confirm that these arguments don't depend on whether Σ is finite.

Trees and proofs

The unsigned trees that we have been exploring were introduced as shorthand for signed trees, a way of setting out arguments in English *about* **PL** inferences. But, as we will now explain, we can also think of them as proofs *in* **PL**.

20.1 Choices, choices ...

Outside the logic classroom, when we want to convince ourselves that an inference is valid, we don't use techniques like the truth-table test. As we pointed out way back in Chapter 5, we usually seek *proofs*: i.e. we try to show that the desired conclusion follows from the premises by chaining together a sequence of obviously acceptable steps into a multi-step argument.

Suppose, then, that we want to develop ways of warranting **PL** inferences by means of proof-reasoning inside **PL**. How should we proceed? (By the way, take talk of '**PL**' in this chapter to cover **PLC** as well.)

It is very important to stress that there is no single 'right' way of setting out proofs. Indeed, in this book we have already touched on no less than *three* different ways of setting out multi-step arguments.

(1) In Chapter 5, we introduced proofs set out linearly, one step below another (though we also added an indentation device to signal when suppositions were being made temporarily, for the sake of argument).

(2) Then in Chapter 8 (see §8.2), we noted that some linear arguments can very naturally be rearranged into tree structures, with the initial assumptions appearing at the tips of branches, and the branches joining as we bring together propositions to derive a consequence. We work down the tree until we are left with the final conclusion of the proof at the bottom of the trunk. (Our examples happened to feature arguments in augmented English about which strings of **PL** symbols are wffs.)

(3) In Chapters 16–18, we found that some other arguments are naturally set out as *inverted* trees. We start with some assumptions at the top of the tree, and work down trying to find contradictions, splitting branches at various choice points as we descend the downwards-branching tree. (Our

examples were again arguments in augmented English, but this time arguments about the tautological validity of **PL** inferences.

Which way of regimenting arguments in (augmented) English should we follow if we want to build a proof system for warranting inferences inside **PL**?

The headline answer is this: if our primary aim is to develop ways of arguing in **PL** that echo the natural contours of ordinary-language arguments, then the first model is the one to emulate. For more advanced technical work, the second model has a lot going for it. But for ease of manipulation by beginners, downwards-branching trees are the thing.

20.2   Trees as arguments in PL

To develop that last point, note next that we in fact don't have to do any new work to get a system for building tree arguments in **PL**. *For our unsigned downwards-branching trees can already be thought of as arguments in* **PL**.

Of course, that is not how we introduced them. Given a **PL** inference

   A $(P \wedge R), \neg(P \wedge Q) \therefore \neg Q$

we set about warranting it by supposing there is a valuation which makes the premises and negation of the conclusion true, and then trying to show that that supposition leads to contradiction(s). That's all metalinguistic reasoning. Here's a corresponding tree in full-dress form, quotation marks and all:

   A′

$$'(P \wedge R)' \Rightarrow T$$
$$'\neg(P \wedge Q)' \Rightarrow T$$
$$'\neg\neg Q' \Rightarrow T$$
$$'P' \Rightarrow T$$
$$'R' \Rightarrow T$$
$$'Q' \Rightarrow T$$

$$'\neg P' \Rightarrow T \qquad '\neg Q' \Rightarrow T$$
$$* \qquad\qquad *$$

Now, we later got into the habit of writing trees in unsigned form for brevity's sake, dropping the quotation marks and then the explicit assignment of truth:

   A″

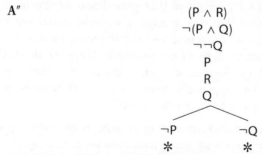

But, at least as originally introduced, a pruned tree like this still abbreviates a way of arguing metalinguistically, i.e. a way of arguing *in English* about whether there is a possible valuation which makes the premisses of the **PL** inference true and conclusion false.

However – and here's the new suggestion – we can equally well look at the pruned tree another way. *It can also be regarded as a multi-step argument framed purely in PL terms* (recall that we quietly announced at the very outset, §8.1, that '✷' belongs to the alphabet of **PL**, so *all* the symbols on the tree really do belong to that language). Imagine again wanting to defend the inference **A**, but this time without straying into meta-linguistic reasoning. We take the premisses '(P ∧ R)', '¬(P ∧ Q)' and the negated conclusion '¬¬Q' as suppositions. Then we argue step-by-step, first to 'P' and 'R', then to 'Q', and then on, via the alternatives '¬P' and '¬Q', to absurdities. Our tree argument reveals that our initial supposition leads to unavoidable contradiction. So that warrants the inference **A** as valid (compare §13.8(i)).

The point generalizes. We can systematically re-read our originally meta-linguistic truth-trees as being arguments in **PL**, arguments that aim to reveal inconsistencies. If the tree headed $A_1, A_2, ..., A_n, \neg C$ closes, i.e. ends up in contradictions, take that as an object-language argument warranting the inference $A_1, A_2, ..., A_n \therefore C$.

20.3 More natural deductions?

Still, it might very reasonably be said, this is certainly not the most natural way of building arguments in **PL**.

Consider this very simple vernacular argument:

B   Popper and Russell both wrote on political philosophy. It isn't the case that both Popper and Quine wrote on political philosophy. So Quine didn't write on political philosophy.

That is obviously valid. But if you need to convince yourself that it *is* valid, here's a natural line of argument, rather laboriously set out in the style of §5.4:

B′ (1) Popper and Russell both wrote on political   (premiss)
         philosophy.
    (2) It isn't the case that both Popper and Quine   (premiss)
         wrote on political philosophy.
             Suppose temporarily, for the sake of argument,
    (3) │   Quine wrote on political philosophy.   (supposition)
    (4) │   Popper wrote on political philosophy   (from 1)
    (5) │   Both Popper and Quine wrote on   (from 3, 4)
         │   political philosophy.
    (6) │   *Contradiction!*   (from 2, 5)
             So the supposition that leads to this absurdity must be wrong.
    (7) Quine didn't write on political philosophy.   (RAA)

Now let's render this natural reasoning into **PL**, using 'P' for 'Popper wrote on political philosophy', etc. The inference **B** gets translated as '(P ∧ R), ¬(P ∧ Q) ∴ ¬Q', i.e. as **A** again. And we might replicate the natural line of reasoning **B′** in

PL along the following lines (the numbering and English side-commentary isn't part of the argument proper).

B″	(1)	(P ∧ R)	(premiss)
	(2)	¬(P ∧ Q)	(premiss)
	(3)	Q	(supposition)
	(4)	P	(from 1)
	(5)	(P ∧ Q)	(from 3, 4)
	(6)	*	(from 2, 5)
	(7)	¬Q	(RAA)

Let's compare this 'natural deduction' in **PL** with the tree proof **A″** warranting the same inference.

One obvious difference is that **B″**, like most real-life arguments, *ends* with the desired conclusion. To be sure, there is a passage of indirect argumentation *en route*, as we make a temporary supposition for the sake of argument: but the overall argument takes us ultimately to the desired conclusion. An **A″**-type tree however always *starts* by assuming the negation of the conclusion along with the premisses, and unpicks the implications of that.

But that's perhaps not the fundamental difference. Thinking about **B″** more, note that at line (4) we move from the premiss '(P ∧ R)' to the interim conclusion 'P'. This takes us from a more complex to a simpler wff by (so to speak) *eliminating* the connective '∧' according to the rule '∧-Elimination':

(∧E)   From a wff of the form (A ∧ B) we can infer A, and equally we can infer B.

But then at line (5) we move from the two previous steps, 'P' and 'Q' to their conjunction '(P ∧ Q)'. Another perfectly valid inference of course! This time, however, we move from a pair of simpler wffs to a more complex one by *introducing* the connective '∧' into a wff according to the rule

(∧I)   From the wffs A, B we can infer their conjunction (A ∧ B).

This is typical of most proof systems. They have (in a phrase) both *elimination* and *introduction* rules for using connectives in inferences.

And now contrast the tree-building rules invoked in tree proofs (§17.1). These rules *always* take us from more complex to simpler wffs. For example, if a conjunction like '(P ∧ Q)' is already on a path, we add its conjuncts; but we never move in the opposite direction (never take a pair of conjuncts and add the conjunction to the tree). *Our tree-building rules are all 'elimination' rules.*

The two noted differences between 'natural deductions' and tree proofs are of course intimately related. If we want to be able to argue directly *to* conclusions, and conclusions can themselves be complex, we'll typically need to be able to build up complex wffs by introducing connectives: so we'll need introduction rules.

However, once we have introduction rules available, proof-building can be trickier than in an elimination-rules-only system. Imagine, for example, that we'd started the proof **B″** in the same way but then got stuck in a groove, applying and reapplying the rule (∧I):

B″	(1)	(P ∧ R)	(premiss)
	(2)	¬(P ∧ Q)	(premiss)
	(3)	⎸Q	(supposition)
	(4)	⎸P	(from 1)
	(5)	⎸((P ∧ R) ∧ P)	(from 1, 4)
	(6)	⎸(¬(P ∧ Q) ∧ ((P ∧ R) ∧ P))	(from 2, 5)
	(7)	⎸(Q ∧ (¬(P ∧ Q) ∧ ((P ∧ R) ∧ P)))	(from 3, 6)
	...	⎸...	

These inferences are all perfectly valid, but take us further and further away from our desired conclusion. They are, no doubt, daft inference moves to make, given we are trying to prove '¬Q'. But that's the point: having introduction rules may well allow us to generate more 'natural' proofs, but it also allows us to drift off into cascades of irrelevancies. By contrast, systems of tree proofs (which only have elimination rules) put very much tighter constraints on what we can do at any point in building a proof.

There's no right choice here. It's a trade-off between 'naturalness' and ease of proof-building. For experience suggests that trees *are* easier for beginners (after all, we've already pretty painlessly slipped into using them). Less thought is required to get a tree proof to work. This is already so for simple propositional logic: but is emphatically the case when we turn to the logic of quantifiers.

Which all goes to explain our choice of a less 'natural' proof-style in this book. (This way, we can still convey the general idea of a formal proof-system, without having to spend too much time talking about strategies for proof-discovery in natural deduction systems.)

20.4 What rules for trees?

So we'll say no more about 'natural deduction' proofs (whether linear proofs like **B″** or rearranged into downward-joining trees as in §8.2). But even if we fix on downward-branching trees as our proof-structure of choice, that doesn't fully determine which rules of proof-building we should adopt for truth-functional logic. In §17.1 we fixed on one set of rules to adopt, but it isn't unique.

The basic rule for tree-building is: extend an open branch by adding to it further wffs which must also be true assuming that wffs already on the branch are true (splitting the tree into two branches when we have to consider alternatives).

Given that basic principle, we could adopt – for example – the following rules for dealing with negated disjunctions and negated conjunctions. Schematically,

$$\begin{array}{cc} \neg(A \lor B) & \neg(A \land B) \\ | & | \\ (\neg A \land \neg B) & (\neg A \lor \neg B) \end{array}$$

These new rules are legitimate because each tells us to add to the tree a wff which must be true if the given wff is true. In fact, we could *replace* our original rules (c) and (e) with these two new rules. For note, when combined with the existing rules (b) and (d) for unpacking conjunctions and disjunctions, our new rules yield

$$\neg(A \vee B)$$
$$|$$
$$(\neg A \wedge \neg B)$$
$$|$$
$$\neg A$$
$$\neg B$$

$$\neg(A \wedge B)$$
$$|$$
$$(\neg A \vee \neg B)$$
$$\nearrow \qquad \diagdown$$
$$\neg A \qquad \neg B$$

which, in two steps, give us back the effects of our original one-step rules.

That kind of possible variation in choice of rules is obviously pretty trivial. Here is a more significant variation. For any C, $(C \vee \neg C)$ is *always* true on any valuation of its atoms. Hence adding a wff of that form to a branch would conform to the general principle that we can extend branches by adding further wffs which are true on the valuation which makes the initial wffs true. So we *could* introduce the rule that we can at any point extend a branch by adding an instance of the law of excluded middle (§12.1; see also Exercise 18B). As you'd expect this makes no difference at all to what we can prove using trees. A major cost of adding such a rule is that it introduces into our tree system the possibility of getting lost adding irrelevancies. But intriguingly, having this additional rule available can also *speed up* proofs radically (at least, that's true when we extend trees to cover quantifier logic, in the style of Chapters 25 onwards). George Boolos has given a pretty example of an inference which can be proved valid in a few pages when this new rule is available, and for which 'the smallest closed tree contains more symbols than there are nanoseconds between Big Bangs' when it isn't! So if we particularly care about maximizing the number of proofs which are *feasible* (can be done in a sensible amount of time) we'll add the suggested new rule – and if we don't, then we won't.

These last remarks, like the discussion of the previous section, again drive home the point that there is no single 'right' proof system for truth-functional propositional logic. Which type of system we prefer to use, and which rules we adopt within that type of system, is a matter of how we weight their various virtues and shortcomings. Don't be fazed when you discover that different logic texts make different choices.

20.5 '⊢' and '⊨', soundness and completeness

This is as good a place as any to introduce another bit of standard symbolism which you could well encounter elsewhere. Suppose we pick on some preferred style of proof layout, and some set of rules of inference for propositional logic. Call that proof system S. If there is some array of wffs which counts as a proof in S with premisses $A_1, A_2, ..., A_n$ and conclusion C, we write $A_1, A_2, ..., A_n \vdash_S C$.

Compare: '⊨' is a metalinguistic symbol, added to English (in §13.7) for abbreviatory purposes, to express the *semantic* relation that holds between some premisses and a conclusion when there is no valuation which makes the former all true and the latter false. '$\vdash_S$' is also a metalinguistic symbol, added to English for abbreviatory purposes, this time to express the *syntactic* relation that holds when we can construct a suitable proof in system S. This is a syntactic relation because whether an array of wffs counts as a proof doesn't depend on their meaning, but only on whether the wffs are related in the ways allowed by the formal proof-building rules of S. (The use of the single turnstile for the syntactic proof relation and the double turnstile for the semantic entailment relation is an absolutely standard notational convention: learn it. If context makes it clear what proof-system S is in question, the subscript can be dropped.)

Obviously, for any proof system worth its keep, we want the following to hold for any A_i and C:

(S)   If $A_1, A_2, ..., A_n \vdash_S C$ then $A_1, A_2, ..., A_n \vDash C$.

Suppose that *didn't* hold. Then we could have a 'proof' in S that C followed from the A_i when in fact C isn't entailed by the premisses. The proof system wouldn't be reliable, or *sound* (to use the usual term).

We'd also like the following:

(C)   If $A_1, A_2, ..., A_n \vDash C$ then $A_1, A_2, ..., A_n \vdash_S C$.

In other words, what we'd like is that, whenever the A_i taken together do in fact entail C, then there is a proof in S which reflects that. We'd like our proof system to be adequate to the task of warranting all the tautologically valid inferences, or to be *complete* (again to use the standard term). There are a large number of variant proof systems for propositional logic on the market which are, as we want, sound and complete.

Returning from generalities to the particular case of tree proofs: let's now write $A_1, A_2, ..., A_n \vdash_T C$ when there is a closed **PL/PLC** tree headed by $A_1, A_2, ..., A_n, \neg C$. Then we indeed have soundness and completeness, i.e.

(SC) $A_1, A_2, ..., A_n \vdash_T C$ if and only if $A_1, A_2, ..., A_n \vDash C$.

For that just neatly sums up what we proved in Chapter 19.

20.6 Summary

- **PL** trees were introduced as (shorthand for) metalinguistic arguments about truth-valuations of **PL** wffs. But closed trees can easily be reinterpreted as object-language proofs (proofs of the inconsistency of the set of wffs $A_1, A_2, ..., A_n, \neg C$, thus warranting the inference $A_1, A_2, ..., A_n \therefore C$).

- Tree proofs are not the most natural proofs, but are perhaps most easily managed by beginners.

- Different proof systems have different pros and cons: there is no 'right' system for truth-functional logic.

Interlude

After propositional logic

In the last seven chapters:

- We have argued that the material conditional is the only candidate for translating ordinary language conditionals by a truth-function, and we have discussed how good (or bad) a translation this is.

- We found it convenient, though not strictly necessary, to add a special symbol to **PL** to express the material conditional, and a symbol for the material biconditional too, giving us the extended language **PLC**.

- We introduced truth-trees as a 'fast track' way of determining whether a **PL** or **PLC** inference is tautologically valid.

- We showed that the tree method does indeed work as advertised (i.e. a correctly constructed tree closes if and only if the argument being tested is valid).

- Finally, we saw how to reconstrue trees (originally introduced as regimenting metalinguistic reasoning *about* **PL/PLC** wffs) as proofs *in* **PL/PLC**.

But put together with what has gone before – now making some twenty chapters and nearly two hundred pages in total – this still hasn't taken us beyond elementary propositional logic.

Which, you might very well think, is really rather too much of a good thing. And, true enough, our careful extended treatment of propositional logic is none too exciting in itself. However, our main reason for dwelling on it is that it *does* now give us a very secure basis from which to move on to tackle something much more substantial, namely *predicate logic* (or *first-order quantification theory*, to use a more technical label).

This theory deals with the logic of inferences involving quantifiers such as 'all' and 'some' and 'no' as well as the connectives. It's arguable that this, together with the logic of identity, in effect covers *all* the deductive reasoning that we need in mathematics and science. But whether or not that bold thesis is right, it is undeniable that quantificational logic is very rich and powerful.

We'll adopt the same two-stage strategy we used in tackling propositional logic. So we design a suitable formal language **QL** for expressing quantified propositions; then we look at how to assess arguments when framed in **QL**.

- We'll find that, while the syntax and semantics of **PL/PLC** was very straightforward, the design of the language **QL**, with its quantifier/variable notation, requires more sophistication.

- We can define a notion of q-validity for **QL** inferences, which is a sort of generalization of the idea of tautological validity for **PL/PLC** inferences.

- But there is no prospect of being able to mechanically test for q-validity in the way that we can mechanically test for tautological validity (in fact, we can *prove* that there is no mechanical test for q-validity).

- That means that when warranting **QL** inferences, proof methods really come into their own. We'll build a proof system on the basis of the now familiar tree proofs for propositional logic. It is a relatively easy extension.

Our first task, then, is to motivate the quantifier/variable notation on which **QL** is built.

Quantifiers

In this chapter, we reflect on some of the vagaries of quantifiers in ordinary English. We find some pointers for the design of an artificial language suitable for exploring the logic of quantifiers.

21.1 Quantifiers in arguments

In Chapters 7 to 20, we were concerned with a very limited class of inferences, those whose validity or invalidity turns solely upon the way that the premises and conclusion are built up using the propositional connectives. So we didn't need to discern any internal structure below the level of whole sentential clauses. We are now going to consider inferences whose validity (or lack of it) *does* depend on a finer level of structure. Take for example:

> Fido is a cat.
> All cats are scary.
> So, Fido is scary.

> The Baron hates anyone who loves Alma.
> Fritz loves Alma.
> So, the Baron hates Fritz.

> Some philosophers admire every logician.
> No philosopher admires any charlatan.
> So, no logician is a charlatan.

These three arguments are all clearly valid, and equally clearly their validity is not a matter of the distribution of propositional connectives (because they don't contain any such connectives!). If we exposed no more than the way that whole sentential clauses are related by any propositional connectives, then these arguments could only be assigned the unrevealing and utterly unreliable inferential form A, B, so C. We therefore need to dig deeper.

What is crucial for these arguments is the occurrence of *quantifying* expressions such as 'all', 'anyone', 'some', 'every', 'no', which are used to form complex terms like 'all men', 'some philosophers' etc. Our target is to understand the logic of such arguments.

21.2 Quantifiers in ordinary language

English expressions of generality behave in tediously complex ways. For example, consider the following as stand-alone claims:

(1)   Every play by Shakespeare is worth seeing.
(2)   Any play by Shakespeare is worth seeing.

They express the same *universal generalization* about all the plays. But compare the way the same sentences embed in wider contexts:

(3)   I don't believe that every play by Shakespeare is worth seeing.
(4)   I don't believe that any play by Shakespeare is worth seeing.
(5)   If every play by Shakespeare is worth seeing, I'll eat my hat.
(6)   If any play by Shakespeare is worth seeing, I'll eat my hat.

The messages expressed in each pair are quite different: so 'every' and 'any' sentences are not generally equivalent.

Next, consider:

(7)   All plays by Shakespeare are worth seeing.
(8)   Each play by Shakespeare is worth seeing.

These also say, it seems, just the same as (1) and (2). And compare, e.g.

(9)   I don't believe that all plays by Shakespeare are worth seeing.
(10)   If all plays by Shakespeare are worth seeing, I'll eat my hat.

(9) is equivalent to (3), and (10) to (5). So sentences containing 'all' typically behave on embedding like sentences with 'every' rather than 'any'. Likewise for 'each' sentences (though they don't seem to embed quite so naturally). So can at least we say that 'all', 'each' and 'every' constructions are just stylistic variants?

Well, here too there are complications. Not every occurrence of 'all' can be simply replaced by 'every' or 'each': consider 'All fresh snow is white'. And some 'all' propositions seem naturally to take the reading 'all, considered as totality', while a corresponding 'each' proposition takes the reading 'each, considered separately'. For example, compare:

(11)   Jack can read all of the plays in a day.
(12)   Jack can read each of the plays in a day.

(11) reports a feat of speed-reading; (12) makes Jack sound a bit slow.

So, given that we are interested in the basic logic of arguments involving generalizations, but are not especially interested in theorizing about the complexities of English usage, it seems very attractive to pursue the strategy of constructing an artificial language with just *one*, simply-behaved, way of forming universal generalizations. Similarly for some other quantifiers. Then, as with propositional logic, we can tackle the assessment of quantificational inferences in two stages. First, we translate the argument into our artificial language, treating the quantifiers in a uniform and tidy way; and then we assess the translated argument for validity.

21.3   Quantifiers and scope

There's another reason for pursuing this strategy of building an artificial language, in addition to wanting to side-step the multiple varieties of English usage. For English expressions of generality are beset by *scope ambiguities* (for introductory remarks on scope ambiguity, see §7.2 and §9.6, where we first encountered the notion in connection with structurally ambiguous sentences of the kind *A or B and C*).

To work up to an example, consider the following pair of sentences:

(1)   Some Republican senators smoked dope in the nineteen sixties.
(2)   Some women died from illegal abortions in the nineteen sixties.

As a shock headline in the newspaper, (1) is naturally read as saying, of some present Republican senators, that they have a dark secret in their past (namely that they dabbled in illegal drugs in the sixties). But (2) is not to be read, ludicrously, as saying of some present women that they have a secret in *their* past (namely that they died of an illegal abortion back in the sixties). The claim is of course that some women *then* existing died from illegal abortions. A natural way of describing the situation is to say that we treat the tense operator 'in the sixties' as having a different *scope* in the two sentences. In the first case we take the tense operator to modify only the verb phrase 'smoked dope', while in the second case it is the whole proposition 'some women died of illegal abortions' which is (so to speak) to be evaluated with respect to the sixties. With a bit of grammatical butchery of tenses, and adding brackets for maximal clarity, we might thus represent the natural readings of the two sentences as follows:

(1′)   (Some Republican senators are such that)(in the sixties it was true that) they smoke dope.
(2′)   (In the sixties it was true that)(some women are such that) they die from illegal abortions.

Here the final verbs are now deemed to be tenseless, and the order of operators shows the relative scopes of the tense operator and the quantifier, i.e. shows how much of the sentence concerns the past.

Consider, however, the following sentence:

(3)   Some senior Republican senators plotted against the party leadership in 1988.

This seems simply ambiguous: are we saying that some of the *then* Republican senators plotted, or rather that some of the *current* lot plotted when they were young turks? We could be saying either

(3′)   (In 1988 it was true that)(some senior Republican senators are such that) they plot;

or else

(3″)   (Some senior Republican senators are such that)(in 1988 it was true that) they plot.

Both readings are fairly naturally available, so the original sentence is ambiguous. And the ambiguity is plausibly diagnosed as a *scope ambiguity*. The issue is: which has wider scope, which governs more of the sentence, the tense operator or the quantifier?

Mixing expressions of generality and *tense* operators is prone to produce scope ambiguities. So too is mixing quantifiers with *modal* operators (i.e. with expressions of necessity and possibility). Consider, for example, the philosophical claim

(4)   All sensory experiences are possibly delusory.

Does this mean that each and every sensory experience is one which, taken separately, is open to doubt, i.e.

(4′)   (Every sensory experience is such that)(it is possible that) it is delusory?

Or is the thought that the whole lot could be delusory all at once, i.e.

(4″)   (It is possible that)(every sensory experience is such that) it is delusory?

These claims are certainly different, and it is evidently philosophically important which reading is meant e.g. in arguments for scepticism about the senses. For even if (4′) is true, (4″) doesn't follow. Compare

(5)   Every competitor might possibly win,

and its two readings:

(5′)   (Every competitor is such that)(it is possible that) he wins.
(5″)   (It is possible that)(every competitor is such that) he wins
        = (It is possible that) every competitor wins.

It could be that (5′) is true because the tournament is fairly run and the competitors well matched, while (5″) is false because it is a knock-out tournament with rules ensuring that only one ultimate winner will emerge.

However, modal logic and the logic of tense operators are both beyond the scope of this book. So, to come to the kinds of case that *do* immediately concern us, let's consider next what happens when we mix expressions of generality with *negation*.

We'll take just one kind of example. Usually, a statement of the kind 'Some *F*s are not *G*s' is heard as entirely compatible with 'Some *F*s are *G*s'. For example, 'Some students are not logic enthusiasts' is no doubt sadly true; but it is compatible with the happier truth 'Some students *are* logic enthusiasts'. But now consider the following case. Jack says 'Some footballers deserve to earn five million a year'. Jill, in response, takes a high moral tone, expostulates about the evils of huge wages in a world of such poverty and deprivation, and concludes: 'So some footballers damned well do *not* deserve to earn five million a year'. Jill certainly doesn't intend to be heard as saying something compatible with Jack's original remark! Thus, though vernacular English sentences of the form 'Some

*F*s are not *G*s' are usually construed as meaning

(6′)   (Some *F*s are such that)(it is not the case that) they are *G*s,

in some contexts the negation can be understood to have wide scope, i.e. to govern the whole sentence, with the resulting message being:

(6″)   (It is not the case that)(some *F*s are such that) they are *G*s.

As in the tensed and modal cases, then, there is a potential for scope ambiguity when we mix quantifiers with negation – though context, emphasis, and plausibility considerations very often serve to privilege one reading and prevent misunderstanding.

Finally, let's consider what can happen when a sentence contains more than one quantifier.

Suppose you discover each of these is true:

(7)   Jack is besotted with some film star;
(7′)   James is besotted with some film star;
(7″)   Jeremy is besotted with some film star.

Here you have found a property shared by each of Jack, James and Jeremy, namely the property of being bewitched by some star or other (not necessarily the same one in each case). You might be tempted to generalize from that sample to the rest of the logic class, and arrive at the rather depressing speculation that

(8)   *Every* man in the class is besotted with some film star,

meaning that every man is like Jack, James and Jeremy in being enthralled by some star or other.

Now a rather different scenario. This time you wonder whether any of these are true:

(9)   Every man in the class is besotted with Cameron Diaz,
(9′)   Every man in the class is besotted with Gwyneth Paltrow,
(9″)   Every man in the class is besotted with Jennifer Lopez.

Here you have predicated the same property of each of Cameron Diaz, Gwyneth Paltrow and Jennifer Lopez, i.e. the property of being warmly admired by all men in the logic class. Now perhaps each of these predications turns out to be false (there is an exceptional man in each case). You might then retreat to the less specific speculation that perhaps

(10)   Every man in the class is besotted with *some* film star

meaning that there is some star who is admired by every man in the group. Here we have the same doubly quantified sentence again as (8), but in the new context it conveys a different message. For there is all the difference between saying that everyone admires someone or other and saying that someone is admired by everyone (just as there is all the difference between saying that every road leads somewhere or other, and saying that there is somewhere where every road leads: compare §4.4). So the ordinary English sentence 'Every man in the class is besotted with some film star' is potentially ambiguous between two readings:

(10′)   (Every man in the class is such that)(there is some film star such that) he is besotted with her.

(10″)   (There is some film star such that)(every man in the class is such that) he is besotted with her.

Again, therefore, we have a potential scope ambiguity, this time when we get certain doubly quantified sentences. Though again we should not exaggerate the problem; context and perhaps a default preference for the first reading normally serve to remove this ambiguity in practice.

Note that these four kinds of scope ambiguity, where quantifiers are mixed with tense, modality and negation and other quantifiers, do *not* occur when proper names are used instead of the quantifier expressions. Consider:

(3°)   John Doe plotted in 1988.

(5°)   Kasparov might possibly win.

(6°)   Michael does not deserve five million a year.

(10°)   Jack is besotted with Gwyneth Paltrow.

Those are all quite unambiguous. As it is sometimes put, genuine names are *scopeless*; whereas with quantifiers issues of relative scope can arise.

So all this points to a key design requirement on an artificial language apt for use when we are exploring the logic of generality. *We must devise our notation so that the scope of a quantifier is always clear and there is no possibility of scope ambiguities.* And this means building in at ground level some way of marking the contrast between quantifiers and scopeless names.

21.4   Expressing quantification unambiguously

How then shall we express generality in a formal language? Well, consider again the linguistic device which we slipped into using above, i.e. the device of using quantifier operators like 'Every man in the class is such that', 'Some Republican senator is such that', ... These quantifier operators are *prefixed* to sentential forms constructed with pronouns, such as 'he wins', 'he plotted', 'he is besotted with her', and so forth. The *order* of the quantifier operators with respect to other quantifiers and negation (or other operators like tenses and modalities) then gives their relative scope according to the obvious rule *operators apply only to what follows them.* This very natural device enables us to express ourselves very clearly and without ambiguity, even if at the expense of a little prolixity. So why not use the very same device in a formal language?

There is just one wrinkle: suppose we want to express the claim that *everyone loves someone* using prefixed quantifiers. Then it won't quite do to regiment the natural reading as

(1)   (Everyone is such that)(there is someone such that) he/she loves him/her.

For that just introduces a new ambiguity. Which quantifiers are the pronouns respectively tied to? Is it, so to speak, everyone or someone who is doing the loving? But it is easy to rephrase: we just need something like

(2) (Everyone is such that)(there is someone such that) the former loves the latter.

This is plainly to be distinguished from the other, unnatural, reading

(3) (There is someone such that)(everyone is such that) the latter loves the former.

The expressions 'the former', 'the latter' function as pronouns linked to particular quantifiers. If we were dealing with more than two quantifiers, we could instead use 'the first', 'the second', 'the third', as in (say)

(4) (Every integer is such that)(every integer is such that)(there is some integer such that) the third is the product of the first and second,

which says that multiplying any two integers yields an integer.

Using pronouns like 'former/latter' or 'the first/second/third' in a formal context, however, would have considerable drawbacks. For when manipulating complex sentences with many quantifiers, what counts as 'the first' or 'the second' etc. will keep changing. So here's a neat alternative trick, borrowed from the mathematicians. *We introduce some new pronouns 'x', 'y', 'z' ...; and then we tag our quantifier prefixes with these new pronouns in order to signal which quantifier goes with which pronoun.* Then we can say 'For any integers x and y there's an integer z such that z is the product of x and y'. Or more stiltedly, we can put it like this:

(4') (Every integer x is such that)(every integer y is such that)(there is some integer z such that) z is the product of x and y.

This use of 'x', 'y' and 'z' of course comes very naturally in this mathematical example: but the same notation could be readily adopted for use much more widely, so that we write, e.g.

(2') (Everyone x is such that)(someone y is such that) x loves y.
(3') (Someone y is such that)(everyone x is such that) x loves y.

There is nothing mysterious about the use of the so-called *individual variables* 'x' and 'y' here: they are simply a snappier replacement for everyday pronouns.

Now for a bit of mere abbreviation. Instead of 'every x is such that', we could use simply the notation '$\forall x$', with the rotated 'A' reminding us that this is also the *all* operator. And instead of 'there is some y such that' let's simply use '$\exists y$', with the rotated 'E' reminding us of another way of expressing that quantifier, namely 'there *exists* some y such that'. Then, at least given a context where it is clear that what is in question are people, our last examples can be abbreviated very briskly, as

(2") $(\forall x)(\exists y) x$ loves y,
(3") $(\exists y)(\forall x) x$ loves y.

Given a different context where it is clear that what is in question are integers, we can equally crisply write our arithmetical proposition

(4″) $(\forall x)(\forall y)(\exists z)(z$ is the product of x and $y)$.

Now, so far, all this could still be treated as just so much augmented, abbreviated English. But, in essence, it exemplifies the notation for generality we will be adopting in **QL**, a version of the standard logicians' language for encoding quantificational arguments. We will use variables corresponding to 'x', 'y', 'z' to do the work of pronouns, and use prefixed quantifiers on the model of '$\forall x$', '$\exists y$' to do the work of 'Every x is such that' and 'Some y is such that'. And that, apart from a few bells and whistles, will give us all we need.

In sum, the design of **QL** is guided by one very simple and compelling insight, namely that propositions involving quantifiers can be unambiguously expressed by using prefixed quantifier operators applied to core clauses involving pronouns. Once that basic idea has been understood, the rest really is plain sailing!

21.5 Summary

- Ordinary English expressions of generality behave in complex ways (consider the availability of 'all', 'every', 'any', and 'each' to express universal generalizations, and compare the behaviour of these quantifiers in complex contexts). And sentences involving such quantifiers are prone to scope ambiguities.

- So (following the 'divide and rule' strategy introduced in §7.3) we want to devise a formal artificial language **QL** that avoids the complexities and scope ambiguities of English. Our guide will be a rather heavy handed and prolix apparatus already available in English, i.e. the use of prefixed quantifier operators of the kind 'everyone is such that', 'someone is such that', which can be used to express messages unambiguously.

- So, **QL** will be supplied with quantifiers like '$\forall x$', '$\exists y$', etc. which operate in a similar way, along with variables like 'x', 'y', etc. to act as corresponding pronouns.

QL introduced

In the last chapter we introduced the idea of the quantifier/variable notation for expressing general propositions. This chapter gives a relaxed, informal sketch of the formal language **QL** which incorporates that idea.

22.1 Names and predicates

We want, for a start, to be able to formalize arguments like 'Fido is a cat; all cats are scary; so, Fido is scary'. We are going to use a quantifier/variable notation for expressing the second, general, premiss. But as a preliminary step, we first need to adopt a standard mode of expressing and revealing the structure of *non-general* propositions such as the first premiss, 'Fido is a cat'. We therefore need, as basic ingredients of **QL**, a stock of *names* and *predicates*.

Since we won't be digging any deeper than the name/predicate/quantifier structure of sentences, the names and predicates in **QL** might as well be as simple as possible, i.e. single letters. Hence, for the *names* (or as we officially call them, *individual constants*) let's for a start use

m, n, o, ...

(i.e. lower case letters, not from the end of the alphabet).

A predicate comes with a number of slots waiting to be filled up with names (or pronouns, etc.) to form sentences. Thus compare the one-place English predicates '... smokes', '... is a philosopher'; the two-place predicates '... is besotted with ...', '... was a pupil of ...'; the three-place predicate '... prefers ... to ...', '... is between ... and ...'; and so forth. We will similarly supply **QL** with

one place predicates: F, G, H, ...
two place predicates: L, M, ...
three place predicates: R, S, ...

and so on (all upper case letters).

Then the rule for constructing the most basic sort of sentence of **QL** is simply that an *n*-place predicate combines with *n* names (not necessarily different ones) to form an *atomic wff*. And just as a matter of convention (of no deep significance at all) we form that combination *by writing the predicate followed by the* *right number of names.* Thus

    Fn, Lnm, Loo, Rnmo, Romo

are sample atomic wffs.

   Of course, when we come to present the official syntax of **QL**, we'll need to be more precise, and not just give open-ended lists of the expressions which are to count as names/constants and predicates. We'll then also add to the class of atomic wffs. But in this chapter we will continue to give an incomplete and informal presentation (diving into the full-dress official story straight away would introduce distracting complications).

   So much, then, for some unofficial syntax for atomic wffs. As for the informal interpretation of our atomic wffs, that is equally straightforward. We fix on some code-book or interpretation for **QL**, for instance the one where

   'm' means *Socrates*
   'n' means *Plato*
   'o' means *Aristotle*

   'F' means ... *is wise*
   'L' means ... *loves* ...
   'R' means ... *prefers* ... *to* ...

Then the atoms above will, of course, get the following interpretation:

   'Fn' means *Plato is wise.*
   'Lnm' means *Plato loves Socrates.*
   'Loo' means *Aristotle loves himself.*
   'Rnmo' means *Plato prefers Socrates to Aristotle.*
   'Romo' means *Aristotle prefers Socrates to himself.*

Note, the order in which names occur when combined with a predicate matters in **QL** just as it does in English.

22.2 Connectives in QL

QL has two ways of building up more complex expressions from simpler ones. The application of quantifiers is one. But first, we will just take over from **PLC** our five propositional connectives (including '⊃' and '≡', and keeping the familiar bracketing conventions). The result of applying these connectives to a wff or pair of wffs of **QL** again yields another wff. So the following are wffs:

   (Fm ∨ Fn)
   ¬(Lnm ∧ Lmn)
   (Rnmo ⊃ (Lnm ∧ ¬Lno))
   ((Fm ∨ Fn) ≡ (Lom ∨ Lon))

The meaning of the connectives remains as before. (Simple exercise: translate these, interpreting the names and predicates as in the previous section.)

   The only point to emphasize here is that the connectives added to **QL** remain essentially *propositional* connectives, i.e. they continue to combine proposition-like clauses. So we must translate

Plato and Aristotle are wise

as

(Fn ∧ Fo)

and *not* as

F(n ∧ o)

A propositional connective cannot connect constants; hence '(n ∧ o)' is simply ill-formed. Similarly, to translate

Plato is wise and handsome

we need something like

(Fn ∧ Gn)

and *not*

(F ∧ G)n

which is also ill-formed; a propositional connective can't connect QL predicates.

22.3   Adding the quantifiers

We are following the design plan of §21.4, so we need to add to our language a stock of *pronouns*, more often called *individual variables*:

x, y, z, ...

(i.e. lower case letters from the end of the alphabet). We also need a couple of symbols to use in forming quantifiers

∀, ∃.

Either of those symbols followed by an individual variable yields a *quantifier*. Thus

∀x, ∀y, ∀z, ...

are *universal quantifiers* meaning, roughly, 'everyone/everything x is such that', etc. We can also read them, more snappily, as 'for all x', 'for all y', etc. Likewise,

∃x, ∃y, ∃z, ...

are *existential quantifiers* meaning, roughly, 'someone/something x is such that', etc. We can read them, more snappily, as 'for some x', 'for some y', etc.

By the way, in our informal remarks before (§21.4), we used bracketing round quantifiers: we wrote, e.g., '(∀x)(∃y)x loves y'. Such brackets aren't needed, though many logic books do retain them. We will henceforth opt for the more modern, bracket-free, notation. You will have no trouble at all reading the bracketed version if you are familiar with our austere version. (Though note that some older books use simply '(x)' instead of '(∀x)'.)

In the end, we want to construct complete quantified sentences, without pronouns dangling free (i.e. without pronouns that aren't tied to associated quantifiers). So a natural first rule for building up **QL** wffs introduces quantifiers and variables *together*. The rule is: Take a wff with one or more occurrences of some constant (i.e. some name); we then '*quantify in*' to the place(s) in the wff occupied by the constant; i.e. we replace all the occurrences of the constant with some variable, and then prefix a quantifier formed from the same variable. Put schematically, the rule is:

> Take a **QL** wff $C(\ldots c \ldots c \ldots)$ containing at least one occurrence of the constant c. Replace all the occurrences of c with some variable v new to C, and prefix the result with $\forall v$ or $\exists v$, so we get $\forall v C(\ldots v \ldots v \ldots)$ or $\exists v C(\ldots v \ldots v \ldots)$. That yields another wff.

The following expressions are then wffs by our rules:

> Rmno
> $\exists z$Rzno
> $\forall x \exists z$Rznx
> $\forall y \forall x \exists z$Rzyx

At each step here, we quantify in to one of the remaining places held by a constant using a new variable. Why do we insist on *new* variables at each step? Well, otherwise we could form e.g.

> Rmno
> $\forall x$Rxno
> $\forall x \forall x$Rxnx
> $\forall x \forall x \forall x$Rxxx

In the last two expressions, the duplicate quantifiers are (at best) redundant.

Here's another longer chain of wffs, built up in stages from atoms by applying connectives or quantifying in:

> Fn
> Rmno
> (Fn $\land$ Rmno)
> $\exists x$(Fn $\land$ Rmnx)
> $\forall y \exists x$(Fn $\land$ Rynx)
> $\neg \forall y \exists x$(Fn $\land$ Rynx)
> (Fn $\supset \neg \forall y \exists x$(Fn $\land$ Rynx))
> $\forall z$(Fz $\supset \neg \forall y \exists x$(Fz $\land$ Ryzx))

Note that, at the last stage, we are quantifying in to all three positions held by 'n'.

Now for the basic rules for interpreting quantified sentences. When we assert '$\forall x$Fx', we are saying 'Everything x is such that x is F' (fill in the appropriate interpretation of 'F'). But what counts as 'everything'? What are we generalizing over? It could be people, or living beings, or stars, or numbers, or .... Until that's fixed, we won't know how to interpret the universal quantification. So, in order to interpret quantified formulae, we need a specification of the *domain of discourse*, where the domain is the collection of objects that we are 'quantifying over', the set of 'everything' we are talking about. (And we'll take it that the domain of discourse isn't empty, i.e. it contains at least one object: for more on this, see §§27.2, 29.3.)

'∀xFx' is then interpreted as saying that *everything in the domain of discourse is F*. Similarly, '∃xFx' says that *something in the domain of discourse is F*.

Extending this to more complex quantified wffs is fairly straightforward. Just bear in mind the equivalence between a universal quantifier ∀v and the English 'everything (in the domain) is such that', and an existential quantifier ∃v and the English 'at least one thing (in the domain) is such that'.

Putting that a bit more carefully, suppose again that C(...c...c...) is a wff containing one or more occurrences of the constant c. This says something more or less complicated about the object picked out by c. In other words, C(...c...c...) attributes to that object some more or less complex property. Then:

> ∀vC(...v...v...) says that *everything* in the domain of discourse has the property which plain C(...c...c...) attributes to the object picked out by c.
>
> ∃vC(...v...v...) says that *at least one thing* in the domain has the property which plain C(...c...c...) attributes to the object picked out by c.

Some illustrations will help. Go back to the interpretation of the names and predicates introduced at the end of §22.1. Then 'Fn' is a wff meaning that Plato is wise. Quantifying in, this is a wff too:

∃xFx

To interpret it we need to know the domain of discourse. Suppose it is people. Then the wff will mean *someone is wise*. Hence

¬∃xFx

is also a wff and says that it isn't true that someone is wise, i.e. it says *no one is wise*. (By the way, this quick availability of a way of expressing 'no one' propositions shows why there would be no point in adding a special quantifier to QL to capture the 'no' quantifier.)

Secondly, '¬Fn' is also a wff (saying that Plato is unwise). By the syntactical building rule, we can quantify in to produce another wff,

∃x¬Fx.

And this (by the interpretation rule) says that someone has the property attributed to Plato by '¬Fn'; i.e. it says *someone is unwise*.

Note then that our last two wffs, which mix quantifiers and negation, are each quite unambiguous. As desired, we lose the potential scope ambiguity of the English 'Someone is not wise'.

Another example: '(Fn ⊃ Lmn)' says that if Plato is wise, then Socrates loves him. In other words, it attributes to Plato the complex property of being, if wise, then loved by Socrates. Replace the 'n' by a variable 'x', and quantify in:

∀x(Fx ⊃ Lmx)

This says of everyone what (Fn ⊃ Lmn) says of Plato; i.e. it says that everyone is such that *if* they are wise, *then* Socrates loves them (note: the connective still connects propositional clauses, though ones involving pronouns). In simpler words, it says: Socrates loves anyone who is wise.

A third example: since 'Lmn' is a wff, so is

$$\exists y Lmy.$$

By the associated semantic rule, this says that Socrates loves someone or other. Applying the syntactic rule for a quantifier building again, we can quantify in to construct another wff

$$\forall x \exists y Lxy.$$

By the semantic rule for reading ∀–propositions, this sentence says that everyone has the property attributed to the bearer of 'm' by '∃yLmy': i.e. it says that *everyone* has the property of loving someone or other (i.e. everyone is such that there is someone they love). Starting again from 'Lmn', we can also form the wff

$$\forall x Lxn$$

which says that Plato is loved by everyone; and then we can go on to construct

$$\exists y \forall x Lxy.$$

By the interpretation rule, this says that someone has the property that '∀xLxn' attributes to Plato; in other words, the last wff says that there is someone such that everyone loves them. So note that '∀x∃yLxy' and '∃y∀xLxy' are each unambiguous, and have quite different interpretations.

Which is exactly as things should be, since the quantifier/variable notation is expressly designed to give us unambiguous ways of representing the different thoughts that can ambiguously be expressed by a single English sentence, 'everyone loves someone'.

22.4 Domains, and named vs. nameless things

Finally, a couple of important points worth emphasizing. First, suppose a wff

$$(1) \qquad Fn$$

is true. Then there is something (whatever it is which is named by 'n'!) which has the property expressed by 'F'. Intuitively, then, (1) entails the corresponding existential quantification

$$(2) \qquad \exists x Fx$$

But, to be a bit more careful, note that this entailment will only hold if the domain of discourse does contain the thing named by 'n'. So when we are interpreting **QL**, we need to ensure that *everything named by some constant is in the domain of discourse*. Else the intuitive inference from (1) to (2) would fail.

With that assumption in play, any wff of the form

$$(3) \qquad C(...c...c...)$$

entails a corresponding existential quantification

$$(4) \qquad \exists v C(...v...v...)$$

But note – our second point – that *(4) might be true, even if no corresponding wff of the form (3) is true.* For there may be items in the domain of discourse that have no name on a particular reading of QL. Hence, it could be that no named item has the property expressed by C (so nothing like (3) is true), but one of the nameless things *does* have that property, making (4) true.

Exactly similarly, if a wff like (3) is *false*, then the corresponding universal quantification

(5)   ∀vC(...v...v...)

will be false. But the converse does not hold. A wff like (5) might be false on some interpretation of QL, even if no corresponding wff of the form (3) is false. The named items might all have the property expressed by C, though one of the nameless things lacks it.

A wff like (1) is said to be a *substitution instance* of the wff (2). More generally a wff of the form (3) is a substitution instance of the corresponding (4) and (5); we arrive at it by stripping off the quantifier and substituting a name for the variable tied to that quantifier. Then the point we've just been making is this. We *can't* in general say that an existential quantification is true if and only if one of its substitution instances is true. There may not be enough named things! Likewise, we can't say that a universal quantification is true if and only if all of its substitution instances is true.

And here we'll stop for the moment. We'll devote Chapters 27 and 28 to setting the full-dress official syntax and semantics of QL: as always, the devil is in the details. When we have done that, we'll be able to develop a sharp notion of q-validity for arguments couched in QL, analogous to our sharp notion of tautological validity for arguments couched in PL. But meanwhile, we have enough of the basics in view to be able to work through a stock of translation examples over the next two chapters, in order to develop a better feel for the language; and then we will informally introduce QL trees. First, however, another quick ...

22.5 Summary

- QL is supplied with quantifiers such as '∀x', '∃y' (analogues of the semi-English 'everything x is such that', 'something y is such that), together with variables 'x', 'y', etc. to act as corresponding pronouns. The rest of the language is as simple as can be: to represent the basic subject–predicate structure of simple claims we use *individual constants*

    m, n, o, ...

    and *predicates* with various numbers of places.

    F, G, ... L, M, ... R, S, ...

    The basic syntax of QL then is as follows: An atomic wff of QL is formed from an *n*-place predicate followed by *n* constants. Complex wffs can be constructed using the familiar connectives and also by 'quantifying in' – i.e. if C(...c...c...) is a wff, so are ∀vC(...v...v...) and ∃vC(...v...v...).

• Interpreting QL sentences involves fixing on a domain of discourse, and explaining the meanings to the basic vocabulary of names and predicates. And then the basic idea is that ∀vC(...v...v...) says that everything in the domain has the property expressed by C, i.e. the property that C(...c...c...) attributes to the object picked out by c. And ∃vC(...v...v...) says that there is at least one thing in the domain which has that property.

Exercises 22

Take the following interpretation of the language QL:

> The domain of discourse is people
> 'm' stands for Socrates
> 'n' stands for Plato
> 'F' means ... *is wise*
> 'G' means ... *is a philosopher*
> 'L' means ... *loves* ...

Check that the following are wffs, and then translate them into natural English:

1. (Fn ⊃ ∃xFx)
2. ∃y(Gy ∧ Fy)
3. ∃x(Gx ∧ Lmx)
4. ∀x(Gx ∧ Lmx)
5. ∀x(Gx ⊃ Lmx)
6. ∃x¬(Fx ∧ Lxn)
7. ¬∃x(Fx ∧ Lxn)
8. (Fn ∧ ∀xLxn)
9. ∃y(Fy ∧ ∀xLxy)
10. ∀z(Lzm ≡ Lnz)
11. (Gn ⊃ ∃zLnz)
12. (Gn ⊃ ∃z(Lnz ∧ Fz))
13. ∀y(Gy ⊃ ∃z(Lyz ∧ Fz))
14. ∃z(Fz ∧ ∀y(Gy ⊃ Lyz))

QL explored

We will delay fussing about the fine details of **QL** until Chapters 26 and 27. This chapter and the next one explore a variety of issues about translating between English and **QL**, the language of predicate logic.

23.1 The quantifiers interrelated

Suppose, as before, that 'F' expresses the property of being wise, and the domain of discourse is all people. Then

∀x¬Fx

says everyone is unwise. So

¬∀x¬Fx

says it isn't the case that everyone is unwise. In other words, it says that someone *is* wise. So it is equivalent to

∃xFx

More generally, a wff of the form

¬∀v¬C(...v...v...)

is true just so long as it isn't the case that C is false of everything, i.e. just so long as

∃vC(...v...v...)

is true. Thus we have a general equivalence between ∃v and ¬∀v¬.
   Similarly

∃x¬Fx

says there is at least one person who is unwise. So

¬∃x¬Fx

says it isn't the case that someone is unwise. In other words, it says that everyone is wise. So it is equivalent to

∀xFx

More generally, a wff of the form

$$\neg \exists v \neg C(...v...v...)$$

is true just when

$$\forall v C(...v...v...)$$

is true. Thus we have a general equivalence between $\forall v$ and $\neg \exists v \neg$.

In short, the quantifiers are interdefinable using negation, just as conjunction and disjunction are interdefinable using negation.

The parallel is not surprising. We can think of '$\exists x Fx$' as like a big disjunction, '$(Fm \lor Fn \lor Fo \lor ...)$' and '$\forall x Fx$' as like '$(Fm \land Fn \land Fo \land ...)$' – at least if we assume that everything has a name (cf. §22.4). Then, slightly abusing '$\equiv$' to abbreviate 'if and only if', and being careless with brackets, we have

$$\exists x Fx \equiv (Fm \lor Fn \lor Fo \lor ...) \equiv \neg(\neg Fm \land \neg Fn \land \neg Fo \land ...) \equiv \neg \forall x \neg Fx$$

We've already met the central type of equivalence, relating disjunctions to negated conjunctions, in §11.8. Analogously

$$\forall x Fx \equiv (Fm \land Fn \land Fo \land ...) \equiv \neg(\neg Fm \lor \neg Fn \lor \neg Fo \lor ...) \equiv \neg \exists x \neg Fx$$

We could therefore do quantificational logic using a language with just one of the quantifiers (universal or existential, take your pick) just as we could do truth–functional logic using just one of conjunction or disjunction (plus negation, see §11.9). But just as it is customary and convenient to use both conjunction and disjunction in **PL**, so it is customary and convenient to use both quantifiers in **QL**.

Note we also have

$$\neg \exists x Fx \equiv \neg \neg \forall x \neg Fx \equiv \forall x \neg Fx$$

That gives us *two* simple translations of 'No-one is wise'. We could have added a third basic style of quantifier to **QL**, e.g. using the symbol '**N**', so that $NvCv$ holds when nothing is C. But that would make the language more complicated, and increase the number of rules needed later for dealing with quantifier arguments. Yet the gain would be trivial, since we can so easily express the 'no' quantifier using '$\exists$' or '$\forall$' with negation. The gain isn't worth the cost.

23.2 Expressing restricted quantifications

We saw that when we put **QL** to use, we need to fix the interpretation of the relevant constants and predicates, and we also have to fix the intended domain of discourse. Further, these interpretations must all be *kept* fixed throughout a particular application. After all, even the inference

$$\forall x Fx \therefore \forall x Fx$$

can have a true premiss and a false conclusion if we are allowed to change the domain in mid-argument.

Suppose we want to translate the claim

Some philosophers are logicians.

We *could* take the domain of discourse to be philosophers; and then we'd be able to translate this by something like '∃xGx' (where 'G' is interpreted as meaning *logician*). Likewise, suppose we want to translate

All logicians are rational.

Taking this just by itself, we could select a different interpretation of QL, with the domain this time being logicians; and then the translation could be simply e.g. '∀xHx'. But suppose we want to encode the *whole* argument

Some philosophers are logicians, all logicians are rational; so some philosophers are rational.

We can't simultaneously take the domain of discourse to be the philosophers and also to be the logicians (for these are distinct, if overlapping, groups). The translation of the whole argument into a single version of QL must use a common domain while rendering all the quantifiers. So how can we do this?

We start by selecting a domain of discourse wide enough to include everything we are interested in. In this case, the whole domain of people will do. We then need some way of expressing the *restrictions* on the generalizations that occur in the premisses and conclusion. But plausibly we have

Some philosophers are logicians ≡ (Someone x is such that)(x is a philosopher and x is a logician).

So, using 'F' for *philosopher*, we might translate the first premiss as

∃x(Fx ∧ Gx).

And arguably we also have

All logicians are rational ≡ (Everyone x is such that)(if x is a logician, then x is rational).

Here, the conditional very plausibly is *just* a material conditional. The claim that all logicians are rational makes no stronger claim about the connection between being a logician and being rational than that, for anyone you choose, either they aren't a logician at all or they are rational. So the second premiss translates as

∀x(Gx ⊃ Hx).

Then the whole argument can be rendered as

∃x(Fx ∧ Gx), ∀x(Gx ⊃ Hx) ∴ ∃x(Fx ∧ Hx)

where now the interpretation of the predicate letters *and* the assignment of the domain of discourse is, as we want, stable throughout.

Note though that one thing is plainly lost in this translation. The English plurals in 'some philosophers are logicians' surely indicate there's *more* than one. The standard translation in QL says only that there is *at least* one. Later, when we come to look at QL=, we'll see how to express 'more than one' with the resources of that richer language. But for the moment, we'll swallow the slight loss of content in the translation; it's worth the gain in simplicity.

Generalizing, the standard way of translating restricted quantifications into QL is as follows:

> *All A are B* translates as a wff of the form $\forall v(Av \supset Bv)$
>
> *Some A is/are B* translates as a wff of the form $\exists v(Av \land Bv)$

Here we are using italicized schematic variables in the obvious double way (with '*A*' re-used on the right for the QL translation of English predicate indicated by '*A*' on the left, and so on).

But why does the translation of the restricted 'all' proposition need a conditional, while the 'some' proposition needs a conjunction? Three observations should help clarify this point:

- Translating (1) 'all logicians are rational' (for example) as (1') '$\forall x(Gx \land Hx)$' would plainly be wrong; (1') says that everyone is a rational logician!

- Translating (2) 'some philosophers are logicians' as (2') '$\exists x(Fx \supset Gx)$' must also be wrong. Suppose that no philosophers are logicians, but also that there is some non-philosopher called 'n' in QL. Then (2) is *false* by hypothesis. But '$(Fn \supset Gn)$' would be true, in virtue of having a false antecedent. So its existential generalization (2') would also be *true* (see §22.4). Hence (2) and (2') aren't true in the same circumstances.

- Thirdly, note we have e.g.

     Not all logicians are irrational ≡ some logicians are rational.

  Compare the QL wffs

    $\neg\forall x(Gx \supset \neg Hx) \equiv \exists x\neg(Gx \supset \neg Hx) \equiv \exists x(Gx \land Hx)$

  Here the first wff is equivalent to the second by the $\forall$–$\exists$ relation we met in §23.1, and the second is equivalent to the third given the way negated '$\supset$'s turn into '$\land$'s. Assume the left-hand wff translates the 'not all' claim. Then we have to translate the 'some' claim by the right-hand wff, the existential quantification of a *conjunction*, if the equivalences are to come out right.

Finally what about restricted 'no' propositions, such as 'no logician is rational'? There are two ways to go, for compare

   No logician is rational ≡ It isn't the case that some logician *is* rational
     ≡ $\neg\exists x(Gx \land Hx)$

   No logician is rational ≡ Every logician is non-rational
     ≡ $\forall x(Gx \supset \neg Hx)$.

Generalizing again, we have

> *No A is B* translates as a wff of the form $\neg\exists v(Av \land Bv)$ or $\forall v(Av \supset \neg Bv)$

23.3 Existential import

Leaving aside that issue about plurals noted in the last section, how far do these standard translations really capture the content of ordinary-language restricted quantifications? Consider these two propositions:

(1)   All logicians are rational.
(2)   Some logicians are rational.

Does (1) logically entail (2)? And compare their translations into QL:

(1′)   ∀x(Gx ⊃ Hx)
(2′)   ∃x(Gx ∧ Hx)

Does (1′) logically entail (2′)?

The second question is easily answered. Suppose it happens that there are no logicians in the domain. Then 'G' is true of nothing in the domain. Then there is nothing of which 'G' and 'H' are jointly true. So (2′) is false. But everyone is such that (he is a logician ⊃ he is rational), because the antecedent of the material conditional is always false and hence the conditional always true. So (1′) is then true. Hence, there is a way of making (1′) true and (2′) false. So (1′) doesn't logically entail (2′).

To be sure, (1′) plus the existential assumption '∃xGx' *will* entail (2′). But the point is that this additional existential assumption is needed.

What, however, of the ordinary language claims? Traditionally it is supposed that (1) *does* entail (2) – i.e. that if it can correctly be said that all logicians are rational, then there must be some logicians to *be* rational. On this view, then, the universal quantification in the vernacular (1) is supposed already to have 'existential import', i.e. to embody the assumption that there are some logicians. And if that is correct, the standard translation which renders (1) by something like (1′) which lacks existential import is inadequate.

Still, we might wonder whether tradition gets it right. Consider for example Newton's second law in the form

(3)   All particles subject to no net force have constant velocity.

Surely to accept this law isn't to commit ourselves to the actual existence of any such particles: doesn't the law remain true even if, as it happens, every particle is subject to some net force? Another example: couldn't the notice

(4)   All trespassers will be prosecuted!

be true even if there are no trespassers; indeed it might be because (4) is true that there *are* no trespassers.

We needn't now argue the issue whether ordinary language *All A are B* propositions are actually ambiguous between uses that have existential import (and so entail the existence of As) and uses that don't. Let's just make it clear that the default translation of such propositions into QL goes against tradition and supposes that they *don't* necessarily have existential import. We can always explicitly add back an explicit existential clause of the form ∃vAv into the translation of *All A are B* if, on a particular occasion, we think it really matters.

23.4 More on variables

What's the relation between '∀xFx', '∀yFy', and '∀zFz'? Well, fix the domain, fix the interpretation of the predicate 'F', and these all say exactly the same thing.

Alphabetic choice of variables is neither here nor there. Why so?

Well, recall that what matters about variables is the way they bind quantifiers to particular slots in predicate expressions. They serve to mark links between quantifiers and places in predicates; it doesn't matter exactly how we mark the links, so long as the patterns of linkage remain the same. Thus take the following wffs:

(i) $\forall x \exists y \, Lxy$
(ii) $\forall y \exists z \, Lyz$
(iii) $\forall y \exists x \, Lyx$
(iv) $\forall y \exists x \, Lxy$

(i), (ii) and (iii) differ only in using different variables to mark exactly the same pattern of linkages between the quantifiers and the slots in the predicate 'L'. In each case, the initial universal quantifier gets tied to the first slot in the predicate, and the existential quantifier to the second slot. The wffs say exactly the same, e.g. that everyone has a beloved. (iv) expresses a quite different proposition, e.g. the thought that everyone is herself a beloved. To represent the common proposition expressed by (i) to (iii), we could in fact use a variable-free notation along the lines of

$$\forall \exists \; L \_ \_$$

and then we can expresses the thought (iv) thus:

$$\forall \exists \; L \_ \_$$

Indeed, as Quine has shown, we can not only do without the variables, but even do without these braces joining quantifiers to slots in predicates. So the standard quantifier/variable notation for expressing generality is not the *only* possible notation that would serve our logical purposes of expressing general propositions without ambiguity. Still, any satisfactory logical symbolism for generality must embody the fundamental insights so perspicuously captured by the standard notation, namely that quantifiers are crucially different from names, and can be thought of as operators applying to simpler expressions, where it is crucial *which order* these operators get applied in.

23.5 'Revealing logical form'

In the next chapter, the translational ideas explored in this chapter will be applied to a series of increasingly complex examples. But first it is worth pausing to make some general methodological comments on what we are up to.

Typical ordinary-language quantifier phrases like 'all men', 'some philosophers', 'none of the women in the logic class' express *restricted* generalizations, running over some explicitly characterized class of things. And the class of things we are generalizing about often shifts between premises and conclusion as we develop an argument. In **QL**, by contrast, our two basic quantifiers are interpreted as running over some domain held fixed throughout an argument (and the domain is not explicitly characterized). To replicate in QL the content of everyday restricted quantifications, we must thus use conditionals and conjunctions to carve out the relevant subclasses from the wider domain.

Here's another divergence between QL and English. Compare 'Socrates is a man', 'Socrates is pale', 'Socrates is walking'. The first proposition involves a term which picks out a kind of being, *man*, and says that Socrates belongs to this kind. It is a response to the question 'What kind of thing is Socrates?'. By contrast, the second proposition isn't a response to *that* question; to be told that Socrates is pale doesn't tell you what kind of being Socrates is (a man, a dog, an elephant, a mountain ...). Rather it is an answer to the question 'What is Socrates like?'. The third proposition is an answer to a different question again, i.e. 'What is Socrates doing?'. And there's a long tradition going back to Aristotle which takes such distinctions marked in ordinary language to track metaphysically important differences between kinds, properties and activities. But QL cheerfully flattens any such distinctions, translating the English propositions by (as it might be) 'Fn', 'Gn', 'Hn'.

Now, when it comes to evaluating QL arguments, we will find that the fact that QL uses unrestricted quantifiers, and flattens distinctions between attributions of membership of a kind and attributions of other properties, all makes for a very neat simplicity. The price we pay is a certain distance from the semantic structures of ordinary English. But given our overall approach, that's not a serious objection.

Some writers suppose that, when representing propositions in a formal language like QL, we are aiming to describe structure that is already there in ordinary language, but just not visible to the naked eye. Early in the last century, we find Bertrand Russell and Ludwig Wittgenstein gripped by the idea that the surface look of ordinary language disguises the true 'logical form' of propositions. They proposed that a central concern of philosopher-logicians should be to 'analyse' propositions to reveal this putative underlying structure. And following their lead, some logicians liked to think (perhaps some still like to think) that the now standard notations of logic perspicuously represent the true semantic structures hidden under the surface complexities of language.

If you take some such line, then it is a real embarrassment that standard QL apparently discerns 'and's and 'if's buried in common-or-garden generalizations (and flattens those distinctions we mentioned). Is it really plausible to suppose, for example, that when we say 'All logicians are rational', we are already meaning to generalize – in some beneath-the-surface sort of way – over conditionals holding true in a wider domain?

However, that kind of question just doesn't arise on our approach. We aren't pretending that PL, QL and the like follow the contours of a logical skeleton underlying ordinary discourse. We are taking a much more modest line. The claim is simply that, for certain purposes, it is good strategy to avoid the messy complexities of English usage (e.g. in respect of expressing generality) by (1) translating arguments into a nicely behaved, unambiguous language, and then (2) evaluating the arguments in their tidily regimented form. For this strategy to work, it is *not* necessary to suppose that the artificial languages reveal structures already somehow present in English. It is enough that wffs in the new language behave in truth-relevant ways which sufficiently respect the contents of the claims made in the original English. By those standards, QL is a pretty resounding success.

Whenever we deploy our overall strategy (whether in propositional logic, quantificational logic, the modal logic of necessity, or elsewhere), we'd ideally like both very direct translations into our relevant artificial language L, *and* a simple L which is easy to manipulate. But the more 'natural' we try to make the translations between English and L, the greater the number of fine semantic distinctions we may need to mark in L, and so the more complex everything will get. And that will mean, in particular, increased complexity when we come to evaluate arguments in L. In practice, then, we are often faced with a trade-off between closeness of translation and ease of logical manipulation. We've seen this before when we adopted the material conditional, giving us the simplicity of truth-functional logic for **PLC**, but at the price of some rather dubious-seeming translations of everyday 'if's. And again, in the present case, there's a trade-off. We buy the relative simplicity of QL at the price of having to shoehorn our everyday restricted quantifications into a language where such quantifications have to be mocked up using conjunctions and conditionals, and at the price too of obliterating some metaphysically significant distinctions between kinds of predication. But the price is right: the translations, although sometimes a little cumbersome, enable us to capture the truth-relevant content of ordinary claims well enough to assess the validity of a great number of arguments relying on the presence of quantifiers.

23.6 Summary

- The quantifiers of QL can be interdefined using negation, but we will continue to deploy both.

- The quantifiers of QL run over some given domain of discourse. To replicate the content of everyday restricted quantifications, we use conditionals and conjunctions to carve out the relevant restricted subclasses from the wider domain. Thus

  All A are B translates as $\forall v(Av \supset Bv)$
  Some A is/are B translates as $\exists v(Av \wedge Bv)$
  No A is B translates as $\neg\exists v(Av \wedge Bv)$ or $\forall v(Av \supset \neg Bv)$

- Since QL is *not* being presented as somehow representing or encoding the 'real' underlying logical structure of ordinary language, the fact that we need to use conditionals and conjunctions in expressing restricted quantifications is no objection.

Exercises 23

A   Suppose 'm' stands for Socrates, 'n' stands for Plato, 'o' stands for Aristotle, 'Fx' means *x is a philosopher*, 'Gx' means *x is wise*, 'Mxy' means *x taught y*. Take the domain of discourse to consist of people. And then translate the following into **QL**:

   1.   Socrates taught Plato and Plato taught Aristotle.
   2.   Aristotle taught neither Socrates nor Plato.
   3.   Plato taught someone.
   4.   Some philosophers are wise.
   5.   Some wise people aren't philosophers.
   6.   No one taught Socrates.
   7.   If Socrates taught Plato, then someone taught Plato.
   8.   Whoever Socrates taught is wise.
   9.   Any philosopher who was taught by Plato taught Aristotle.
   10. No wise philosopher was taught by Aristotle.

B   Which of the following pairs of wffs are equivalent (i.e. imply each other), and why? When they aren't equivalent, give interpretations to illustrate the non-equivalence.

   1.   ∃x∀y∃zRyxz; ∃z∀y∃xRyzx
   2.   ∃x∀y∃zRyxz; ∃z∀x∃yRxyz
   3.   (∀xFx ⊃ Fn); (∀zFz ⊃ Fn)
   4.   (∀xFx ⊃ ∀xFx); (∀zFz ⊃ ∀yFy)
   5.   ∃x∃yLxy; ∃y∃xLxy
   6.   ∀x∀yLxy; ∀y∀xLxy
   7.   ∀x(Fx ∧ Gx); (∀xFx ∧ ∀xGx)
   8.   ∀x(Fx ∨ Gx); (∀xFx ∨ ∀xGx)
   9.   ∃x(Fx ∧ Gx); (∃xFx ∧ ∃xGx)
   10. ∃x(Fx ∨ Gx); (∃xFx ∨ ∃xGx)

C   We can render 'Plato and Aristotle are philosophers' by e.g. '(Fm ∧ Fn)'. Why can't we render 'Plato and Aristotle are classmates' by something like '(Gm ∧ Gn)'? Consider other cases of predicates *F* where we can't render something of the form 'Plato and Aristotle are *F*' by something of the type '(Fm ∧ Fn)'. What can be learnt from such cases about the expressive limitations of QL?

More QL translations

The best way to get really familiar with the language **QL** is to tackle some more translation exercises. That's the business of this chapter. Some of the examples get fairly tricky. We rarely need to tangle with such complex cases in practice: the point is just to show how the translations can be done with a bit of care.

24.1   Translating English into QL

Let's adopt the following interpretations:

'm' means *Maldwyn*
'n' means *Nerys*
'o' means *Owen*
'F' means *is a man*
'G' means *is a woman*
'L' means … *loves* …
'M' means … *is married to* …
'R' means … *prefers* … *to* …
The domain of discourse is to be all people.

And we'll translate the following reports of Welsh affairs:

1.   Whoever is loved by Owen is loved by Maldwyn too.
2.   Every woman who loves Maldwyn is loved by Owen.
3.   Maldwyn loves some woman who loves Owen.
4.   No man who loves Nerys loves Owen or Maldwyn.
5.   Every man loves someone.
6.   Everyone Owen loves is loved by someone Nerys loves.
7.   No woman loves every man.
8.   No woman loves any man.
9.   If everyone loves Nerys then Owen does.
10.   If anyone loves Nerys then Owen does.
11.   Someone who is married loves Nerys.
12.   Anyone who is married loves someone they aren't married to.
13.   A married man only loves women.

14.   Not every married man loves any woman who loves him.

15.   Nerys loves any married men who prefer her to whomever they are married to.

Do pause here to see how many of these you can translate given our explanations so far. But don't worry if that's not a lot at this stage.

We'll take the examples in turn, and proceed fairly systematically.

- *Stage one*   We re-express the claim in English using prefixed-quantifiers-plus-variables (along the lines of §21.4). Typically we'll need to use restricted quantifiers – either simple ones such as 'some man x is such that', 'every woman y is such that', or else ones involving relative clauses like 'everyone x who is a man is such that' or 'some woman y who loves Owen is such that'.

- *Stage two*   We then translate these restricted quantifiers into QL following §23.2. So for the simpler cases we use e.g. the following patterns:

(Every man x is such that)$Cx \mapsto \forall x(Fx \supset Cx)$
(Some woman y is such that)$Cx \mapsto \exists y(Gy \wedge Cx)$
(No man z is such that)$Cx \mapsto \forall z(Fz \supset \neg Cx)$ or $\neg \exists z(Fz \wedge Cx)$.

Here and below, our new type of arrow '$\mapsto$' can be read as 'translates as'; and we are again using the same schematic letter for double duty in the obvious way.

And what about the cases with relative clauses? 'Everyone x who is F is such that ...' is equivalent to 'Everyone x is such that, if x is F, then ...', and gets translated the same way. 'Some woman y who loves Owen' is equivalent to 'Someone y who-is-a-woman-*and*-loves-Owen': so we can express this more complex quantification by using a conjunctive restrictive clause.

Note, as we work through the second stage en route to the final result, we will allow ourselves to write down some unholy mixtures of English and QL (rather than always doing the translation in one fell swoop). Purists will hate this. Let them fuss away! These bilingual intervening stages are a useful heuristic, a very helpful ladder which can (to borrow a phrase) be thrown away once climbed. Such intervening stages can be skipped as soon as you can do without them.

1.   Whoever is loved by Owen is loved by Maldwyn too
       $\mapsto$ (Every x who is loved by Owen is such that) Maldwyn loves x
       $\mapsto$ (Every x who is loved by Owen is such that) Lmx
       $\mapsto \forall x(Lox \supset Lmx)$

2.   Every woman who loves Maldwyn is loved by Owen
       $\mapsto$ (Every woman x who loves Maldwyn is such that) x is loved by Owen
       $\mapsto$ (Every woman x who loves Maldwyn is such that) Lox
       $\mapsto \forall x((Gx \wedge Lxm) \supset Lox)$

for x is a-woman-who-loves-Maldwyn just if x is a woman *and* x loves Maldwyn.

3. Maldwyn loves some woman who loves Owen
   ↦ (Some woman x who loves Owen is such that) Maldwyn loves x
   ↦ (Some woman x who loves Owen is such that) Lmx
   ↦ ∃x((Gx ∧ Lxo) ∧ Lmx)

4. No man who loves Nerys loves Owen or Maldwyn
   ↦ (No man x who loves Nerys is such that) x loves Owen or
      Maldwyn
   ↦ (No man x who loves Nerys is such that)(Lxo ∨ Lxm)
   ↦ ¬∃x((Fx ∧ Lxn) ∧ (Lxo ∨ Lxm))
      or ∀x((Fx ∧ Lxn) ⊃ ¬(Lxo ∨ Lxm))

5. Every man loves someone
   ↦ (Every man x is such that)(someone y is such that) x loves y
   ↦ (Every man x is such that)(someone y is such that) Lxy
   ↦ (Every man x is such that) ∃yLxy
   ↦ ∀x(Fx ⊃ ∃yLxy)

The translation '∀x∃y(Fx ⊃ Lxy)' would in fact do as well (see §3 below on the thorny topic of moving quantifiers around). The other, very much less natural, reading of the English can be rendered

   ↦ (Someone y is such that)(every man x is such that) x loves y
   ↦ (Someone y is such that)(every man x is such that) Lxy
   ↦ (Someone y is such that) ∀x(Fx ⊃ Lxy)
   ↦ ∃y∀x(Fx ⊃ Lxy)

6. Everyone Owen loves is loved by someone Nerys loves
   ↦ (Everyone x whom Owen loves is such that)(someone y who
      is loved by Nerys is such that) x is loved by y
   ↦ (Everyone x whom Owen loves is such that)(someone y who
      is loved by Nerys is such that) Lyx
   ↦ (Everyone x whom Owen loves is such that) ∃y(Lny ∧ Lyx)
   ↦ ∀x(Lox ⊃ ∃y(Lny ∧ Lyx))

That at any rate captures the more natural reading. The other possible reading will come out as

   ∃y(Lny ∧ ∀x(Lox ⊃ Lyx))

7. No woman loves every man
   ↦ (No woman x is such that)(every man y is such that) x loves y
   ↦ (No woman x is such that)∀y(Fy ⊃ Lxy)
   ↦ ¬∃x(Gx ∧ ∀y(Fy ⊃ Lxy))
      or ∀x(Gx ⊃ ¬∀y(Fy ⊃ Lxy))

8. No woman loves any man
   ↦ (No woman x is such that)(some man y is such that) x loves y
   ↦ (No woman x is such that)∃y(Fy ∧ Lxy)
   ↦ ¬∃x(Gx ∧ ∃y(Fy ∧ Lxy))
      or ∀x(Gx ⊃ ¬∃y(Fy ∧ Lxy))

Or at least, that's the more obvious reading. But consider the conversation 'Nerys seems to love *any* man at all!', 'No! No woman (not even Nerys) loves *any* man'. So (8) *can* sometimes be read as saying the same as (7).

9.   If everyone loves Nerys then Owen does
     ↦ If (everyone x is such that) x loves Nerys, then Owen loves Nerys
     ↦ If ∀xLxn, then Lon
     ↦ (∀xLxn ⊃ Lon)

Note the way this is bracketed: the proposition we are translating is overall a *conditional*, which has a universally quantified antecedent. Compare and contrast:

     ∀x(Lxn ⊃ Lon)

This wff is true in different circumstances. For suppose that Maldwyn loves Nerys and Owen doesn't. Then '∀xLxn' and 'Lon' are both false, so the conditional '(∀xLxn ⊃ Lon)' is *true*. However, (Lmn ⊃ Lon) is false: so its universal quantification '∀x(Lxn ⊃ Lon)' is *false* too.

10. In most contexts, 'If anyone loves Nerys then Owen does' says the same as 'If there's someone or other who loves Nerys, then Owen loves her'. So:

If anyone loves Nerys then Owen does
     ↦ If (someone x is such that) x loves Nerys, then Owen loves Nerys
     ↦ If ∃xLxn, then Lon
     ↦ (∃xLxn ⊃ Lon)

But consider the context 'Anyone loves Nerys, and if anyone loves Nerys, then Owen does.' Here it seems that (10) can be read as expressing the same message as (9).

11. Someone who is married loves Nerys
     ↦ (Someone x who is married is such that) x loves Nerys
     ↦ (Someone x who is married is such that) Lxn
     ↦ ∃x(x is-married ∧ Lxn)

Now, we have in our QL vocabulary the *two-place* predicate 'M' meaning ... is married to .... But how do we translate the *one-place* predicate '... is married'? Well, to be married is to be married to someone. So 'x is married' can be translated '∃yMxy'. So we can finish the translation thus:

     ↦ ∃x(∃yMxy ∧ Lxn)

12. Anyone who is married loves someone they aren't married to
     ↦ (Everyone x who is married is such that)(there is someone y, who
        x isn't married to, such that) x loves y
     ↦ (Everyone x who is married is such that) ∃y(¬Mxy ∧ Lxy)
     ↦ ∀x(x is-married ⊃ ∃y(¬Mxy ∧ Lxy))
     ↦ ∀x(∃zMxz ⊃ ∃y(¬Mxy ∧ Lxy))

It is very good policy, when introducing new variables into a wff, always to use letters that don't already appear in the wff. That way, tangles can't arise. But in fact, as the official syntax in Chapter 26 will make clearer, at the last step it *would* have been permissible to use 'y' again and write

$\mapsto \forall x(\exists y Mxy \supset \exists y(\neg Mxy \land Lxy))$

Why? Roughly speaking, because these two existential quantifications are isolated from each other, their scopes do not overlap, and the local cross-linking of variables and quantifiers is quite unambiguous. More about that anon.

13. A married man only loves women
    $\mapsto$ (Every man x who is married is such that)(x loves only women)
    $\mapsto$ (Every man x who is married is such that)(everyone y is such that) if x loves y, y is a woman.

On the natural reading, 'a married man' is universally generalizing; and we are reading 'x only loves women' as saying anyone x loves is a woman, leaving it open as to whether there *is* anyone x loves. So, continuing,

$\mapsto$ (Every man x who is married is such that)$\forall y(Lxy \supset Gy)$
$\mapsto \forall x((Fx \land x\text{ is-married}) \supset \forall y(Lxy \supset Gy))$
$\mapsto \forall x((Fx \land \exists z Mxz) \supset \forall y(Lxy \supset Gy)).$

14. Not every married man loves any woman who loves him
    $\mapsto$ It is not the case that (every man x who is married is such that) (every woman y who loves x is such that) x loves y
    $\mapsto$ It is not the case that (every man x who is married is such that) (every woman y who loves x is such that)Lxy
    $\mapsto \neg$(Every man x who is married is such that)$\forall y((Gy \land Lyx) \supset Lxy)$
    $\mapsto \neg\forall x((Fx \land \exists z Mxz) \supset \forall y((Gy \land Lyx) \supset Lxy)).$

15. A final example for the moment (and don't fret too much about this kind of contorted case: it is here to illustrate the reach of our step-by-step translation techniques, and not because we'll frequently need to be assessing arguments with premisses this complex!)

Nerys loves any married men who prefer her to whomever they are married to
    $\mapsto$ (Every x who is-a-married-man and who prefers-Nerys-to-whomever-x-is-married-to is such that) Nerys loves x.

The quantifier is restricted here by two conjoined conditions. The first one is now familiar, so we get to

$\mapsto \forall x(\{(Fx \land \exists z Mxz) \land [x \text{ prefers-Nerys-to-whomever-}x\text{-is-married-to}]\} \supset Lnx)$
$\mapsto \forall x(\{[Fx \land \exists z Mxz] \land \forall y[\text{if } x \text{ is married to } y \text{ then } x \text{ prefers Nerys to } y]\} \supset Lnx)$
$\mapsto \forall x(\{[Fx \land \exists z Mxz] \land \forall y[Mxy \supset Rxny]\} \supset Lnx).$

And that, surely, is enough examples of English-to-QL translation to be going on with (there are plenty more in the exercises!).

24.2   Translating from QL

Translating *from* an unfamiliar language tends to be a lot easier than translating *into* that language. Once we have learnt to spot the devices QL uses for expressing restricted quantifications, it is usually pretty easy to decode the message. And if the worst comes to the worst, and we are faced with a dauntingly complex wff, we can always simply reverse the step-by-step procedure that we have just been using. To illustrate, now go back and read through each of the examples in the last section again, but this time starting with the QL wff, and working back up to its rendition into English.

Here's just one more example. Consider

$$\forall x(\exists y(Mxy \land Lyx) \supset \neg\exists y(Lxy \land \forall z(Mxz \supset Rxyz)))$$

and let's suppose the QL-to-English dictionary is as in §24.1. Overall, this wff has the form $\forall x(A(...x...) \supset \neg B(...x...))$, so says *No x which is A is such that x is B*. As a first stage, then, we have

No x who is such that $\exists y(Mxy \land Lyx)$ is such that $\exists y(Lxy \land \forall z(Mxz \supset Rxyz))$

which is readily interpreted as

No x who is married to someone who loves them is such that $\exists y(Lxy \land \forall z(Mxz \supset Rxyz)))$

which in turn decodes as

No one x, who is married to someone who loves them, is such that there's someone y they love who is such that $\forall z(Mxz \supset Rxyz)$.

That says

No one x, who is married to someone who loves them, is such that there's someone y they love who is such that x prefers y to anyone x is married to.

Or in unaugmented (though not quite unambiguous) English

No one who is married to someone who loves them loves someone they prefer to whomever they are married to.

24.3   Moving quantifiers

Leave aside the case of translating 'no' sentences, where we always have (at least) two equally good options. Leave aside too the point that the alphabetical choice of variables – as in '$\forall x Fx$' versus '$\forall y Fy$' – is arbitrary. There are other cases too where there can be distinct but equally good translations. To take a very simple example, suppose we want to translate 'Someone loves someone'. Then there is nothing to choose between '$\exists x \exists y Lxy$' and '$\exists y \exists x Lxy$'. Similarly, there is nothing to choose between '$\forall x \forall y Lxy$' and '$\forall y \forall x Lxy$'. True, we can't unrestrictedly swap quantifiers around; indeed, the whole rationale of our notation is that it allows us to unambiguously mark the difference between the messages expressed by e.g. '$\forall x \exists y Lxy$' and '$\exists y \forall x Lxy$'. However, *immediately adjacent quantifiers of the same kind can be interchanged.*

To labour the point, consider the case where we have neighbouring existential quantifiers. Reading **QL** as in §24.1, '∃yLny' attributes to Nerys the property of loving someone. When we quantify in, we get '∃x∃yLxy', which says that *someone* has the property that '∃yLny' attributes to Nerys (cf. §22.3), i.e. the property of loving someone. So, '∃x∃yLxy' holds just if there is a pair of people in the domain (not necessarily distinct) such that the first loves the second. Likewise '∃xLxn' attributes to Nerys the property of being loved by someone. Quantifying in, we get '∃y∃xLxy', which says that *someone* has the property that '∃xLxn' attributes to Nerys, i.e. the property of being loved by someone. So '∃y∃xLxy' holds just if there is a pair of people in the domain (not necessarily distinct) such that the first loves the second. That's the same truth-condition as for '∃x∃yLxy'. And the point generalizes to any pair of wffs of the form $\exists v \exists w C(...v...w...)$, $\exists w \exists v C(...v...w...)$. Similarly for pairs of universal quantifiers.

Consider next another sort of case. Suppose we want to translate 'Nerys is a woman everyone loves'. The following will in fact do equally well:

(Gn ∧ ∀xLxn) *or* ∀x(Gn ∧ Lxn)

The latter holds just in case everyone x is such as to make it true that (Nerys is a woman and x loves her), which holds just in case Nerys is a woman loved by everyone, which is what the first says. More generally (borrowing '≡' to express equivalence), we have

$$(A \wedge \forall v B(...v...)) \equiv \forall v (A \wedge B(...v...))$$

if the variable v doesn't occur in A. The order of the conjuncts of course doesn't matter. Similarly, we have

$$(A \wedge \exists v B(...v...)) \equiv \exists v (A \wedge B(...v...))$$

if the variable v doesn't occur in A. Return to example (11) in §1, 'Someone who is married loves Nerys'. We gave this translation:

∃x(∃yMxy ∧ Lxn)

Now we see it would have been equally legitimate to write

∃x∃y(Mxy ∧ Lxn)

given the variable 'y' does not occur in the second conjunct.

And we can do similar manipulations with disjunctions. Suppose we want to translate 'Either Owen loves Nerys or nobody does'. We could equally well write

(Lon ∨ ∀x¬Lxn) *or* ∀x(Lon ∨ ¬Lxn)

More generally, we have

$$(A \vee \forall v B(...v...)) \equiv \forall v (A \vee B(...v...))$$

where v doesn't occur in A, and the order of the disjuncts doesn't matter. Likewise, we have

$$(A \vee \exists v B(...v...)) \equiv \exists v (A \vee B(...v...))$$

The case where we really have to be careful involves conditionals. We have, as you might expect,

$$(A \supset \forall v B(...v...)) \equiv \forall v(A \supset B(...v...))$$
$$(A \supset \exists v B(...v...)) \equiv \exists v(A \supset B(...v...))$$

However, note very carefully the following:

$$(\forall v B(...v...) \supset A) \equiv \exists v(B(...v...) \supset A)$$
$$(\exists v B(...v...) \supset A) \equiv \forall v(B(...v...) \supset A)$$

Extracting a universal quantification from the antecedent of a conditional (when the variable doesn't occur in the consequent) turns it into an existential quantification. This shouldn't be surprising when you recall that antecedents of conditionals are like negated disjuncts – remember $(A \supset B)$ is equivalent to $(\neg A \vee B)$ – and recall too that when quantifiers tangle with negation they 'flip' into the other quantifier. Consider, for example, the following chain of equivalences:

$$(\forall x Fx \supset Fn) \equiv (\neg\forall x\, Fx \vee Fn) \equiv (\exists x\neg Fx \vee Fn) \equiv \exists x(\neg Fx \vee Fn)$$
$$\equiv \exists x(Fx \supset Fn).$$

But having noted the various equivalences in this section for the record, do be very, very cautious in using them (careless application can lead to trouble)!

24.4 Summary

- Translation into QL can proceed semi-automatically, if (as a half-way house) we transcribe vernacular sentences into an augmented English deploying quantifier operators of the kind 'Every/some/no A, v, which is B is such that', and then use the schemas:

  (Every A, v, which is B is such that)$Cv \mapsto \forall v((Av \wedge Bv) \supset Cv)$
  (Some A, v, which is B is such that)$Cv \mapsto \exists v((Av \wedge Bv) \wedge Cv)$
  (No A, v, which is B is such that)$Cv \mapsto \forall v((Av \wedge Bv) \supset \neg Cv)$
  $\qquad\qquad\qquad\qquad$ or $\neg\exists v((Av \wedge Bv) \wedge Cv)$

- The order of neighbouring quantifiers can be swapped if the quantifiers are of the same kind. And quantifiers inside conjunctions and disjunctions can, in certain circumstances, be 'exported' outside the conjunction/disjunction.

Exercises 24

A   Suppose 'm' denotes Myfanwy, 'n' denotes Ninian, 'o' denotes Olwen, 'Fx' means *x is a philosopher*, 'Gx' means *x speaks Welsh*, 'Lxy' means *x loves y*, and 'Rxyz' means that *x is a child of y and z*. Take the domain of discourse to consist of human beings. Translate the following into QL:

1.   Ninian is loved by Myfanwy and Olwen.
2.   Neither Myfanwy nor Ninian love Olwen.
3.   Someone is a child of Myfanwy and Ninian.
4.   No philosopher loves Olwen.
5.   Myfanwy and Ninian love everyone.
6.   Some philosophers speak Welsh.
7.   No Welsh-speaker who loves Myfanwy is a philosopher.
8.   Some philosophers love both Myfanwy and Olwen.

9.   Some philosophers love every Welsh speaker.
10.   Everyone who loves Ninian is a philosopher who loves Myfanwy.
11.   Some philosopher is a child of Olwen and someone or other.
12.   Whoever is a child of Myfanwy and Ninian loves them both.
13.   Everyone speaks Welsh only if Olwen speaks Welsh.
14.   Myfanwy is a child of Olwen and of someone who loves Olwen.
15.   Some philosophers love no Welsh speakers.
16.   Every philosopher who speaks Welsh loves Olwen.
17.   Every Welsh-speaking philosopher loves someone who loves Olwen.
18.   If Ninian loves every Welsh speaker, then Ninian loves Myfanwy
19.   No Welsh speaker is loved by every philosopher.
20.   Every Welsh speaker who loves Ninian loves no one who loves Olwen.
21.   Whoever loves Myfanwy, loves a philosopher only if the latter loves Myfanwy too.
22.   Anyone whose parents are a philosopher and someone who loves a philosopher is a philosopher too.
23.   Only if Ninian loves every Welsh-speaking philosopher does Myfanwy love him.
24.   No philosophers love any Welsh-speaker who has no children.

B   Take the domain of quantification to be the (the positive whole) numbers, and let 'n' denote the number one, 'Fx' mean *x is odd*, 'Gx' means *x is even*, 'Hx' means *x is prime*, 'Lxy' means *x is greater than y*, 'Rxyz' means that *x is the sum of y and z*. Then translate the following into natural English:

1. $\neg\exists x(Fx \land Gx)$
2. $\forall x\forall y\exists z\,Rzxy$
3. $\forall x\exists y\,Lyx$
4. $\forall x\forall y((Fx \land Ryxn) \supset Gy)$
5. $\forall x\forall y((Gx \land Rxyn) \supset Fy)$
6. $\forall x\exists y((Gx \land Fy) \land Rxyy)$
7. $\forall x\forall y(\exists z(Rzxn \land Ryzn) \supset (Gx \supset Gy))$
8. $\forall x\forall y\forall z(((Fx \land Fy) \land Rzxy) \supset Gz)$
9. $\forall x(Gx \supset \exists y\exists z((Hy \land Hz) \land Rxyz))$
10. $\forall w\exists x\exists y(((Hx \land Hy) \land (Lxw \land Lyw)) \land \exists z(Rzxn \land Ryzn))$

C   Which of the following pairs are equivalent, and why?

1. $\forall x(Fx \supset Gx)$; $(\forall xFx \supset \forall xGx)$
2. $\exists x(Fx \supset Gx)$; $(\exists xFx \supset \exists xGx)$
3. $\exists x(Fx \supset Gx)$; $(\forall xFx \supset \exists xGx)$
4. $\forall x(Fx \supset Gx)$; $(\exists xFx \supset \forall xGx)$

The claim that, e.g., that a wff of the form $(A \lor \exists xFx)$ is equivalent to one of the form $\exists x(A \lor Fx)$ depends on our stipulation that the domain of quantification isn't empty. Why? Which other equivalences we stated in §24.3 above also depend on that stipulation?

Introducing QL trees

Our discussion of the language **QL** has so far been fairly informal and introductory. We'll need to nail down the syntax and semantics much more carefully, beginning in the next chapter. But in this chapter, as an encouragement to keep going, we'll continue our informal discussions, and show how to develop the tree methods we used in **PL** to prove that some **QL** inferences are valid.

25.1 The ∀-instantiation rule

We introduced **PL** trees as a fast-track, working-backwards, way of looking for a 'bad line' on a truth-table. However, we soon dropped the explicit talk of valuations, and used unsigned trees. And as we noted in Chapter 20, these unsigned trees (initially introduced as regimenting metalinguistic arguments) can also be thought of as reductio proofs couched in **PL**.

Here's the main idea again. We are given an inference

$$A_1, A_2, ..., A_n \therefore C$$

and want to know whether it is valid or not. Equivalently, we want to know whether the set of wffs

$$A_1, A_2, ..., A_n, \neg C$$

is or isn't inconsistent. We can show that this set *is* inconsistent (and so the original inference is valid) by constructing a closed tree with the wffs in question as its trunk. The basic principle which guides extending the tree is this: we extend an open branch *by adding to it further wffs which must also be true assuming that wffs already on the branch are true,* splitting the tree into two branches when we have to consider alternative cases (see §17.1).

We are now going to apply the very same idea to quantificational inferences. Except this time, we are going to *start* by thinking of trees the second way – i.e. not as metalinguistic arguments about valuations but as object-language arguments revealing that a bunch of wffs is inconsistent. (We have to do it this way round if we are going to introduce trees now, for we haven't yet met the appropriate official story about valuations for **QL** wffs. In what follows we rely on our grip on what **QL** wffs mean, and hence when they are true, given an interpretation of their basic vocabulary.)

Take our old friend, the argument 'All philosophers are egocentric; Jack is a philosopher; hence Jack is egocentric' (§1.2). Translated into **QL**, using the obvious code book, and taking the domain of quantification to be people, we have:

    A   Fn, ∀x(Fx ⊃ Gx) ∴ Gn

(We'll take it that the comma and the inference marker officially belong to **QL** as they do to **PL**.) We show this inference is valid by showing that the premisses and the negation of the conclusion form an inconsistent set. We start by setting down those wffs as the trunk of a tree:

    (1)   Fn
    (2)   ∀x(Fx ⊃ Gx)
    (3)   ¬Gn

Now, how should we extend the tree? Assuming that those wffs *are* true together, what else has to be true? Well, if *everything* in some domain satisfies some condition, then any particular named thing in that domain satisfies that condition. So if something of the form ∀*v*C(...*v*...*v*...) is true, then C(...*c*...*c*...) will be true, given *c* names something in the domain. In this case, we are assuming 'Fn' is true, so 'n' certainly should name something in the domain (see §22.4). So the following ought also to be true:

    (4)   (Fn ⊃ Gn)

And now we can just press on, applying the usual tree-building rule for wffs of the form (A ⊃ B); and the tree closes.

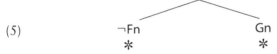

    (5)   ¬Fn   Gn
                          *   *

The supposition that the premisses and negated conclusion of **A** are all true has been revealed to lead to contradiction. Which shows that the argument **A** is indeed valid.

Another simple example. Consider the inference

    B   ∀x(Fx ⊃ Gx), (Fn ∧ ¬Hn) ∴ ¬∀x(Gx ⊃ Hx)

which renders e.g. 'All logicians are philosophers; Russell is a logician who isn't wise; so not all philosophers are wise.' Again, we assume that the premisses and negation of the conclusion are true, and set down those assumptions as the beginning of a tree.

    (1)   ∀x(Fx ⊃ Gx)
    (2)   (Fn ∧ ¬Hn)
    (3)   ¬¬∀x(Gx ⊃ Hx)

Now we can apply the familiar rules for unpacking the connectives:

    (4)   Fn
    (5)   ¬Hn
    (6)   ∀x(Gx ⊃ Hx)

So far, so straightforward. To proceed further, we evidently need to extract some more 'n'-relevant information. (1) says that any thing *x* satisfies the condition '(Fx ⊃ Gx)'. So this will be true in particular of what 'n' denotes. Whence

(7)   (Fn ⊃ Gn)

Similarly from (6) we have

(8)   (Gn ⊃ Hn)

Now it is easy to extract an inconsistency. Using the standard rule for the material conditional twice, we have:

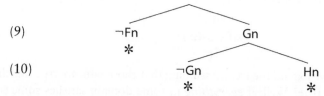

And we are done. The assumption that the premisses and the negated conclusion of **B** are all true together leads to contradictions however we try to work through the consequences of that assumption; so the argument is valid.

The new rule for tree-building which we are using here is the ∀-*instantiation rule* (we'll call it '(∀)' for short). The basic idea is that, given a universal quantification, then any substitution instance of it must also be true: i.e. any wff which is got by stripping off the quantifier and replacing the remaining occurrences of the associated variable with some name of an object in the domain must also be true. But any constant we are already using *is* supposed to name something in the domain; so that motivates the following schematic statement of the rule:

> (∀) If ∀*v*C(...*v*...*v*...) appears on an open path, we can add C(...*c*...*c*...) to the foot of the path, where *c* is any constant which already appears on the path.

Note very carefully that – basically to keep our trees neatly pruned – this rule doesn't insist that we add *all* such instances of the quantified statement to *every* relevant path whenever we deploy it: *the rule simply permits us to draw out implications as and where needed.* And because we don't (and indeed can't) 'use up' all the content of the quantified wff in a single application of (∀), we *don't* check off the quantified wff after applying the instantiation rule.

Here's a slightly more complex example of our new rule in operation, which illustrates the last point. Consider the argument 'Every great logician is also a great philosopher; either Moore or Russell is a great logician; so either Moore or Russell is a great philosopher.' Translating this into QL, and keeping the domain of quantification to be people, the inference is

C   ∀x(Fx ⊃ Gx), (Fm ∨ Fn) ∴ (Gm ∨ Gn)

This is plainly valid. Still, let's *prove* that by supposing, in the usual way, that the premisses are true and the conclusion false, and aiming for contradictions.

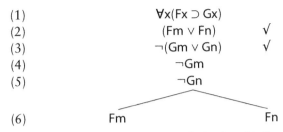

(1)   ∀x(Fx ⊃ Gx)
(2)   (Fm ∨ Fn)   √
(3)   ¬(Gm ∨ Gn)   √
(4)   ¬Gm
(5)   ¬Gn

(6)   Fm   Fn

We can now use (∀) to extract more 'm'-related information, instantiate (1) with the name 'm' and continue the left-hand branch:

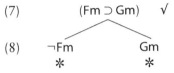

(7)   (Fm ⊃ Gm)   √

(8)   ¬Fm   Gm
          *   *

Next, consider the right-hand branch. We could have added the wff '(Fm ⊃ Gm)' on this branch too, but that won't do any useful work. What we need to do, in order to develop the right-hand branch, is extract 'n'-related information from (1) using (∀) again. Now we see the importance of *not* having checked off (1), and marked it as 'used up'. We hadn't squeezed all the useful juice out of that quantified wff; we need to revisit it and extract more implications. This second application (∀) enables us to finish the tree.

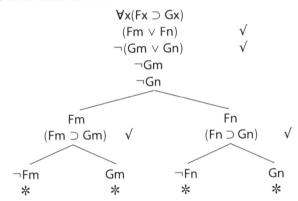

∀x(Fx ⊃ Gx)
(Fm ∨ Fn)   √
¬(Gm ∨ Gn)   √
¬Gm
¬Gn

Fm   Fn
(Fm ⊃ Gm)   √   (Fn ⊃ Gn)   √

¬Fm   Gm   ¬Fn   Gn
 *   *   *   *

The tree closes, and the inference C is shown to be valid.

25.2   Rules for negated quantifiers

We've stated the tree-building rule for deducing implications from a wff of the form ∀*v*C(...*v*...*v*...). What about a tree-building rule for dealing with a negated universal quantification, i.e. a wff of the form ¬∀*v*C(...*v*...*v*...)?

Well, ¬∀*v*C(...*v*...*v*...) is true just so long as ∃*v*¬C(...*v*...*v*...) is true (see §23.1). Hence a possible rule for treating negated universal quantifications is to transform them into equivalent existential quantifications, and pass them over to be dealt with by the rule (∃) for dealing with existential wffs – whatever that rule later turns out to be!

Similarly, note that $\neg \exists v C(...v...v...)$ is true just so long as $\forall v \neg C(...v...v...)$ is. Hence a possible rule for treating negated existential quantifications is to transform them into equivalent universal quantifications, and pass them back to be dealt with by the rule for ($\forall$).

We'll adopt these rules for **QL** tree-building. Given a wff that *starts* with a negated quantifier, we can move the initial negation sign past the quantifier, swapping '$\forall$' with '$\exists$' as it passes. Since that transformation extracts all the content from the original wff – it takes us from one wff to an exactly equivalent wff – this time we *can* check it off after the rule is applied. So schematically:

($\neg\forall$)   If $\neg \forall v C(...v...v...)$ appears on an open path, then we add the corresponding wff $\exists v \neg C(...v...v...)$ to each open path that contains it, and check it off.

($\neg\exists$)   If $\neg \exists v C(...v...v...)$ appears on an open path, then we add the corresponding wff $\forall v \neg C(...v...v...)$ to each open path that contains it, and check it off.

Let's review two quick examples of these new rules at work. Consider

D $\neg \exists x(Fx \wedge Gx)$, Fm $\therefore \neg Gm$

(Perhaps that translates 'No logicians are wise; Russell is a logician; so Russell isn't wise.') The completed tree looks like this:

(1)	$\neg \exists x(Fx \wedge Gx)$	√
(2)	Fm	
(3)	$\neg \neg Gm$	
(4)	$\forall x \neg (Fx \wedge Gx)$	
(5)	$\neg (Fm \wedge Gm)$	√

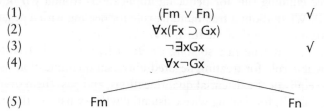

(6) $\neg$Fm $\neg$Gm
           *   *

At (4) we have applied the new ($\neg\exists$) rule to (1); and then we applied ($\forall$) at the next step, and the tree almost immediately closes.

Here's another intuitively valid argument:

E   (Fm $\vee$ Fn), $\forall x(Fx \supset Gx) \therefore \exists x Gx$

(Compare: 'Either Jack or Jill is a student; all students are poor; hence there is someone who is poor.'). The tree starts in the obvious way

(1)	(Fm $\vee$ Fn)	√
(2)	$\forall x(Fx \supset Gx)$	
(3)	$\neg \exists x Gx$	√
(4)	$\forall x \neg Gx$	

(5)   Fm   Fn

We now proceed as on the tree for **C**. First we want to instantiate the two universal quantifications (2) and (4) with 'm' to get 'm'-related information useful for extending the left-hand branch. But there would be no point in adding the resulting wffs to the right-hand branch too (though it wouldn't be *wrong* to do so, just redundant). Rather, we need to extract 'n'-related information useful for the right-hand branch. So we continue:

(6)	(Fm ⊃ Gm)	(Fn ⊃ Gn) (from 2)
(7)	¬Gm	¬Gn (from 4)

Then applying the rule for the conditional twice will yield:

(8)	¬Fm	Gm	¬Fn	Gn
	*	*	*	*

25.3 The ∃-instantiation rule

Consider next the following argument

    F   ∃xFx, ∀x(Fx ⊃ Gx) ∴ ∃xGx

(Perhaps: 'Someone is a great logician; all great logicians are cool; hence someone is cool'). Assume that the premises and the negation of the conclusion are all true, and then apply the rule (¬∃), and we get

(1)	∃xFx	
(2)	∀x(Fx ⊃ Gx)	
(3)	¬∃xGx	√
(4)	∀x¬Gx	

where we have applied the (¬∃) rule to get to line (4).

  (1) tells us that at least one thing in the domain satisfies 'F'. Now, there can't be any harm in picking out one of these things and arbitrarily *dubbing* it. If the original wffs including (1) are consistent, no contradiction can ensue just from this act of selecting and naming. Let's use a constant that isn't currently being used for other purposes – e.g. 'a' (for 'arbitrary'!). So by hypothesis we have

(5)	Fa

What else do we know about the thing now dubbed 'a'? Given the universal claims (2) and (4), we can deduce particular claims about it as follows:

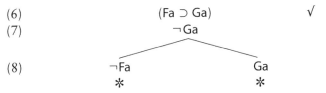

(6)	(Fa ⊃ Ga)	√
(7)	¬Ga	
(8)	¬Fa Ga	
	* *	

The tree closes. To repeat, the mere act of dubbing one of the things which are 'F' can't have created the contradiction. So the contradiction must have been in the original assignments of truth to each of (1) to (3). Hence there can be no way of making (1) to (3) all true, and so the inference **F** is valid.

Another quick example. Take the evidently valid syllogism 'All logicians are philosophers; all philosophers are wise; so all logicians are wise.' Using the obvious translation we get,

> **G**   ∀x(Fx ⊃ Gx), ∀x(Gx ⊃ Hx) ∴ ∀x(Fx ⊃ Hx)

We start the tree in the standard way

(1)	∀x(Fx ⊃ Gx)
(2)	∀x(Gx ⊃ Hx)
(3)	¬∀x(Fx ⊃ Hx)

and apply the (¬∀) rule to get

(4)	∃x¬(Fx ⊃ Hx)

(4) tells us that there is at least one thing in the domain such that the condition expressed by '¬(Fx ⊃ Hx)' holds of it. Pick one, and let's dub it with a name not otherwise being used in this argument, say 'a' again. Then we have

(5)	¬(Fa ⊃ Ha)

Applying the (∀) rule to (1) and (2), we get the following further information about this thing dubbed 'a':

(6)	(Fa ⊃ Ga)
(7)	(Ga ⊃ Ha)

Then applying the standard propositional rules to (5) to (7) in turn yields

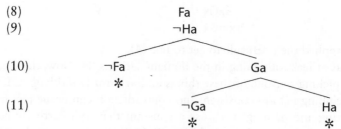

So the tree closes. Again, the contradictions here cannot have been created by the mere act of selecting and then dubbing something with a name not otherwise in use. Therefore the trouble must be in the original supposition that (1) to (3) are all true; so argument **G** is valid.

Note that absolutely crucial qualification: contradictions cannot be created by the mere act of dubbing something *with a name not otherwise in use*. Of course, if a name has already been recruited for some other purpose, then using it again to dub something may very well cause trouble, as we shall now see.

Consider the following very simple case. The inference 'Someone is a great philosopher: hence Derrida is a great philosopher' is obviously bogus. Similarly the inference

> **H**   ∃xFx ∴ Fn

is equally invalid. Now suppose we start a tree, and then – at line (3) – instantiate the existential premiss with the name already in use, to get

(1)   ∃xFx
(2)   ¬Fn
(3)   Fn
                                 ✳

We have then absurdly 'proved' that the fallacious argument **H** is valid!

Here's another example of the same mistake getting us into trouble. Consider the inference:

I   ∃xFx, ∃xGx ∴ ∃x(Fx ∧ Gx)

Suppose we set out to build a quantifier tree, starting:

(1)   ∃xFx
(2)   ∃xGx
(3)   ¬∃x(Fx ∧ Gx)

We apply the (¬∃) rule to get

(4)   ∀x¬(Fx ∧ Gx)

Now, if there is at least one thing which satisfies 'F', it can do no harm to pick one and dub it 'a', so we could write

(5)   Fa

Likewise, if there is at least one thing which satisfies 'G', it can do no harm to pick one and dub it with some name. But suppose we offend by wrongly using the same name again, and add

(6)   Ga

to the tree. Given (4), we could continue

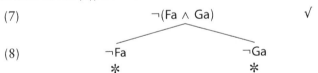

The tree then closes, giving the verdict that argument **I** is valid, which it isn't! Compare 'Someone is very tall; someone is very short; hence someone is very tall and very short.'

It's evidently step (6) that caused the trouble. To repeat, although (2) tells us that something satisfies 'G', and it can do no harm to pick out such a thing and dub it, *we mustn't suppose that it is the same thing that we have already dubbed* '*a*'. At that step (6), we are only entitled to add something like

(6)   Gb

We could continue

(7)   ¬(Fa ∧ Ga)

as before, and also add

(8)   ¬(Fb ∧ Gb)

But this plainly isn't going to get the tree to close.

In summary, then, all this motivates the following ∃-*instantiation rule*:

> (∃) Given an existential quantification of the form ∃vC(...v...v...) on an
> open path, then we can add C(...c...c...) to each open path that it is
> on, where c is a 'new' constant, i.e. one which *hasn't yet appeared on
> any of those paths*; we then check off the original wff.

∃vC(...v...) tells us that something – and it may be only one thing – is C. So
picking such a thing (and dubbing it) exhausts the claim's commitments: that's
why we need to check it off once the rule (∃) is applied.

Our two proposed rules for un-negated quantifiers might look confusingly
similar, so let's point up the essential difference loud and clear.

- The rule (∀) says that you can instantiate a universal quantifier with any
  name *already in use (on the path)*.

- The (∃) rule says that you can instantiate an existential quantifier with any
  name *that isn't already in use (on the relevant paths)*.

In Chapter 28, we'll return to motivate these quantifier tree-building rules much
more carefully (but we'll need a proper account of the semantics for QL first).
However, our quick-and-dirty explanations will suffice for the moment.

25.4 More examples

To help fix ideas, let's show that the following simple arguments are all valid by
translating into QL and using trees (try these examples before reading on):

J   Some chaotic attractors are not fractals; all Cantor sets are fractals;
     hence some chaotic attractors are not Cantor sets.

K   If there is a logician who isn't a philosopher, then Jones isn't a philos-
     opher. Jones *is* a philosopher. So every logician is a philosopher.

L   Only if Owen is happy does he love someone. If anyone loves Nerys
     then Owen does. Maldwyn loves Nerys. So Owen is happy.

The first should raise no translational difficulties:

J′   ∃x(Fx ∧ ¬Gx), ∀x(Hx ⊃ Gx) ∴ ∃x(Fx ∧ ¬Hx)

and the finished tree looks like this (still adding line numbers for ease of reference):

(1)	∃x(Fx ∧ ¬Gx)	√
(2)	∀x(Hx ⊃ Gx)	
(3)	¬∃x(Fx ∧ ¬Hx)	√
(4)	∀x¬(Fx ∧ ¬Hx)	
(5)	(Fa ∧ ¬Ga)	√
(6)	(Ha ⊃ Ga)	√
(7)	¬(Fa ∧ ¬Ha)	√

(8)
(9)

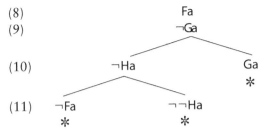

How was this tree produced? The first three lines are automatic, and applying the
(¬∃) rule immediately gives (4). That yields an existentially quantified wff and
two universally quantified wffs unchecked.

Now, a useful rule of thumb is *where possible apply the (∃) rule before the (∀)*
rule. Why? Because instantiating an existential quantification always introduces
a *new* name, and you'll probably need to go back to the universal quantifications *afterwards* to extract any needed information about the named object. But
in fact, in this case, we don't get a choice anyway. The only rule we can apply
once we've got to line (4) is the (∃) rule to (1). So we pick one of the things
which makes '∃x(Fx ∧ ¬Gx)' true, and dub it 'a', yielding (5), and checking off
(1). We then extract 'a'-relevant information from (2) and (4) by use of the (∀)
rule. With (6) and (7) in place, the rest is straightforward.

We can translate the next argument thus:

   K′   (∃x(Fx ∧ ¬Gx) ⊃ ¬Gn), Gn ∴ ∀x(Fx ⊃ Gx)

Note carefully the bracketing in the translation of the first premiss. The tree then
starts in the standard way, applying the rule (¬∀) to (3):

(1)   (∃x(Fx ∧ ¬Gx) ⊃ ¬Gn)   √
(2)   Gn
(3)   ¬∀x(Fx ⊃ Gx)   √
(4)   ∃x¬(Fx ⊃ Gx)

(5)   ¬∃x(Fx ∧ ¬Gx)   ¬Gn
                 | ✱

The left-hand branch now continues

(6)   ∀x¬(Fx ∧ ¬Gx)
(7)   ¬(Fa ⊃ Ga)

applying first the (¬∃) rule to (5), and then the (∃) rule to (4). Then, after applying the (∀) rule to (6), we quickly finish:

(8)   ¬(Fa ∧ ¬Ga)
(9)   Fa
(10)   ¬Ga

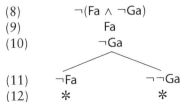

Finally, here's a translation of the third argument (why the *existential* quantifiers for 'anyone'? – compare §24.1, Example 10):

   L′   (∃xLox ⊃ Ho), (∃xLxn ⊃ Lon), Lmn ∴ Ho

The tree is then fairly straightforward:

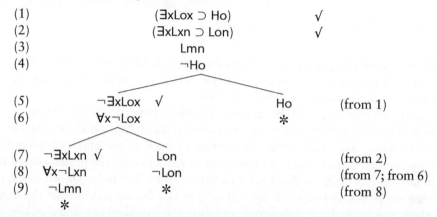

(1)	(∃xLox ⊃ Ho) √	
(2)	(∃xLxn ⊃ Lon) √	
(3)	Lmn	
(4)	¬Ho	
(5)	¬∃xLox √ Ho	(from 1)
(6)	∀x¬Lox *	
(7)	¬∃xLxn √ Lon	(from 2)
(8)	∀x¬Lxn ¬Lon	(from 7; from 6)
(9)	¬Lmn *	(from 8)
	*	

Note that on the right-hand branch at (8), we could have instantiated the universal quantifier at (6) by any of the names 'm', 'n', or 'o': our choice, however, is the sensible one, as it makes the right-hand branch close. Similarly at (9), we make the obvious choice of how best to instantiate the quantifier at (8).

25.5   Open and closed trees

We've introduced the language QL. We've explained four rules that we need in addition to the rules for connectives in order to construct QL tree arguments to show various inferences are valid. There's a little work to be done tidying up the syntax for our formal language; and there's rather a lot of work to be done developing the semantics. We need to explore trees further, and to discuss whether any more tree-building rules are necessary to give us a 'complete' system (compare §20.4). In fact, we'll see that we do need to slightly augment the rules we've so far stated. But that is all to come later. For the moment, let's concentrate on emphasizing a crucial point which should be made straight away.

Predicate logic really comes into its own when we are dealing with inferences involving propositions that involve more than one quantifier. We'll look at a range of examples in later chapters; but here's just one to be going on with (and to set the scene for the point we need to make). Consider the argument 'Everyone loves someone who is kind; no cruel person is kind; Caligula is cruel. Still, it follows that not everyone is cruel'. That's evidently valid, if we read the first premiss as meaning everyone loves at least one kind person. For take Caligula. Although he is cruel, he loves someone who is kind. And that person, being kind, isn't cruel. So not everyone is cruel.

Let's warrant this argument with a tree proof. Here's a translation:

   M   ∀x∃y(Fy ∧ Lxy), ∀x(Gx ⊃ ¬Fx), Gn ∴ ¬∀xGx

The tree starts in the obvious way.

(1)	∀x∃y(Fy ∧ Lxy)	
(2)	∀x(Gx ⊃ ¬Fx)	
(3)	Gn	
(4)	¬¬∀xGx	√
(5)	∀xGx	

We now have three universal quantifiers to instantiate, using the rule (∀). But we'll tackle (1) first, so we can 'get at' the embedded existential quantifier. At the moment, we only have one constant in play, so we have to instantiate (1) with that, to yield

(6)	∃y(Fy ∧ Lny)

Now we'll follow the heuristic rule we noted before: where possible, we apply the (∃) rule before the (∀) rule. We have to instantiate (6) with a name *new* to the argument:

(7)	(Fa ∧ Lna)	√
(8)	Fa	
(9)	Lna	

We can now continue by extracting more 'a'-relevant information by instantiating the universal quantifications (2) and (5):

(10)	(Ga ⊃ ¬Fa)	√
(11)	Ga	

(12) ¬Ga	¬Fa
✳	✳

And once more we are done.

So far, so good. But note we could have taken a wrong turning after line (7). For we might be struck by the fact that we now have the new name 'a' in play, and we could instantiate (1) again with this new name, thus ...

(8′)	∃y(Fy ∧ Lay)

We then have a new existential quantification which we can instantiate in turn with a constant new to the argument:

(9′)	(Fb ∧ Lab)

Now suppose we (unwisely, but quite legitimately) use the name 'b' to instantiate (1) again to infer

(10′)	∃y(Fy ∧ Lby)

and instantiate *this* with respect to a new name:

(11′)	(Fc ∧ Lbc)

Evidently we could go on, and on, and on without end. The unwise strategy of returning to (1) whenever we can instantiate it with a new name could prevent the tree from ever closing! (Query: but what entitles us to suppose that there are all those different things in the domain to be named 'a', 'b', 'c', ...? Maybe the domain only contains one thing! Reply: the question is based on a misunderstanding. We are introducing a sequence of new *names* one after another; but maybe the names refer to the same *thing*, maybe to different things: that is left open.)

So here we encounter a very deep and important difference between **PL** and **QL** trees. In the propositional case it doesn't matter (except for the aesthetics) in which *order* the available rules are applied. If a tree closes, then any legally constructed tree with the same initial trunk will also close (see §§16.4, 19.3). *In the quantificational case, by contrast, the order in which rules are applied matters.* Our first tree for **M** closed: but we've seen that a second tree, equally constructed in accordance with the rules, could be extended without limit, without closing.

Now, finding *one* well-constructed closed tree for a **QL** argument is enough to prove that the argument is valid. If *one* line of reasoning from the assumption that the premisses and the negation of the conclusion are true leads to contradictions all round, that settles it: if the premisses are true, the conclusion must be true too. This result isn't subverted by the fact that some *other* line of reasoning trails off in the wrong direction and fails to reveal a contradiction.

How can **QL** trees have branches going on for ever when that can't happen for **PL** trees? Because our rules for unpacking complex **PL** wffs systematically 'use up' wffs, always taking us to shorter and shorter wffs: and – assuming there are only a finite number of premisses to begin with – eventually we will simply run out of rules to apply. We must end up with a completed tree, either open or closed. In the quantificational case, however, we can get into cycles like the one that takes us from (6) to (8') to (10') to ..., where the wffs don't become any simpler, and so the tree never finishes.

Of course, our second, un-ending, tree for **M** is constructed according to a pretty crazy strategy. And in fact, there is a decently systematic strategy which will always deliver us a closed tree for an argument *if there is one*. But intriguingly, there is no mechanical test which will show us (in finite time) whether applying this strategy will eventually deliver a closed tree; there is no mechanical test for whether a quantified argument is valid or not. More on this anon (see Chapter 30).

But having seen where all this development of **QL** is taking us, and in particular having had a glimpse of how we can develop a tree method for warranting inferences, we really need now to return to basics, to tidy up the details. That's the business for the next few chapters.

25.6 Summary

- We have introduced four new rules for quantifier tree-building, which we've labelled (∀), (¬∀), (∃), (¬∃), and given some motivation for them. We'll state the rules for **QL** trees more carefully in Chapter 29.

- QL trees give us *sound* proofs of validity: i.e. when a properly constructed tree closes, the argument whose premises and negated conclusion start the tree is indeed valid. But we noted that valid arguments can also generate (infinite) open trees.

- QL trees are not a *mechanical method* for demonstrating the validity or invalidity of arguments. There can't be one.

Exercises 25

A   Show the following simple arguments are valid by translating into **QL** and using trees.

1.   Everyone is rational; hence Socrates is rational.
2.   No-one loves Angharad; hence Caradoc doesn't love Angharad.
3.   No philosopher speaks Welsh; Jones is a philosopher; hence Jones does not speak Welsh.
4.   Jones doesn't speak Welsh; hence not everyone speaks Welsh.
5.   Socrates is rational; hence someone is rational.
6.   Some philosophers speak Welsh; all Welsh speakers sing well; hence some philosophers sing well.
7.   All electrons are leptons; all leptons have half-integral spin; hence all electrons have half-integral spin.
8.   All logicians are philosophers; all philosophers are rational people; no rational person is a flat-earther; hence no logician is a flat-earther.
9.   If Jones is a bad philosopher, then some Welsh speaker is irrational; but every Welsh speaker is rational; hence Jones is not a bad philosopher.

B   Consider the following rule

($\neg\exists'$) If $\neg\exists v C(...v...v...)$ appears on an open path, then we can add $\neg C(...c...c...)$ to that path, where c is any constant which already appears on the path.

Show informally that this rule would do as well as our rule ($\neg\exists$). What would be the analogous rule for dealing with negated universal quantifiers without turning them first into existential quantifiers?

C   Suppose we had set up predicate logic with a *single* quantifier formed using the symbol 'N', so that $N v C v$ holds when nothing is C. Show that the resulting language would be expressively equivalent to our now familiar two-quantifier language **QL**. What would be an appropriate set of rules for tree-building in a language with this single quantifier?

The syntax of QL

Over the last four chapters, we have explored the language **QL** in an informal way, and we have introduced the tree method for warranting quantifier arguments once they are regimented into **QL** form. In this chapter, we pin down the syntactic details of **QL** much more carefully, as a preparation for getting the semantics into clear focus in the next chapter. We start with a couple of preliminary sections, motivating the full-dress syntactic story in §26.3.

26.1 How not to run out of constants, predicates or variables

In **PL**, we supplied ourselves with an unlimited number of atomic wffs. That's because we didn't want to put a limit on the complexity of arguments that we can express using that language. For exactly the same reason, we will supply ourselves with an unlimited number of constants and predicates in **QL**. And we don't want to run out of variables either, because we want to be able to express propositions with arbitrary numbers of quantifiers.

Consider predicates first. Each predicate of **QL** comes with a fixed number of slots to be filled up with names (or individual variables). We won't put any fixed limit on the number of places a predicate can have. Why shouldn't we be able to argue with five-place or twenty-five-place predicates?

So we want an unlimited supply of n-place predicates, for each n. Probably the neatest way of securing this is to help ourselves to the numerals. Assume 'P' (for 'predicate') is in the alphabet of **QL**; then we can adopt the rule that any occurrence of this letter with superscript and subscript numerals j, k (where j, $k \geqslant 0$) counts as a predicate. The *super*script fixes the number of places the predicate has; and the *sub*script distinguishes the various predicates which share that number of places. To be sure, it won't look very pretty having predicates like

$$P^3_{14} \, , \; P^3_{15} \, , \; P^9_{254} \, , \; ...$$

but at least we are not going to run out of supplies!

We could produce the same effect more in the style of **PL** by using two sequences of primes (one sequence to indicate the number of places, and another to distinguish among predicates with the same number of places): but that would be typographically even messier.

We'll adopt the same device of using numerical subscripts to give us an unlimited supply of names and variables. Let's assume that 'c' (for 'constant') is in the alphabet of **QL**, and have the rule that any occurrence of 'c' with a subscript numeral $k \geq 0$ counts as a constant. Similarly, let's assume 'v' (for 'variable') is in the alphabet; there shouldn't be any confusion with the '∨' sign for disjunction. The corresponding rule to generate an unlimited number of variables is that any occurrence of 'v' with a subscript numeral counts as a variable.

By using the subscript trick, we now know we are not going to run out of constants, predicates and variables.

On the other hand, we most certainly *don't* want to be continually writing down ugly wffs like

$$P^3_{14}c_7c_3c_9 \, , \quad \forall v_5 P^3_{14}c_7v_5c_9 \, , \quad \ldots$$

or worse. Nor do we want retrospectively to rule out all our examples over recent chapters from counting as official wffs of **QL**. So, although it complicates things, let's also add to our vocabulary all the unadorned letters for names, predicates, and variables that we've met before, plus a few extra.

One last comment. A three-place predicate like P^3_{14} takes three names to form an atomic wff of **QL**. What about a zero-place predicate like P^0_2 (for note, we did allow the superscript indicating the number of places to take the value zero)? By parity of reasoning, a zero-place predicate is an expression that takes zero names to make an atomic wff – *in other words, it already is an atomic wff!* So a zero-place predicate functions exactly like a propositional atom of **PL**. This has the kind of neatness that mathematically inclined logicians just *love*. Instead of saying that our language **QL** has both predicates of all degrees from *one* upwards plus propositional atoms as well, we say more crisply that the language has predicates of all degrees from *zero* upwards. (It could be convenient, however, to have a few simple letters for propositional atoms in addition to the zero-superscripted instances of 'P'. Since 'P' and 'R' are already recruited for other uses, let's add 'A', 'B', 'C' for this purpose. By this stage in the game, no-one is going to confuse these object-language sans-serif letters with the italic letters '*A*', '*B*', '*C*' we use as metalinguistic schematic variables when generalizing about wffs).

26.2 How to introduce quantifiers

In Chapter 22, we showed how we can build up quantified wffs by the operation of *quantifying in*. We take a wff with a constant in it, and then – in one go, so to speak – we replace the constant with a variable and also prefix a quantifier formed using that same variable (thereby 'binding' the variable to the quantifier). To take the simplest kind of example, we can start from the atomic wff 'Fm', and quantify in to form the wff '∃x Fx'.

Here's an alternative way of proceeding. We could say that atomic wffs are formed by combining a predicate with the right number of *constants and/or variables*. So, by this more generous rule for atomic wffs, 'Fx' is itself a wff. Then the rule for adding quantifiers is: given a wff $C(\ldots v \ldots v \ldots)$ involving one or more occurrences of the variable v, but *not* already containing the quantifier $\forall v$ or $\exists v$, then we form another wff by prefixing one of those quantifiers. So since 'Fx' is a wff by our first new rule, '$\exists$xFx' is a wff by the new second rule. (This second rule says we can only prefix quantifiers that link to variables that aren't already tied to quantifiers. That prevents us going on to form expressions like '$\exists$x$\exists$xFx' or '$\exists$y$\exists$xFx'.)

In Chapter 22, we quickly demonstrated that '$\forall$y$\forall$x$\exists$zRzyx' is a wff by setting out a possible way of constructing it by three operations of quantifying in, with the following sequence of steps:

Rmno
$\exists$zRzno
$\forall$x$\exists$zRznx
$\forall$y$\forall$x$\exists$zRzyx

Note, by the way, that we could equally well have followed the steps

Rnom
$\exists$zRzom
$\forall$x$\exists$zRzox
$\forall$y$\forall$x$\exists$zRzyx

So on our previous story, wffs don't have a unique 'constructional history' (cf. §8.2). But given our new rules, to construct the same wff, we instead start from 'Rzyx' and simply add the quantifiers like this:

Rzyx
$\exists$zRzyx
$\forall$x$\exists$zRzyx
$\forall$y$\forall$x$\exists$zRzyx

and this constructional history is unique under our new rules. That's one differece between the old and new rules. But the key difference between our two ways of proceeding is that on the new one, expressions with variables 'dangling free', like 'Rzyx' or '$\forall$x$\exists$zRzyx' count as wffs, which they didn't before. *Every old-style wff is a new-style wff, but not vice versa.*

Quantifiers in QL need variables, and the point of variables is to be tied to quantifiers; so our approach in Chapter 22, where they get introduced in tandem is in some ways the more natural one. But the more liberal two-step approach, which also allows wffs with 'free' variables, has some presentational advantages when we turn to technicalities. And perhaps more to the point, it is still the more common approach (to the extent that it would be odd to study quantificational logic and never hear of wffs with free variables). We'll follow the common convention and adopt the more liberal rules for forming wffs from now on.

26.3   The official syntax

We'll now put together all the ideas about syntax from Chapter 22, and from §§26.1, 26.2 above, into one composite story. This is done for convenience and

---

The official syntax for QL

The alphabet for **QL** is as follows:

> *for forming constants*: a, b, c, m, n, o
> *for forming predicates*: A, B, C, F, G, H, L, M, P, R, S
> *for forming variables*: v, w, x, y, z
> *for superscripts and subscripts*: the arabic numerals
> *for constructing complex wffs*: $\land$, $\lor$, $\neg$, $\supset$, $\forall$, $\exists$, (,)
> *for expressing arguments*: $\therefore$, $*$, *plus also the comma.*

The basic elements of **QL** are then:

> *individual constants*: a, b, c, m, n, o, c_k
> *individual variables*: w, x, y, z, v_k
> *0-place predicates (propositional atoms)*: A, B, C, P_k^0
> *1-place predicates*: F, G, H, P_k^1
> *2-place predicates*: L, M, P_k^2
> *3-place predicates*: R, S, P_k^3
> *j-place predicates*: P_k^j

where the indices *j*, *k* run over all values $\geqslant 0$.

We add a standard definition:

> A *term* is an individual constant or individual variable.

Then the rules for forming atomic wffs are straightforward:

(A1)   For all $j \geqslant 0$, a *j*-place predicate followed by *j* terms is an atomic wff of **QL**.

(A2)   Nothing else is an atomic wff.

Finally, we can state the rules for the full class of wffs:

(W1)   Any atomic wff of **QL** is a wff.

(W2)   If *A* is a wff, so is $\neg A$.

(W3)   If *A* and *B* are wffs, so is $(A \land B)$.

(W4)   If *A* and *B* are wffs, so is $(A \lor B)$.

(W5)   If *A* and *B* are wffs, so is $(A \supset B)$.

(W6)   If *A* is a wff, and *v* is an individual variable which occurs in *A*, (while neither $\forall v$ nor $\exists v$ occurs in *A*), then $\forall v A$ is also a wff.

(W7)   If *A* is a wff, and *v* is an individual variable which occurs in *A*, (while neither $\forall v$ nor $\exists v$ occurs in *A*), then $\exists v A$ is also a wff.

(W8)   Nothing is a wff that cannot be shown to be so by applications of rules (W1) to (W7).

---

for the sake of completeness; *there are no new ideas here*. We state the rules using metalinguistic variables like '*A*' and '*v*' in the now familiar way.

The story looks complicated; but divide it into two parts. The second part, comprising the definition of a term, and the A-rules and W-rules, is clean and simple. So the complication is all in the first part where we give the basic elements for our version of QL. To repeat, the mess is *there* so we don't always have to write messy wffs with letters bedecked with superscripts and subscripts.

Let's quickly show that the following (see §24.1, Example 14) is a wff by our official rules:

$$\neg \forall x((Fx \land \exists z Mxz) \supset \forall y((Gy \land Lyx) \supset Lxy))$$

First, laboriously, we'll give an annotated proof in the style of §§5.2, 8.2, listing the principles of construction that are being appealed to at each stage:

(1)	'Fx' is a wff	(by A1)
(2)	'Mxz' is a wff	(by A1)
(3)	'∃zMxz' is a wff	(from 2 by W7)
(4)	'(Fx ∧ ∃zMxz)' is a wff	(from 1, 3 by W3)
(5)	'Gy' is a wff	(by A1)
(6)	'Lyx' is a wff	(by A1)
(7)	'(Gy ∧ Lyx)' is a wff	(from 5, 6 by W3)
(8)	'Lxy' is a wff	(by A1)
(9)	'((Gy ∧ Lyx) ⊃ Lxy)' is a wff	(from 7, 8 by W5)
(10)	'∀y((Gy ∧ Lyx) ⊃ Lxy)' is a wff	(from 9, by W6)
(11)	'((Fx ∧ ∃zMxz) ⊃ ∀y((Gy ∧ Lyx) ⊃ Lxy))' is a wff	(from 4, 10 by W5)
(12)	'∀x((Fx ∧ ∃zMxz) ⊃ ∀y((Gy ∧ Lyx) ⊃ Lxy))' is a wff	(from 11 by W6)
(13)	'¬∀x((Fx ∧ ∃zMxz) ⊃ ∀y((Gy ∧ Lyx) ⊃ Lxy))' is a wff	(from 12 by W2)

But this chain of inferences is, of course, more perspicuously represented in the form of a construction tree in the manner of §8.2:

```
                                        Gy        Lyx
                    Mxz                 ─────────────
        Fx      ∃zMxz                   (Gy ∧ Lyx)         Lxy
        ─────────────                   ──────────────────────
        (Fx ∧ ∃zMxz)                    ((Gy ∧ Lyx) ⊃ Lxy)
                                        ∀y((Gy ∧ Lyx) ⊃ Lxy)
        ───────────────────────────────────────────────────
        ((Fx ∧ ∃zMxz) ⊃ ∀y((Gy ∧ Lyx) ⊃ Lxy))
        ∀x((Fx ∧ ∃zMxz) ⊃ ∀y((Gy ∧ Lyx) ⊃ Lxy))
        ¬∀x((Fx ∧ ∃zMxz) ⊃ ∀y((Gy ∧ Lyx) ⊃ Lxy))
```

where, recall, every expression on the tree is to be understood to come within an unwritten frame '"..." is a wff'. Here's another quick example, the wff

$$\forall x(((Fx \land \exists z Mxz) \land \forall y(Mxy \supset Rxny)) \supset Lnx)$$

(which is the translation for Example 15, §24.1) has the tree

```
            Mxz                 Mxy         Rxny
        Fx      ∃zMxz           ─────────────
        ─────────────          (Mxy ⊃ Rxny)
        (Fx ∧ ∃zMxz)           ∀y(Mxy ⊃ Rxny)
        ((Fx ∧ ∃zMxz) ∧ ∀y(Mxy ⊃ Rxny))                Lnx
        ────────────────────────────────────────────────
        (((Fx ∧ ∃zMxz) ∧ ∀y(Mxy ⊃ Rxny)) ⊃ Lnx)
        ∀x(((Fx ∧ ∃zMxz) ∧ ∀y(Mxy ⊃ Rxny)) ⊃ Lnx)
```

Given our new construction rules, construction trees are unique – at least, if we stick to the convention adopted in §8.2 and, when branches join, put the wffs above the inference line in the same left-to-right order in which they occur as components in the wff below the line.

One more construction tree, this time for

$$\forall x(\exists y Mxy \supset \exists y(\neg Mxy \wedge Lxy))$$

(the second version of the translation for Example 12, §24.1). The variable 'y' in the end gets quantified twice; we want to show that this is legitimate.

$$
\begin{array}{cc}
& Mxy \\ \cline{2-2}
& \neg Mxy \qquad Lxy \\ \cline{2-2}
Mxy & (\neg Mxy \wedge Lxy) \\ \cline{1-1}\cline{2-2}
\exists y Mxy & \exists y(\neg Mxy \wedge Lxy) \\ \hline
\multicolumn{2}{c}{(\exists y Mxy \supset \exists y(\neg Mxy \wedge Lxy))} \\ \hline
\multicolumn{2}{c}{\forall x(\exists y Mxy \supset \exists y(\neg Mxy \wedge Lxy))}
\end{array}
$$

Note that the two '∃y' quantifiers get introduced on separate branches of the tree. At each introduction, left and right, the quantifier is applied to a wff where the variable 'y' occurs and hasn't yet been quantified. So the construction follows our rules. Compare

$$\forall x(Fx \supset \exists x(Gx \wedge Lxn))$$

which isn't a wff by our rules. If we try to produce a tree, we could get as far as showing that 'Fx' and '∃x(Gx ∧ Lxn)' are wffs, and so

$$(Fx \supset \exists x(Gx \wedge Lxn))$$

is a wff. But we can't use (W6) to add the quantifier '∀x', because that last wff already contains one of the quantifiers '∀x' or '∃x'.

26.4 Some useful definitions

We'll define (or redefine) some useful syntactic notions. First, as in **PL**,

- A wff *A* is a *subformula* of a wff *B* if *A* appears anywhere on the construction tree for *B*.

And, similarly to **PL**,

- The *main logical operator* (quantifier or connective) of *A* is the operator introduced at the final step in the construction tree.

Next, we want to capture the idea of an expression being within the scope of an operator (quantifier or connective). As in §8.2, a neat way of doing this is to define some wff as *being* the scope of an operator, and then any part of this wff will count as being *in* the scope of the operator:

- The *scope* of a quantifier in a **QL** wff is the subformula that starts with that quantifier. In other words, it is the wff that occurs at the step on the construction tree where the quantifier is introduced. Take for example the wff '∀x(∃yMxy ⊃ ∃y(¬Mxy ∧ Lxy))':

the scope of the first '∃y' is '∃yMxy';
the scope of the second '∃y' is '∃y(¬Mxy ∧ Lxy)';
the scope of '∀x' is the whole wff.

- The scope of a *connective* is likewise the wff on the construction tree where the connective is introduced.

Now let's tidy up the intuitive notion of a 'free' variable. The idea is that a variable dangles free when it isn't tied down by a matching quantifier. So let's say

- An occurrence of a variable v in a wff A is *bound* if and only if it is in the scope of some matching quantifier ∀v or ∃v that occurs in A.

- An occurrence of a variable v in a wff A is *free* if and only if it is not bound.

Thus consider the wffs

(Mxy ⊃ Rxny)
∀y(Mxy ⊃ Rxny)
((Fx ∧ ∃zMxz) ∧ ∀y(Mxy ⊃ Rxny))

In the first, both variables are free. In the second, 'y' is bound and 'x' is free. In the third, 'y' and 'z' are both bound, but 'x' still remains free. Again, in

(Mxy ⊃ ∃y(¬Mxy ∧ Lxy))

which counts as a wff by our rules, the first occurrence of 'y' is free but the others are bound (so we can't quantify the whole wff using '∀y' again).

With that terminology now to hand, the rules (W6) and (W7) come to this: If A is a wff, and v is an individual variable which only occurs free in A, then ∀vA and ∃vA are also wffs.

We need a term for those wffs without any occurrences of variables remaining free:

- A *sentence* of QL (alternatively, a *closed wff*) is a wff with no free occurrences of variables.

- If a wff isn't closed it is *open*.

If 'F' means *... is wise*, then (remembering that variables operate like pronouns) we might read an open wff like 'Fx' as meaning something like *he is wise*. But taken as a stand-alone claim, without anything to tell us who 'he' is referring to, this doesn't express a determinate proposition: it isn't yet either true or false. By contrast, closed wffs can be true or false.

Finally, we define again the notion of a *substitution instance* of a quantified wff (§22.4):

- Suppose a wff $C(...c...c...)$ results from ∀$vC(...v...v...)$ or ∃$vC(...v...v...)$ by stripping off the initial quantifier, and then substituting some individual constant c for all the occurrences of the associated variable v. Then $C(...c...c...)$ is said to be an *instance* of the quantified sentence.

26.5 Summary

- We have now taken a semi-permissive line on what is to count as an official wff of **QL**. We are allowing wffs with free variables, though we are still ruling out expressions with redundant or duplicated quantifiers.

- We have given ourselves an unlimited supply of predicates, names and variables, using numerical subscripts and superscripts.

- The key building rule for constructing quantified sentences is: if A is a wff, and v is an individual variable which only occurs free in A, $\forall v A$ and $\exists v A$ are also wffs.

- Demonstrations that an expression is indeed a wff according to our rules can be set out as construction trees; looking at how the tree is built up will reveal the scope of the various logical operators, and which occurrences of variables are free at which stage of the construction.

Exercises 26

Which of the following are wffs of **QL**? Which are sentences, i.e. closed wffs? In the open wffs, which occurrences of which variables are free? What is the main logical operator of each wff?

1. ∃z∀y(Myz ∨ ¬∀y(Mxy ∧ Lxy))
2. ∃z(∀yMyz ∨ ¬(Mxy ∧ Lxy))
3. ∃z(∀yMyz ∨ ¬∀y(Mxy ∧ Lxy))
4. ¬∃z(∀yMyz ∨ ¬∀x∀y(Mxy ∧ Lxy))
5. ∃z∀x(∀yMyz ∨ ¬∀y(Mxy ∧ Lxy))
6. ∀x(Gx ⊃ ∃y∃z(Hy ∧ Hz) ∧ Rxyz))
7. (Gx ⊃ ∃x∃z((Hy ∧ Hz) ∧ Rxyz))
8. ∀x∃y(Gx ⊃ ∃y¬∃z((Hy ∧ Hz) ∧ Rxyz))
9. ∀x(Gx ⊃ ∃y∀x((Hy ∧ Hz) ∧ Rxyz))
10. (∀x(Gx ⊃ ∃z((Hy ∧ Hz) ∧ Rxyz)) ∨ ∃zHz)
11. ∀x∀y∀z((Fx ∧ Fy) ∧ Rzxy) ⊃ Gz)
12. ¬∀x∀y(((Fx ∧ Fy) ∧ ∀zRzxy) ⊃ Gz)

Q-valuations

Having now tidied up the syntax, in this chapter we give the official story about the semantics of **QL**. In particular, we define the key notion of a q-valuation of **QL** wffs (the analogue of a truth-valuation for **PL**).

27.1 Q-valuations vs. interpretations

When describing the semantics for **PL** in Chapter 9, we quickly moved on from discussing *interpretations* of wffs to discussing *valuations*. A valuation, recall, starts with an assignment of truth-values to some of the building blocks of **PL** (i.e. some atomic wffs). This initial assignment then generates a valuation of all the molecular wffs that can be built up from these atomic wffs using the logical resources of the language (i.e. the connectives). We also defined the notion of tautological validity – where an inference is tautologically valid if every valuation which makes the premisses true makes the conclusion true. Tautological validity implies plain validity; but there are valid arguments (e.g. a **PL** translation of 'Jo is a sister, hence Jo is female') which are not tautologically valid.

We now proceed similarly, by introducing the notion of a *q-valuation*. An interpretation of (some) **QL** wffs tells us what the relevant wffs *mean*; a q-valuation fixes which of these wffs are *true*. A q-valuation must start by fixing the domain of discourse; then it needs to assign truth-relevant values to some of the building blocks of **QL** (i.e. some constants and predicates). This initial assignment then generates a valuation of all the wffs that can be built up from these ingredients using the logical resources of the language. In particular it tells us which closed wffs (sentences, wffs without free variables) are true. We can then define the key notion of *q-validity*, where an inference is q-valid if every q-valuation which makes the premisses all true makes the conclusion true. Q-validity implies plain validity; but there are valid arguments (e.g. a **QL** translation of 'Jo is a sister, hence Jo is female') which are not q-valid.

Valuations are crucial for propositional logic, thought of as the study of arguments that are valid because of the way that the truth-functional connectives distribute among premisses and conclusion. Q-valuations are crucial for quantificational logic, thought of as the study of arguments that are valid because of the way connectives and quantifiers distribute among premises and conclusion. Irritatingly and confusingly, you will find q-valuations are elsewhere usually called 'interpretations'. But an interpretation (in the intuitive sense of the term, which accords with our use in this book) gives the *meaning* or *sense* of some expressions: by contrast, a q-valuation just specifies the features of these expressions which fix the truth-values of wffs.

27.2 Q-valuations defined

Given a bunch of **QL** wffs, let's say that their *vocabulary* is the set V of constants and predicates that appear in those wffs. So a q-valuation for a set of wffs with vocabulary V needs to (1) fix a domain D for the quantifiers to run over, and then assign truth-relevant properties to (2) the constants and (3) the predicates in V.

(1) We'll allow the domain D to be any collection of objects, big or small. The domain might be a 'natural' set like all people, or all numbers. But any old gerry-mandered collection, such as the set {The Eiffel Tower, the number four, Bill Clinton, Mount Everest} – or even the tiny set {Jennifer Lopez} containing just one item, namely Jennifer – can also serve as a domain of discourse.

Could we allow the domain to contain absolutely *everything*? It isn't clear that this idea makes sense. For take a domain as big as you like. Then isn't there always a more inclusive collection of things containing all the objects in the original domain plus something else, namely the *set* of those objects? But we won't tangle here with that kind of issue. We'll just allow domains to be any determinate collection of objects, however big that might possibly be.

To go to the other extreme, could we allow the domain to be *empty*, to have *nothing* in it? There are pros and cons to the proposal that we should allow empty domains. Tradition is on the side of requiring that any domain contains at least one object; and, as we'll see, some technicalities do go more smoothly if we follow tradition. *So we'll take the standard line and require domains to be non-empty.* More on this in §29.3.

(2) Consider an ordinary-language sentence like 'Bill is wise'. What is required for this to be true? A certain man has to be wise. And the role of the name 'Bill' is to pick out the individual who has to be wise if the sentence is to be true. In other words, what matters about the name here, as far as truth is concerned, is to whom it refers.

To be sure, there can be differences in associations, tone, or sense between names that have the same reference, like 'Bill' and 'Mr Clinton'. But as far as the truth of what we say goes, it *normally* doesn't matter which name we use. For example, 'Bill kissed Monica' and 'Mr Clinton kissed Monica' attribute the same action to the same man, and hence are true or false together.

Constants are the **QL** analogues of ordinary-language names. So similarly, *what matters about a constant, as far as truth is concerned, is which object it refers to*. Or as we'll say: the q-value of a constant is its *reference*. Of course, this object should be in the domain of discourse; for the reference of a constant c should be one of the things that is in question when we talk of 'everything'!

(3) The third component of a q-valuation deals with predicates. Again, we take our lead from elementary reflections about their ordinary-language correlates.

'Bill is wise' is true just if the man picked out by the name 'Bill' is one of the wise. That way of putting it immediately suggests a simple way of looking at what matters about '... is wise' as far as truth is concerned. The predicate is correlated with a set of things – the things that satisfy the condition of being wise. If Bill is in that set, the sentence in question is true; if not, not.

The set of objects that satisfy a predicate is standardly called the *extension* of the predicate. So another way of putting things is this: a sentence like 'Bill is wise' is true just if the object referred to by the name is in the extension of the predicate. (Maybe *no one* is wise: in that case the extension of the predicate is empty.)

Moving from ordinary language to **QL**, we again can say that (as far as truth is concerned) what matters about a one-place predicate is its extension, the set of objects of which the predicate is true. These objects should be in the domain D; the things of which a predicate is true should again be among the things which count as 'everything'. So the q-value of a one-place predicate F is a (possibly empty) set of objects from the domain. And an atomic wff of the form Fc will be true if and only if the q-value of c (i.e. the object which c refers to) is in the q-value of F (i.e. is in the extension of F).

What about evaluating two-place predicates? Well, 'Romeo loves Juliet' is true just so long as the *pair* of people ⟨Romeo, Juliet⟩ is in the set of pairs such that the first in the pair loves the second. So what matters about a relational term like 'loves' is the set of *pairs* of things it picks out: this set of pairs is the extension of the relation. Likewise for two-place relations in **QL**. The q-value of a two-place relation R is its extension, a set of pairs of things from the domain. And a wff of the form Rmn is true if and only if the q-values of the constants m and n taken in that order form a pair that belongs to the extension of R. (When we talk about pairs here, we of course mean *ordered* pairs, so that we can indeed sensibly talk of a first and second member.)

Similarly, the q-value of a three-place relation S is a set of ordered triples (namely, the triples that satisfy S). A wff of the form $Smno$ is true if the q-values of the names taken in order form an ordered triple which is in the extension of the predicate. And so it goes.

It is useful to introduce some standard jargon here. We've spoken of individual objects, ordered pairs or couples, and ordered triples. What comes next, of course, are ordered quadruples, quintuples, and so on. So let's refer to all such items as *tuples*: tuples with n members are, of course, *n-tuples*. We can then snappily say: for $n > 0$, the q-value of an n-place predicate is a set of n-tuples. (That works, at any rate, if we stipulate that '1-tuples of objects' are just the objects.)

But remember, we are also allowing *zero*-place predicates in **QL** (§25.1). Such expressions are already atomic wffs in their own right, just like the propositional variables in **PL**. If there are any such expressions in the vocabulary V, then we will give them their own distinctive kind of values on a q-valuation, namely truth-values.

So putting all that together, we have the following definition:

A q-valuation defined over vocabulary *V*:

(1) specifies a (non-empty) collection of objects to be the domain *D*;

(2) it assigns to any constant in *V* some object in *D* as its q-value;

(3) it assigns a truth-value to any 0-place predicate in *V*; and it assigns to any *n*-place predicate in *V*, where $n > 0$, a (possibly empty) set of *n*-tuples of objects from *D*.

27.3 The semantics of quantifiers: a rough guide

A q-valuation defined over the vocabulary *V* fixes the truth-values for *atomic sentences* built from constants and predicates in *V* in the obvious way. An atomic sentence formed from an *n*-place predicate followed by *n* constants is true just in case the *n*-tuple formed by taking the q-values of the *n* constants in order is in the extension of the *n*-place predicate (except for the special case where $n = 0$, when the q-valuation directly assigns the 0-place predicate its own truth-value).

Further, we'll let **QL**'s truth-functional connectives look after themselves for the moment in the obvious way (so it will still be the case that a conjunction is true if both the conjuncts are true, and so on and so forth).

So the key question is how to deal semantically with quantified wffs. For example, when is the wff '∀xFx' true?

- We remarked before on a *wrong* answer. We can't say that '∀xFx' is true just when all its substitution instances in the given vocabulary *V* are true, i.e. when 'Fm', 'Fn', 'Fo', ... are all true. There may not be names in *V* for every object in the domain; so even if all the instances we can construct are true, '∀xFx' could still be false, because of an uncooperative unnamed object (see §22.4).

- To introduce a *correct* answer, note that 'Every man is such that he is mortal' is true just so long as it doesn't matter which man in the domain we choose, he is mortal – i.e. if 'He is mortal' is true, *whoever we use the pronoun 'he' to point to in the domain of men*. We can say something similar about '∀xFx' (for recall that we introduced variables in **QL** to do the work of pronouns, and remember too that we are now allowing expressions with free variables to count as wffs). So, '∀xFx' is true if 'Fx' is true when 'x' is treated as a pronoun pointing to an object in the domain, *no matter which object in the domain the pronoun is taken to be picking out*.

The rejected answer wrongly tries to explain when '∀xFx' is true by generalizing about *names*. Our new, correct, suggestion explains when it is true by generalizing about *objects* in the domain.

Following that lead, consider three wffs of the forms

(a) $\forall v C(...v...v...)$

(b) $\exists v C(...v...v...)$

(c) $C(...v...v...)$

Then the key idea is that (a) is true if the corresponding wff (c) comes out true when *v* is treated as a pronoun, *no matter what* object in the domain the pronoun *v* is taken to be 'pointing to'. Likewise (b) is true when there is *some* object in the domain such that (c) comes out true when *v* is treated as a pronoun picking out that object.

This straightforward and entirely intuitive idea is the key that unlocks the semantics of quantifiers. Everything that follows in the rest of this chapter just develops this one idea a bit more formally; it should be easy enough to grasp the official version as long as you don't lose sight of this underlying motivation.

So, to develop the key idea, we now allow q-valuations to be *extended* to valuations which assign temporary references to one or more variables. Take the simplest case where just one variable *v* is involved:

- We extend a given valuation *q* by assigning an object *o* from the domain for *v* to 'pick out' when the variable occurs free. Call this extension q^+. We can then treat wffs with *free* occurrences of *v* as if the variable is a temporary name for the object *o* assigned to the variable by q^+. So, $\forall v C(...v...v...)$ is true on *q* just if $C(...v...v...)$ is true on *every* possible extension q^+ giving a reference to *v* – i.e. no matter what object *o* in the domain is assigned to *v*, $C(...v...v...)$ still comes out true when *v* is treated as a pronoun referring to *o*. Similarly, $\exists v C(...v...v...)$ is true on *q* if $C(...v...v...)$ is true on *some* extension q^+.

Let's take a simple example. Suppose we are interpreting items in the vocabulary $V = \{\text{'m'}, \text{'F'}, \text{'L'}\}$ to mean, respectively, *Socrates*, ... *is a philosopher*, and ... *loves* ..., and we understand the quantifiers as running over all people (past and present). Then the corresponding q-valuation *q* will say:

(1)   the domain of discourse is the set of all people;
(2)   'm' refers to Socrates (i.e. has Socrates as its q-value);
(3)   the extension of 'F' is the set of philosophers; and the extension of 'L' is the set of pairs of people such that the first loves the second.

Helping ourselves to the obvious rules for the connectives, we can now work out the value of the wff

(a)   ∃x(Fx ∧ Lmx)

(translation: *Socrates loves some philosopher*). This is true so long as there is some extension q^+ of *q*, which assigns an object in the domain to 'x', which makes

(a*)   (Fx ∧ Lmx)

come out true. Which there is. Take the extension q^+ which assigns Plato to 'x'. Then, with the variable/pronoun treated as referring expression picking out Plato, 'Fx' is true since Plato is in the set of philosophers; and 'Lmx' is true since the pair ⟨Socrates, Plato⟩ is famously in the set of pairs such that the first member of the pair loves the second. Hence the conjunction '(Fx ∧ Lmx)' is true on q^+.

Now compare

(b) $\forall x(Fx \supset Lmx)$

(translation: *Socrates loves every philosopher*). This comes out false. For (b) is true only if

(b*) $(Fx \supset Lmx)$

is true on *every* possible extension q^+ of q. But consider the extension q^+ which assigns Bertrand Russell to 'x'. Russell is in the set of philosophers, so 'Fx' is true on q^+. 'Lmx' is false, because the pair ⟨Socrates, Russell⟩ is presumably not in the set of pairs such that the first member of the pair loves the second. So this q^+ makes the antecedent of (b*) true and its consequent false, so makes (b*) false. Hence (b*) isn't true on all possible extensions of q; so (b) is indeed false on the original q-valuation.

So far, so good. Things get only a little more complicated when we have to deal with wffs involving more than one quantifier. We then have to consider extended valuations which assign references to more than one variable. Take

(c) $\exists y \exists x(Fx \wedge Lyx)$

(translation: *There's someone who loves a philosopher*). (c) is true on our valuation q so long as

(c*) $\exists x(Fx \wedge Lyx)$

is true on *some* possible extension q^+ which assigns a reference to the variable 'y'. One such extension q^+ assigns Socrates to 'y'. So let's ask, is (c*) true on q^+? Well, (c*) will be true on that valuation if there is some *further* extension q^{++} which as well as assigning Socrates to 'y' also assigns someone to 'x', where

(c**) $(Fx \wedge Lyx)$

is true on q^{++}. And there *is* such an extension – assign Plato to 'x' again!

What about

(d) $\forall y \exists x(Fx \wedge Lyx)$

(translation: *Everyone loves some philosopher*)? This is true on q if for *every* extension q^+ which assigns a reference to 'y' there is a further extension q^{++} which assigns a reference to 'x' which makes (c**) true.

And so it goes. The main ingredients for our official story about the semantics of **QL** are now in place. And it should already seem plausible that, however complex a closed **QL** wff W is, a q-valuation for its vocabulary suffices to fix W's truth-value (just as a valuation of the atoms in a complex **PL** wff fixes the truth-value of that wff). And generalizing: a q-valuation defined over the vocabulary V of some set of closed wffs Σ will similarly fix the truth-values of every wff in Σ.

The rest of this chapter just wraps up these ideas into an official story presented rather more formally. But if you find that kind of more abstract presentation hard going, just skim and skip, or even omit. *Our informal presentation so far is all you really need to know.*

27.4   The official semantics

To summarize so far. The standard truth-valuational story about **PL** gives rules for working out the truth-values of more complex wffs in terms of the values of their simpler constituent subformulae (§9.4). We now want to give similar rules for working out the values of complex **QL** wffs in terms of the q-values of *their* constituents. En route, we will want to say how the value of the wff 'VxFx' relates to valuations of its subformula 'Fx', and similarly how the value of 'Vy∃x(Fx ∧ Lyx)' relates to valuations of *its* subformulae, ultimately 'Fx' and 'Lyx'. But of course, those subformulae don't have a truth-value on a q-valuation just of constants and predicates. In order to do systematic semantics, going via these subformulae with free variables, we are going to have to follow the idea of the last section and consider *extended* valuations which treat variables as if they have temporary references. And the truth of 'Fx' on *one* extended valuation won't ensure the truth of 'VxFx'. As we saw, we have to *generalize* over extended valuations to give the right truth-conditions for quantified wffs.

So first, let's augment the definition in the previous box (§27.2) by defining extended valuations. That's easy:

> An *extended q-valuation* defined over vocabulary *V* is a q-valuation defined over *V*, augmented by an assignment of objects as q-values to one or more variables.

Henceforth, we'll use '(extended) valuation' to cover both original q-valuations and extended ones. We are also going to need the following definition:

> An extended q-valuation *q'* is a *v-variant* of the (extended) valuation *q* if *q'* assigns some object *o* as q-value to the variable *v* and otherwise agrees exactly with *q*.

If *q* doesn't assign any object to *v* already, then a *v*-variant *q'* will of course extend *q* by giving a value to *v*. If *q* does already assign an object to *v*, then a *v*-variant *q'* may (but needn't) change that assignment. To revisit an example from the last section, suppose *q* is the q-valuation defined over {'m', 'F', 'L'} that we met before; q_1 is just the same, except that it also assigns Socrates to 'x'; and q_2 is the same as *q* except it assigns Plato to 'x'. Then q_1 is an 'x'-variant of *q* (and of q_2), and q_2 is an 'x'-variant of q_1 (and of *q*).

With that notion to hand we can neatly give the rules that determine the truth value of any wff built from vocabulary *V*, on some valuation *q* which covers *V*. First, we need the rule (Q0) to deal with the atomic wffs on valuations extended or otherwise (recall, 'term' means 'constant or variable', see §26.3):

> (Q0) An atomic wff, i.e. an *n*-place predicate followed by *n* terms, is true on the (extended) valuation *q* if either *n* = 0, and the *q*-value of the 0-place predicate is *true*, or else *n* > 0, and the *n*-tuple formed by taking the *q*-values of the terms in order is in the extension which is the *q*-value of the *n*-place predicate. Otherwise it is false.

Second, here are the clauses to cover the various ways in which the truth-values of more complex wffs depend on the values of simpler ones. For brevity, we use '$\Rightarrow_q$ T' to mean 'is true on the (extended) q-valuation q' (dropping quotation marks before '$\Rightarrow$'; see §16.3). Each clause is intended to apply to any valuation q, and so can be read as if prefixed by 'For all (extended) valuations q, ...':

> (Q1) For any wff A, $\neg A \Rightarrow_q$ T if $A \Rightarrow_q$ F; otherwise $\neg A \Rightarrow_q$ F.
>
> (Q2) For any wffs A, B: $(A \wedge B) \Rightarrow_q$ T if both $A \Rightarrow_q$ T and $B \Rightarrow_q$ T; otherwise $(A \wedge B) \Rightarrow_q$ F.
>
> (Q3) For any wffs A, B: $(A \vee B) \Rightarrow_q$ F if both $A \Rightarrow_q$ F and $B \Rightarrow_q$ F; otherwise $(A \vee B) \Rightarrow_q$ T.
>
> (Q4) For any wffs A, B: $(A \supset B) \Rightarrow_q$ F if both $A \Rightarrow_q$ T and $B \Rightarrow_q$ F; otherwise $(A \supset B) \Rightarrow_q$ T.
>
> (Q5) For any wffs A, B: $(A \equiv B) \Rightarrow_q$ T if either both $A \Rightarrow_q$ T and $B \Rightarrow_q$ T, or both $A \Rightarrow_q$ F and $B \Rightarrow_q$ F; otherwise $(A \equiv B) \Rightarrow_q$ F.
>
> (Q6) For any wff $C(...v...v...)$ with variable v free, $\forall v C(...v...v...) \Rightarrow_q$ T if $C(...v...v...) \Rightarrow_{q^+}$ T for every q^+ which is a v-variant of q; otherwise $\forall v C(...v...v...) \Rightarrow_q$ F.
>
> (Q7) For any wff $C(...v...v...)$ with variable v free, $\exists v C(...v...v...) \Rightarrow_q$ T if $C(...v...v...) \Rightarrow_{q^+}$ T for some q^+ which is a v-variant of q; otherwise $\exists v C(...v...v...) \Rightarrow_q$ F.

Clauses (Q1) to (Q5) preserve the familiar truth-functional reading of the basic connectives; compare (P1) to (P3) in §9.4, and (P4) and (P5) in §14.5. So that confirms that we are indeed justified in carrying over the old connectives to their new use. (Q6) and (Q7) neatly capture the key idea about quantifiers.

It should be clear that, given a closed wff with vocabulary V and a valuation q defined over V, the rules taken together *do* in principle fix the truth-value of the wff on q. But to convince you of this, it might help to work through ...

27.5 A toy example

Take, then, the following q-valuation q:

    (1)   The domain of discourse is {Socrates, Plato, Aristotle}

    (2)   Constants are assigned references as follows:

            'm' $\Rightarrow$ Socrates

            'n' $\Rightarrow$ Plato

    (3)   Predicates are assigned extensions as follows:

            'F' $\Rightarrow$ {Socrates, Aristotle}

            'G' $\Rightarrow$ { }

            'L' $\Rightarrow$ {⟨Socrates, Plato⟩, ⟨Plato, Aristotle⟩, ⟨Plato, Socrates⟩, ⟨Aristotle, Aristotle⟩}

Here, curly brackets are used (as is absolutely standard) to form expressions to denote sets; the members of the set are as listed. In particular, 'G' is being assigned the *empty* set; i.e. *nothing* satisfies that predicate. We are again using the equally standard convention of using angled brackets to indicate tuples; so '⟨Socrates, Plato⟩' denotes the ordered pair whose first member is Socrates and whose second member is Plato, and '⟨Aristotle, Aristotle⟩' denotes the pair whose first member is Aristotle and whose second member is Aristotle too.

A   Consider first '∀xFx'. Then, we have by (Q6)

'∀xFx' $\Rightarrow_q$ T just if 'Fx' $\Rightarrow_{q^+}$ T for every q^+ which is an 'x'-variant of q.

Each of the three objects in the domain is available to be assigned to 'x' by an extension to q, so there are three 'x'-variants of q. Take q_2 to be the second of these, so q_2 is the same as q except it also assigns 'x' the value Plato. By (Q0),

'Fx' $\Rightarrow_{q_2}$ T just if the q-value of 'x' on q_2 is in the q-value of 'F' on q_2, i.e. just if Plato is in {Socrates, Aristotle}.

So, on q_2, 'Fx' is false. Hence '∀xFx' $\Rightarrow_q$ F.

B   Consider next '∀y¬(Fy ∧ Gy)'. By (Q6), and then (Q1)

'∀y¬(Fy ∧ Gy)' $\Rightarrow_q$ T just if '¬(Fy ∧ Gy)' $\Rightarrow_{q^+}$ T for every q^+ which is a 'y'-variant of q; i.e. just if '(Fy ∧ Gy)' $\Rightarrow_{q^+}$ F for every q^+ which is a 'y'-variant of q.

Now consider any 'y'-variant of q. Inspection shows that '(Fy ∧ Gy)' is indeed false on all of them because 'Gy' is always false. So that means '∀y¬(Fy ∧ Gy)' is true on q.

C   Thirdly, consider '∀x∃yLxy'. By (Q6) again,

'∀x∃yLxy' $\Rightarrow_q$ T just if '∃yLxy' $\Rightarrow_{q^+}$ T for every q^+ which is an 'x'-variant of q.

There are again three 'x'-variants to consider. Take the first, q_1, which assigns Socrates to 'x'. Then by (Q7)

'∃yLxy' $\Rightarrow_{q_1}$ T just if 'Lxy' $\Rightarrow_{q^{++}}$ T for some q^{++} which is a 'y'-variant of q_1.

Inspection shows that there *is* such an q^{++}. Take the 'y'-variant of q_1 which assigns Plato to 'y'. On this valuation, 'Lxy' is true, since the pair ⟨the q-value of 'x', the q-value of 'y'⟩, i.e. ⟨Socrates, Plato⟩, is in the extension of 'L'. So '∃yLxy' is true on q_1. Likewise, '∃yLxy' comes out true on the two other 'x'-variants of q. So '∀x∃yLxy' $\Rightarrow_q$ T.

These calculations are all more than a bit laborious (and just imagine if the domain had been a bigger one!). But tracking through this sort of toy example should convince you that the values of shorter subformulae – together with their values taken across families of appropriate variant valuations – will at least in principle always determine the values of more complex wffs in the intended way. The rules do indeed work as advertised!

27.6 Five results about (extended) q-valuations.

In the next few chapters, we'll invoke a number of elementary and intuitive results about (extended) q-valuations. It's convenient to explain them here. But in fact it is enough if you read through claims (V1) to (V5) in their italicized informal versions, and are happy to accept them as intuitively true. You needn't worry too much about the arguments which show that they *are* indeed true.

(V1) *If a valuation makes a universal quantification true, it makes its instances true too* (so long as we use names the valuation 'knows about'). For example, suppose q is an (extended) valuation defined over vocabulary V including the name 'n', and '∀xFx' is true on q. Then 'Fn' must also be true on q. (Proof: Let the reference of 'n' according to q be the object o. By assumption, 'Fx' is true on every 'x'-variant of q. So, in particular, 'Fx' must be true on the variant which assigns object o to 'x'. So o has to be in the extension of 'F'. That means 'Fn' must be true.)

Relatedly, suppose that q is an (extended) valuation defined over vocabulary V which includes no constants and '∀xFx' is again true on q. Then there is an extension q^+ of q to cover the vocabulary V plus the new name 'a', such that 'Fa' is true on q^+. (Proof: By definition there is at least one object o in the domain of q. And by assumption, 'F' is true of everything, so is true of o. Define q^+ as q plus the assignment of o to the new name 'a'. That will make 'Fa' true on q^+.)

(V2) *If a valuation makes an existential quantification true, it has an extension which assigns a q-value to a new* name *which makes an instance of the quantification true too.* For example, suppose q is defined over vocabulary V which does *not* include the name 'a', and suppose '∃xFx' is true on q. Then there is an extension q^+ of q to cover the vocabulary V plus the new name 'a', such that 'Fa' is true on q^+. (Proof: By assumption, 'Fx' is true on some 'x'-variant of q. So there is some object o in the domain that can be assigned to 'x' which makes 'Fx' true. Define q^+ as q plus the assignment of this object o to the new name 'a'. That will make 'Fa' true on q^+.)

(V3) *Valuations respect the equivalence between* ¬∀v *and* ∃v¬. For example, suppose q is a an (extended) valuation which makes '¬∀xFx' true. Then it makes '∃x¬Fx' true. (Proof: By assumption q must make '∀xFx' false. So some 'x'-variant of q makes 'Fx' false. So this variant makes '¬Fx' true. So q has an 'x'-variant which makes '¬Fx' true. So q makes '∃x¬Fx' true.)

(V4) *Valuations respect the equivalence between* ¬∃v *and* ∀v¬. For example, suppose q makes '¬∃xFx' true. Then q makes '∀x¬Fx' true. (Proof: Similar to proof for (V3).)

(V5) *Extending a valuation to deal with a new constant doesn't affect the values of wffs not containing that constant.* (Proof: Inspection of the rules shows that when we calculate the truth-value of a wff W on an (extended) valuation q, we only need to consider the specification of the domain, and the q-values of the names, free variables, and predicates actually in W. Suppose, for example, the name 'a' doesn't appear in *W*; then it won't matter what *q* assigns it. And if this wff *W* is true on some *q* that doesn't evaluate 'a', it will remain true on an extension of *q* which assigns some reference to that irrelevant name.)

Summing up, then, and putting things a little more formally:

(V1) If an (extended) valuation *q* makes a wff of the form $\forall v C(...v...v...)$ true, then (assuming *q* assigns a value to the constant *c*), *q* makes $C(...c...c...)$ true. And if *q* doesn't assign a value to *c*, there is an extension of *q* to q^+ which gives a q-value to *c* and makes $C(...c...c...)$ true.

(V2) If *q* makes a wff of the form $\exists v C(...v...v...)$ true, and *c* is a constant which *q* does *not* evaluate, there is an extension of *q* to q^+ which gives a q-value to *c* and makes $C(...c...c...)$ true.

(V3) If *q* makes a wff of the form $\neg \forall v C$ true, then it also makes true the corresponding wff of the form $\exists v \neg C$.

(V4) If *q* makes a wff of the form $\neg \exists v C$ true, then it also makes true the corresponding wff of the form $\forall v \neg C$.

(V5) Suppose *q* is defined over vocabulary *V* which doesn't include the constant *c*; and suppose q^+ extends *q* by assigning some object in the domain to *c*. Let *W* be some wff using vocabulary from *V*. Then if *q* makes *W* true, so does q^+.

27.7 Summary

- A q-valuation defined over **QL** vocabulary *V* fixes a set of objects to be the domain of discourse *D*; it assigns individual constants in *V* objects in *D* as references; it assigns a truth-value to any 0-place predicate in *V*; and when $n > 0$, it assigns an *n*-place predicate in *V* a (possibly empty) set of *n*-tuples of objects in *D*.

- The key to understanding the semantics of the quantifiers is the following idea: $\forall v C(...v...v...)$ is true if $C(...v...v...)$ would come out true, no matter which object we took *v* to pick out, if we regard *v* here as a pronoun. Similarly $\exists v C(...v...v...)$ is true if $C(...v...v...)$ would come out true, for some possible reference for *v*.

- To implement this idea, we introduce the idea of an *extended* q-valuation, which assigns references to one or more variables as well as individual constants. Then we can say, roughly, that $\forall v C(...v...v...)$ is true if $C(...v...v...)$ is true no matter which object the extended valuation assigns to *v*, i.e. is true on all relevant extended valuations. And $\exists v C(...v...v...)$ is true if $C(...v...v...)$ is true on some such extended valuation.

- Together with the natural rules for the propositional connectives, the rules for evaluating quantified wffs ensure that a q-valuation defined over vocabulary *V* fixes a truth-value for any closed wff built up from that vocabulary.

Exercises 27

Take the following q-valuation:

The domain is {Romeo, Juliet, Benedick, Beatrice}

Constants are assigned references as follows:
'm' ⇒ Romeo
'n' ⇒ Juliet

Predicates are assigned extensions as follows:
'F' ⇒ {Romeo, Benedick}
'G' ⇒ {Juliet, Beatrice}
'L' ⇒ {⟨Romeo, Juliet⟩, ⟨Juliet, Romeo⟩, ⟨Benedick, Beatrice⟩,
⟨Beatrice, Benedick⟩, ⟨Benedick, Benedick⟩}

Then what are the truth values of the following wffs?

1.   ∃x Lmx
2.   ∀x Lxm
3   (∃x Lmx ⊃ Lmn)
4.   ∀x(Fx ≡ ¬Gx)
5.   ∀x(Gx ⊃ (Lxm ∨ ¬Lmx))
6.   ∀x(Gx ⊃ ∃yLxy)
7.   ∃x(Fx ∧ ∀y(Gy ⊃ Lxy))

Now take the following q-valuation:

The domain is {4, 7, 8, 11, 12}

Constants are assigned references as follows:
'm' ⇒ 7
'n' ⇒ 12

Predicates are assigned extensions as follows:
'F' ⇒ the even numbers in the domain
'G' ⇒ the odd numbers in the domain
'L' ⇒ the set of pairs ⟨m, n⟩ where m and n are in the domain and
m is less than n

What are the truth values of the wffs (1) to (7) now?

Q-validity

Using the idea of a q-valuation introduced in the last chapter, we can now define the idea of a q-valid argument (i.e. one whose conclusion must be true on a q-valuation if its premisses are). Unfortunately, we can't mechanically test for q-validity in the way we can for tautological validity. But we will introduce various ways of establishing arguments to be q-valid or q-invalid.

28.1 Q-validity defined

Back in Chapter 13, we met the notion of a tautological entailment:

> The **PL** wffs A_1, A_2, ..., A_n tautologically entail the wff C if and only if there is no valuation of the atoms involved in the premisses A_i and the conclusion C which makes A_1, A_2, ..., A_n simultaneously all true and yet C false.

We can now introduce an analogous definition for *q-entailment*:

> The **QL** closed wffs A_1, A_2, ..., A_n q-entail the closed wff C if and only if there is no q-valuation of the vocabulary V involved in the premisses A_i and the conclusion C which makes A_1, A_2, ..., A_n simultaneously all true and yet C false.

Note we've defined q-entailment only for *closed* wffs – i.e. wffs with no variables dangling free, and which thus have determinate truth-values on a q-valuation of their vocabulary. When certain premisses tautologically entail a conclusion we say that the inference in question is tautologically valid; similarly, when a bunch of premisses q-entail a conclusion, we will say that the inference is *q-valid*.

Here are some more definitions:

- First, if a q-valuation makes some collection of closed wffs all true, then the valuation is said to *satisfy* those wffs or, equivalently, to be a *model* of those wffs. So another way of defining q-validity is this: an inference is q-valid just if there is no model of the premisses and negation of the conclusion.

- Second, if a q-valuation does make the premisses of some inference all true while making the conclusion false, then we will say that the valuation is a *countermodel* for the inference. So we can also say: an inference is q-valid just if it has no countermodels.

- Third, we introduce the **QL** analogue of the **PL** notion of a tautology. A **PL** wff is a tautology if it is true on all valuations of its atoms. Similarly, we will say that a closed **QL** wff is a *q-logical truth* if it is true on all q-valuations of its vocabulary.

However, we should frankly signal that this last bit of terminology is entirely non-standard. In fact, the customary term for a q-logical truth seems to be 'valid wff' – which is a rather perverse bit of nomenclature given the logician's loud insistence elsewhere, as in §2.5, on emphatically distinguishing validity (a property of *inferences*) from truth (a feature of wffs or propositions)!

28.2 Some simple examples of q-validity

Take the argument

  A   ∀x(Fx ⊃ Gx), Fn ∴ Gn

which we might read as the **QL** rendition of 'All philosophers are egocentric; Jack is a philosopher; hence Jack is egocentric' (§§1.2, 25.1).

The inference here is certainly valid; and intuitively it is also q-valid, for it surely depends just on the way the quantifier and the conditional works, and not on the particular interpretation of the predicates. But how can we prove this? After all, we can't directly check through each and every permissible q-valuation to see what happens to the premisses and conclusion in each case: there are just too many possibilities. The first step in specifying a q-valuation is fixing the domain of discourse: but this can be *any* non-empty set you care to choose, however enormous. And having chosen the domain, a one-place predicate can then have as its extension *any* arbitrarily selected set of objects from the domain. The possibilities here are literally endless. So how can we proceed?

Let's try an indirect approach. Suppose there *is* a q-valuation q of the vocabulary {'n', 'F', 'G'} which makes the premisses and the negation of the conclusion of **A** all true. We'll use '$\Rightarrow_q$ T' again as shorthand for 'is T on valuation q', dropping quotation-marks before '⇒' (as we did in §16.1). Then we are supposing:

  (1)   Fn $\Rightarrow_q$ T
  (2)   ∀x(Fx ⊃ Gx) $\Rightarrow_q$ T
  (3)   ¬Gn $\Rightarrow_q$ T

At the end of the last chapter, we noted the obvious principle (V1): if the q-valuation q makes a wff of the form ∀vC(...v...v...) true, then q makes C(...c...c...) true, for any constant c which q evaluates. So, given (2), this principle tells us

  (4)   (Fn ⊃ Gn) $\Rightarrow_q$ T

Now we are in familiar territory with a familiar connective, and can continue:

  (5)   ¬Fn $\Rightarrow_q$ T   Gn $\Rightarrow_q$ T

But the left branch of (5) contradicts (1), and the right branch contradicts (3). *Hence there can be no such valuation q which is a countermodel to* **A**. Which proves that the argument is indeed q-valid.

For a second example, consider

B $\exists xFx, \forall x(Fx \supset Gx) \therefore \exists xGx$

which renders, as it might be, 'Someone is a great logician; all great logicians are cool; hence someone is cool'.

Again, we need to proceed indirectly and assume for the sake of argument that there is a q-valuation q which makes the premisses true and conclusion false:

(1) $\exists xFx \Rightarrow_q T$
(2) $\forall x(Fx \supset Gx) \Rightarrow_q T$
(3) $\neg \exists xGx \Rightarrow_q T$

But if a valuation q makes '$\neg \exists xGx$' true, it will make '$\forall x \neg Gx$' true (as we noted in §27.6, V4). So we must also have

(4) $\forall x \neg Gx \Rightarrow_q T$

Now, if a valuation q makes '$\exists xFx$' true, then some object in the domain must be in the extension of 'F': so we can construct a q^+ which extends q by assigning that object to be the referent of the new name 'a' (§27.6, V2). And then

(5) $Fa \Rightarrow_{q^+} T$

What else holds on q^+? If a wff not involving 'a' is true on a valuation q, then it remains true on any extension q^+ to deal with that constant (see §27.6, V5). So (2) implies that $\forall x(Fx \supset Gx) \Rightarrow_{q^+} T$ too; and applying (V1) again, we get

(6) $(Fa \supset Ga) \Rightarrow_{q^+} T$

Similarly (4) implies

(7) $\neg Ga \Rightarrow_{q^+} T$

But unpacking (6) immediately generates contradictions:

(8) $\neg Fa \Rightarrow_{q^+} T$ $Ga \Rightarrow_{q^+} T$
                 *   *

There can be no q^+ which makes (5) to (7) all true; hence there is no q that is a countermodel to **B**. So this argument is q-valid too.

Let's take just one more example. Consider the inference

C $(\exists x(Fx \wedge \neg Gx) \supset \neg Gn), Gn \therefore \forall x(Fx \supset Gx)$

We again start by assuming that this is *not* q-valid. That is, we suppose that there is a q-valuation q such that:

(1) $(\exists x(Fx \wedge \neg Gx) \supset \neg Gn) \Rightarrow_q T$
(2) $Gn \Rightarrow_q T$
(3) $\neg \forall x(Fx \supset Gx) \Rightarrow_q T$

Immediately, by (V3), we can infer

(4) $\qquad\qquad\qquad \exists x \neg (Fx \supset Gx) \Rightarrow_q T$

Applying to (1) the familiar rule for dealing with conditionals gives

(5) $\qquad \neg \exists x (Fx \wedge \neg Gx) \Rightarrow_q T \qquad\qquad\qquad \neg Gn \Rightarrow_q T$
$\qquad\qquad\qquad\qquad\qquad\qquad\qquad\qquad\qquad\qquad\quad *$

We can apply the principle (V4) to continue the left-hand branch

(6) $\qquad \forall x \neg (Fx \wedge \neg Gx) \Rightarrow_q T$

And now looking at (4), we can apply (V2) again: if a valuation q makes '$\exists x \neg (Fx \supset Gx)$' true, and 'a' is a new constant that q doesn't evaluate, then there must be an extension q^+ of q to cover the new name 'a', such that

(7) $\qquad \neg (Fa \supset Ga) \Rightarrow_{q^+} T$

We again remember that wffs *not* involving 'a' that are true on q must remain true on an extended valuation that assigns some object to 'a'. That holds in particular of (6). So '$\forall x \neg (Fx \wedge \neg Gx)$' remains true on q^+. And applying (V1) we get (8), and then can quickly finish:

(8) $\qquad \neg (Fa \wedge \neg Ga) \Rightarrow_{q^+} T$
(9) $\qquad\qquad Fa \Rightarrow_{q^+} T$
(10) $\qquad\quad \neg Ga \Rightarrow_{q^+} T$

(11) $\neg Fa \Rightarrow_{q^+} T \qquad \neg\neg Ga \Rightarrow_{q^+} T$
(12) $\qquad * \qquad\qquad\qquad\qquad *$

We hit contradictions again. There can be no q^+ which makes (7) and (8) both true; hence no q is a countermodel to C, and so the inference is q-valid.

But of course, we've seen trees very like these before! In fact, strip off the explicit assignments of values and these are *exactly* the trees we met informally in Chapter 25 (Examples **A, F, K**). So what is going on here?

28.3 Thinking about trees again

Recall what we said about **PL** trees. We first introduced these as a way of presenting 'working backwards' arguments in English about valuations. We then dropped the explicit assignments of truth-values, for the sake of brevity. Still, the trees remained ways of presenting *metalinguistic* arguments. In Chapter 20, however, we pointed out that these same unsigned trees could equally well be read as *object-language* arguments (i.e. arguments *in* **PL**, not arguments *about* **PL**). Read that way, they represent inferences within **PL**, revealing the commitments of an initial bunch of assumptions. And a closed tree shows that this bunch of assumptions is inconsistent.

In Chapter 25, we introduced **QL** trees. We didn't then have the notions of a q-valuation and of q-validity to hand. So at that point, we couldn't think of the trees metalinguistically, as arguments about **QL** valuations. We had to think of them in the second way, as representing inferences within **QL**, with a closed tree showing that an initial set of assumptions is inconsistent. And the tree inferences were then justified intuitively, using an informal notion of validity.

But now we *do* have formal semantic notions in play. So we can now construct metalinguistic tree arguments about possible q-valuations.

Looking at our three quick examples of metalinguistic trees, we see that the way they are constructed exactly parallels the object-language arguments we met before. And the principles (V1) to (V5) of §27.6 justify analogues of the intuitive rules for the quantifiers that we adopted before. In particular, (V1) tells us that we can extend a tree argument about the q-valuation q by instantiating a universal quantification with respect to a constant that q already 'knows about'. And (V2) tells us that if q makes an existential quantification true, then there will be an extension q^+ which makes true an instantiation of it by a new name. So we can extend a tree argument about the q-valuation q by instantiating an existential quantification with respect to a new constant, to get something true on q^+.

We will tidy up these observations in §30.1. But what they suggest is that the unsigned trees we introduced in Chapter 25 (as arguments to establish the intuitive validity of certain **QL** inferences) can also be viewed as shorthand for arguments about q-valuations, with closed trees establishing not just the validity but the *q-validity* of inferences.

28.4 Validity, q-validity, and 'quantification logical form'

The idea of tautological entailment captures the idea of an inferential move's being valid by virtue of the distribution of the truth-functional propositional connectives among the premisses and conclusion (i.e. being valid irrespective of the meanings of the atoms the connectives connect). Similarly, the idea of q-validity captures the idea of an inference which is valid in virtue of 'quantificational logical form', i.e. in virtue of the distribution of connectives and quantifiers among the premisses and conclusion (so valid irrespective of the meanings of the names and predicates that are involved in the argument).

Compare our remarks about logical form in §§3.6 and 13.6. While the notion of logical form in general is murky, the notion of a **QL** argument's being valid in virtue of its quantificational logical form is in perfect order and can be formally defined. It is just the notion of q-validity.

An inference can be valid without being tautologically valid; and it can be valid without being q-valid either. For example, consider the reading of **QL** where the domain of discourse is humans, the predicate 'F' means *father*, and 'G' means *male*. Given those interpretations, the little argument

D   ∃xFx ∴ ∃xGx

is evidently valid (if there are fathers, there must be males). But it isn't valid in virtue of quantificational logical form. There are q-valuations of the vocabulary which make the premiss true and the conclusion false: e.g., let the extension of 'F' be the set of great Cambridge philosophers, and let the extension of 'G' be empty. Of course, this q-valuation doesn't conform to the interpretation that we

just gave 'F' and 'G'. That's allowed, though, exactly as we allowed **PL** truth-valuations which do not respect the meanings of atoms (see §13.2). The q-valuations of wffs are intended to generate assignments of truth-values which are
allowed *so far as the meanings of the connectives and quantifiers are concerned*.

However, although there can be arguments which are valid without being tautologically valid, we saw that any **PL** argument which *is* tautologically valid is
plain valid (see §13.2 again). Similarly, although there can be arguments which are
valid without being q-valid, any **QL** argument which is q-valid is plain valid. Or
to put it the other way about, if an argument isn't valid, it can't be q-valid. Why?

Take the single premiss case for simplicity (the generalization is trivial). If the
inference from *A* to *C* is *not* valid, there must be some possible situation which
would make *A* true and *C* false. But when *A* and *C* are wffs of **QL**, a situation
can only make *A* true and *C* false by fixing the domain and imposing certain q-
values on the constituent names and predicates, since these values are all that
matters for determining the truth-values of **QL** wffs. Hence if the inference from
A to *C* is indeed invalid, there must be a q-valuation of their vocabulary which
makes *A* true and *C* false – and that means that *A* does not q-entail *C*. (For a
cautionary remark, however, see §34.4.)

Summing that up: q-validity stands to **QL** arguments as tautological validity
stands to **PL** arguments.

28.5 The undecidability of q-validity

We saw that we can *mechanically decide* whether a **PL** argument is tautologically valid by exhaustively looking at all the possible different valuations of the
atoms that occur in the argument (that's a finite number, of course), and asking
whether any valuation makes the premisses true and the conclusion false. As we
remarked before, we can't similarly test whether a **QL** argument is q-valid by
exhaustively looking at all the possible different q-valuations of the vocabulary
that occur in the argument. There are just too many possibilities.

Still, although we can't do a brute-force search through all possible q-valuations of some vocabulary, couldn't there be some *alternative* mechanical method
of determining whether any argument is q-valid?

There *are* some special cases where there are mechanical tests: the simplest one
is where all the predicates that appear in the argument are one-place. Construct
a **QL** tree, applying the rules as often as possible. Either it closes, and the argument is proved valid. Or it must terminate with no more rules left to apply (once
we've instantiated any universal quantifier with every name in sight), but leaving
an open branch. And then we can read off a countermodel to show the argument
is invalid: see §29.3.

However, as soon as there is even a single two-place predicate featuring in an
argument, trees needn't terminate, and there won't in general be a mechanical
method for deciding q-validity. That result is mathematically demonstrable.
However, we can't demonstrate it here. For a start, we would have to carefully
investigate the general theory of 'mechanical methods', which is in effect the

general theory of what computers can do: and *that* is evidently a non-trivial business which would take us too far afield. (The undecidability result belongs to a family of exciting 'limitative' results about what can be done by mechanical methods, discovered in the 1930s by Kurt Gödel, Alan Turing and others.)

If we can't in general test for q-validity by anything akin to a truth-table test, we *have* to resort to the likes of tree proofs to demonstrate validity. And as we'll see, our tree system can be beefed up to become a *complete* proof system in the sense that, if an argument is valid, we can always show it to be valid using trees. But that point doesn't conflict with the point about undecidability: claiming that we can always prove an argument is q-valid by some approved method if it *is* q-valid doesn't mean that there is a mechanical way of telling whether there is a proof or not.

In short, then, to explore the logic of **QL** arguments further we *need* to consider methods of proof, such as trees. So we return to them in the next chapter.

28.6   Countermodels and invalidity

But before moving on to consider tree proofs again, let's consider how we might directly show that a q-invalid argument is q-invalid.

How do we do that? By finding q-valuations that work as countermodels, of course. Later, we'll see how we can sometimes do this systematically by reading valuations off open trees (§§29.4, 30.4). Here we'll just work very informally through a few examples.

Consider the inference:

E $\forall x(Fx \supset Hx), \forall x(Gx \supset Hx) \therefore \forall x(Fx \supset Gx)$

Suppose we interpret the predicates so this translates as 'All cats are mammals; all dogs are mammals; hence all cats are dogs'. We immediately see that E is invalid by the invalidity principle (§2.3). And this interpretation can be parlayed into a formal countermodel to show the q-invalidity of E. Just put

> Domain of discourse is the set of animals
> Extension of 'F' = the set of cats
> Extension of 'G' = the set of dogs
> Extension of 'H' = the set of mammals

and we get a q-valuation which makes the premises true and the conclusion false.

However, we just picked that counterexample/countermodel out of thin air. Can we proceed a little bit more methodically?

We are looking for a q-valuation which makes '$\forall x(Fx \supset Gx)$' false. That requires there to be an object in the domain which is in the extension of 'F' but not in the extension of 'G' (why?). Let's insert an object into the domain to play this role – anything will do, e.g. Gwyneth Paltrow, Brad Pitt, ... or the number 0. Take the latter (as it is less distracting!). And let's try to be minimalist, and start off with nothing else in the domain, so we'll try to build up our q-valuation starting as follows:

Domain = {0}, i.e. the set containing just 0
Extension of 'F' = {0}
Extension of 'G' = { }, i.e. the empty set

This evidently makes the conclusion of E false; so far, so good. Next, to make
'∀x(Fx ⊃ Hx)' true, anything which is in the extension of 'F' must be in the extension of 'H' (why?). So, let's put

Extension of 'H' = {0}

and the first premiss comes out true. And as it happens, that's enough to make
the second premiss true too. That's because '∀x(Gx ⊃ Hx)' is true on this valuation
so long as '(Gx ⊃ Hx)' comes out true on every extension which assigns an object
in the domain to 'x'. The only object in the domain is 0; and '(Gx ⊃ Hx)' is true
when 'x' is treated as referring to *that*, since the antecedent of the conditional
will be false and the conclusion true. Therefore our mini-model for the premisses
is indeed another countermodel for the inference.

Take a second example:

F   ∀x(Fx ∨ Gx) ∴ (∀xFx ∨ ∀xGx)

This is q-invalid (and plain invalid for most readings of 'F' and 'G'). Compare:
every number is even or odd; but it doesn't follow that either every number is
even or every number is odd. Hence, for a countermodel, suppose the domain is
the set of natural numbers, the extension of 'F' is the even numbers, and the
extension of 'G' is the odd numbers.

Again, we can proceed more methodically. We are looking for a valuation
which makes '(∀xFx ∨ ∀xGx)' false, i.e. makes each of '∀xFx' and '∀xGx' false.
So there must be something in the domain which isn't in the extension of 'F' and
also something in the domain which isn't in the extension of 'G'. Can we get a
countermodel with a one-element domain? Well, if the domain is {0}, then the
extensions of 'F' and 'G' will have to be empty to make '∀xFx' and '∀xGx' false.
But then '∀x(Fx ∨ Gx)' will be false too, so we haven't got a countermodel.

So let's try a two-element domain, say {0, 1}. A bit of experimenting shows
that e.g.

Domain = {0, 1}
Extension of 'F' = {0}
Extension of 'G' = {1}

will do the trick and provide a countermodel.

If we only have monadic predicates in play, things stay fairly simple (again, see
§29.3). But if an inference involves even one dyadic predicate, then we may need
an *infinite* countermodel to show that it is invalid. Thus consider

G   ∀x∃yLxy, ∀x∀y∀z((Lxy ∧ Lyz) ⊃ Lxz) ∴ ∃xLxx

Here, the first premiss says that everything stands in the relation *L* to something
or other; and the second premiss says that *L* is a *transitive* relation (a relation *L*
is transitive if when *a* has *L* to *b*, and *b* has *L* to *c*, then *a* has *L* to *c*: for example

'being an ancestor of' or 'being taller than' are transitive). Now, suppose that these premisses are satisfied in a *finite* domain. Start from some element o_1 in the domain. Then by the first premiss, this initial element is related by L to some element o_2 in the domain, and o_2 is related by L to some element o_3, and so on and so forth. And if there are only a finite number of elements to choose from, we must in the end revisit some element o_n we've met before. In short, we will have a finite L-related chain of elements such that

$$o_1 \to o_2 \to \ldots \to o_n \to o_{n+1} \to \ldots \to o_n$$

But, by the second premiss, if o_n is L to o_{n+1}, and o_{n+1} is L to o_{n+2}, then o_n is L to o_{n+2}; and if o_n is L to o_{n+2} and o_{n+2} is L to o_{n+3}, then o_n is L to o_{n+3}. And so on. In other words, by the transitivity premiss, o_n is L-related to everything further down the chain. So in particular, o_n is L-related to o_n itself. Which makes '$\exists x Lxx$' true. In short, if the premisses of **G** are true in a finite domain, so is the conclusion!

Yet the argument **G** is evidently invalid. Compare 'Take any number, there is some number it is less than; if a number is less than a second number, and the second is less than a third, then the first is less than the third; hence there is a number which is less than itself', which has true premiss and a false conclusion. That's a counterexample to the validity of **G**, and the corresponding infinite countermodel is

> Domain = {0, 1, 2, ...}, i.e. the infinite set of *all* positive integers
> Extension of 'L' = the set of pairs $\langle m, n \rangle$ such that $m < n$.

28.7   Summary

- A q-valid inference is a **QL** inference which is such that, on any q-valuation of the vocabulary V in the premisses and the conclusion, if the premisses are all true together, the conclusion is true too.

- There is no generally applicable mechanical method for determining whether an argument is q-valid (contrast the truth-table test for mechanically settling the tautological validity of **PL** arguments).

- We can establish a q-valid argument to be q-valid by showing that the idea that there is a countermodel (a q-valuation which makes the premisses true and conclusion false) leads to contradiction. Such a proof is again naturally set out as a tree.

- We can establish a q-invalid argument to be q-invalid if we can find a countermodel. But in the general case, countermodels may need infinite domains.

Exercises 28

A   Let's use '$\vDash$' now to abbreviate 'q-entails' (compare §13.7). Which of the following claims are true? Provide arguments in the style of §§28.2 and 28.6 to defend your answers.

1. ∀x(Fx ⊃ Gx) ⊨ ∀x(Gx ⊃ Fx)
2. ∀x(Fx ⊃ Gx) ⊨ ∀x(¬Gx ⊃ ¬Fx)
3. ∀x∃yLxy ⊨ ∀y∃xLyx
4. ∀x((Fx ∧ Gx) ⊃ Hx) ⊨ ∀x(Fx ⊃ (Gx ⊃ Hx))
5. (∀xFx ∨ ∀xGx) ⊨ ∀x(Fx ∨ Gx)
6. ∀x(Fx ⊃ Gx), ∀x(¬Gx ⊃ Hx) ⊨ ∀x(Fx ⊃ ¬Hx)
7. ∃x(Fx ∧ Gx), ∀x(¬Hx ⊃ ¬Gx) ⊨ ∃x(Fx ∧ Hx)
8. ∀x∃y(Fy ⊃ Gx) ⊨ ∀y∃x(Gx ⊃ Fy)
9. ∀x∀y(Lxy ⊃ Lyx) ⊨ ∀xLxx
10. ∀x(∃yLxy ⊃ ∀zLzx) ⊨ ∀x∀y(Lxy ⊃ Lyx)

B   How can the techniques of §§28.2 and 28.6 be used to show that certain
     wffs are or are not q-logical truths? Which of the following claims are q-
     logical truths? Defend your answers.

1. ∀x((Fx ∧ Gx) ⊃ (Fx ∨ Gx))
2. ∃x((Fx ∧ Gx) ⊃ (Fx ∨ Gx))
3. (∀x∀y(Lxy ⊃ Lyx) ⊃ ∀xLxx)
4. (∃x∀yLxy ⊃ ∀y∃xLxy)
5. (∀y∃xLxy ⊃ ∃x∀yLxy)
6. (∀x∀y(Lxy ⊃ ¬Lyx) ⊃ ∀x¬Lxx)

C   (For enthusiasts!) Show that the following claims are true:

1. Some **QL** wffs of the form (A ∨ ¬A) are *not* q-logical truths.
2. Any **QL** sentence of the form (A ∨ ¬A) *is* a q-logical truth.
3. If W is a **PL** wff involving just the **PL** atom A, and if B is a **QL**
   atomic sentence, then the result of systematically replacing every
   occurrence of A in W with an occurrence of B yields a **QL** sentence.
4. If W is a **PL** tautology involving just the **PL** atom A, and if B is a
   QL atomic sentence, then the result of systematically replacing
   every occurrence of A in W with an occurrence of B yields a q-
   logical truth of **QL**.
5. If W is a **PL** tautology involving just the **PL** atoms $A_1, A_2, ..., A_n$,
   and if $B_1, B_2, ..., B_n$ are **QL** atomic sentences, then the result of
   systematically replacing every occurrence of A_i in W with an
   occurrence of the corresponding B_i yields a q-logical truth of **QL**.
6. If W is a **PL** tautology involving just the **PL** atoms $A_1, A_2, ..., A_n$,
   and if $B_1, B_2, ..., B_n$ are any **QL** sentences (atomic or molecular),
   then the result of systematically replacing every occurrence of A_i
   in W with an occurrence of the corresponding B_i yields a q-logical
   truth of **QL**.

More on QL trees

This chapter begins by gathering together the rules for **QL** trees introduced in Chapter 25. We look at some more complex examples using closed trees to demonstrate q-validity. We then explain a needed small addition to one of the rules. Finally, we consider what can be learnt from looking at trees that *don't* close.

29.1 The official rules

In the boxes over the next two pages, we summarize the rules for building the unsigned **QL** trees that we've already encountered. Rules (a) to (i) are the familiar rules for the connectives, trivially re-ordered. Then we add the current four rules for quantifiers. How are these rules to be interpreted and applied?

First, a quick reminder of what we mean when we say that a wff W is e.g. of the form $(A \supset B)$ or $\forall v C(...v...v...)$. We mean, in the first case, that W can be constructed from the template '$(A \supset B)$' by replacing 'A' and 'B' by **QL** wffs: in that case, '$\supset$' is the main logical operator of W (see §26.4). In the second case, we mean that W can be constructed by prefixing the universal quantifier $\forall v$ to a wff where the variable v has only *free*, i.e. unbound, occurrences: the quantifier $\forall v$ is the main logical operator of W.

In each case, then, the tree-building rules tell us how to extend a tree depending on the *main* operator of the wff W which is being processed. Given the wff

$$(\forall x(Fx \wedge Gx) \supset Fn)$$

we can apply the appropriate tree-splitting rule (g), to deal with '$\supset$'; but we *cannot* apply the rule ($\forall$), as '$\forall x$' is not the main operator. Likewise, given the wff

$$\exists x \neg \neg (Fx \wedge Gx)$$

we can apply the quantifier rule ($\exists$) but not the double-negation rule (a). And so on.

Given a **QL** inference (involving only closed wffs, as we said before) we build a corresponding tree in the following stages, just as with a **PL** inference:

(1) Start off the trunk of the tree with the premisses A_1, A_2, ..., A_n and the negation of conclusion C of the inference under consideration.

(2) Inspect any not-yet-closed path down from the top of the tree to see whether it involves an explicit contradiction, i.e. for some wff W it

---

Rules for building QL trees

Suppose that *W* is a non-primitive wff on an open branch. We can then apply the corresponding rule:

(a)   *W* is of the form ¬¬*A*: add *A* to each open path containing *W*. Check off *W*. Schematically,

(b)   *W* is of the form (*A* ∧ *B*): add both *A* and *B* to each open path containing *W*. Check off *W*.

(c)   *W* is of the form ¬(*A* ∨ *B*): add ¬*A* and ¬*B* to each open path containing *W*. Check off *W*.

(d)   *W* is of the form ¬(*A* ⊃ *B*): add *A* and ¬*B* to each open path containing *W*. Check off *W*.

(*A* ∧ *B*) √   ¬(*A* ∨ *B*) √   ¬(*A* ⊃ *B*) √
| | |
A   ¬*A*   *A*
B   ¬*B*   ¬*B*

(e)   *W* is of the form (*A* ∨ *B*): add a new fork to each open path containing *W*, with branches leading to the alternatives *A* and *B*. Check off *W*.

(f)   *W* is of the form ¬(*A* ∧ *B*): add a new fork to each open path containing *W*, with branches leading to the alternatives ¬*A* and ¬*B*. Check off *W*.

(g)   *W* is of the form (*A* ⊃ *B*): add a new fork to each open path containing *W*, with branches leading to the alternatives ¬*A* and *B*. Check off *W*.

(h)   *W* is of the form (*A* ≡ *B*): add a new fork to each open path containing *W*, leading to the alternatives *A*, *B* and ¬*A*, ¬*B*. Check off *W*.

(i)   *W* is of the form ¬(*A* ≡ *B*): add a new fork to each open path containing *W*, leading to the alternatives *A*, ¬*B* and ¬*A*, *B*. Check off *W*.

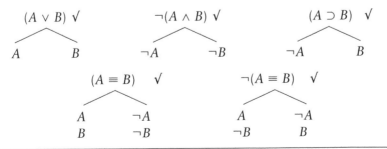

---

*Rules for building QL trees, continued**

(¬∀)   W is of the form ¬∀vC: add ∃v¬C to each open path on which W
       appears. Check off W.

(¬∃)   W is of the form ¬∃vC: add ∀v¬C to each open path on which W
       appears. Check off W.

$$\begin{array}{ccc}
\neg\forall vC \quad \surd & & \neg\exists vC \quad \surd \\
| & & | \\
\exists v\neg C & & \forall v\neg C
\end{array}$$

(∀)   W is of the form ∀vC(...v...v...): add C(...c...c...) to an open path
       on which W appears, where c is a constant on that path (which
       hasn't already been used to instantiate W on that path). W is *not*
       checked off.

$$\begin{array}{c}
\forall vC(...v...v...) \\
| \\
C(...c...c...) \quad [c \text{ old}]
\end{array}$$

(∃)   W is of the form ∃vC(...v...v...): add C(...c...c...) to each open
       path on which W appears, where c is a constant *new* to the paths.
       Check off W.

$$\begin{array}{c}
\exists vC(...v...v...) \quad \surd \\
| \\
C(...c...c...) \quad [c \text{ new}]
\end{array}$$

---

contains both W and also the corresponding formula ¬W. If it does, then
we *close* that path with a star as an absurdity marker.

(3) If we have run out of rules to apply to unchecked non-primitive formulae
left on any *open* path, then we have to stop!

(4) Otherwise, we choose some unchecked non-primitive formula W that sits
on an *open* path. We then apply the appropriate rule, and extend the tree
by adding more (closed) wffs in the described way. We can then check off
W, except after an application of rule (∀).

(5) Loop back to step (2).

As we saw (§25.5), this process may not terminate in the case of QL trees – and
it is possible for injudicious choices at step (4) of the tree-building process to lead
to a non-terminating tree even though *other* choices *would* lead to a closed tree.
Why do we check off a wff W when we apply the appropriate tree-building rule,
except when W is universally a quantified wff and we are applying the (∀) rule?
Because in the other cases, the unpacking rules effectively exhaust the content of
the wff we are unpacking (and there is no point in doing that twice over).

---

* But NB the extended version of the (∀) rule in §29.3!

By contrast, a universal quantified wff tells us that *everything* in the domain satisfies a certain condition. We can of course infer that a particular named object in the domain satisfies that condition. But there is more content to the original quantified claim than is given by that particular implication of it.

And why does the (∀) rule tell us not to re-instantiate a universally quantified wff using the same constant again on the same path? Because *that* would be quite redundant. It would just put a duplicate of the same wff on the same path.

29.2 Further examples of closed trees

Let's consider more inferences involving wffs with multiple quantifiers. Take first:

A   Some people are boastful. No one likes anyone boastful. So some
       people aren't liked by anyone.

To confirm the argument is valid, translate it into **QL**, e.g. by

$$∃xFx, ∀x∀y(Fy ⊃ ¬Lxy) ∴ ∃x∀y¬Lyx$$

where the translation manual we are using is plain. Then the tree starts:

(1)   ∃xFx
(2)   ∀x∀y(Fy ⊃ ¬Lxy)
(3)   ¬∃x∀y¬Lyx

Before we continue, let's set down a couple of handy rules of thumb for applying quantifier tree-building rules in a sensible order:

- At each step, deal with any negated quantifiers first.

- Instantiate existential quantifiers before universals (see §25.4).

Applying these in turn, the tree will continue …

(4)   ∀x¬∀y¬Lyx

Note, by the way, that the rule (¬∃) applied to (3) only pushes the negation sign past the *initial* quantifier. Next, instantiating (1) with a new constant, we have

(5)   Fa

Now how shall we proceed? We eventually want to instantiate the *inner* universal quantifier of (2), to get another occurrence of 'Fa'; but rule (e) does not allow us to instantiate inner quantifiers; the tree-building rules insist we proceed step by step, from the outside in. So let's instantiate (4) instead:

(6)   ¬∀y¬Lya

We've now got another negated quantifier, so deal with that next …

(7)   ∃y¬¬Lya

This gives us a new existential quantification, so the second rule of thumb enjoins us now to deal with *that*, instantiating (7) with a new name, of course:

(8)   ¬¬Lba

Now return to instantiate the initial quantifier in (2) with 'b'; that yields another universally quantified wff, we can instantiate by 'a':

(9)   ∀y(Fy ⊃ ¬Lby)
(10)   (Fa ⊃ ¬Lba)

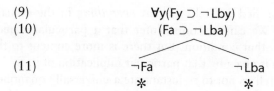

(11)   ¬Fa   ¬Lba
                         *   *

And we are done. The inference is q-valid, and hence valid. (Make sure you see why it was sensible to instantiate (2) with 'b' rather than 'a'.)
    Consider next

B   Everyone admires some logician; whoever admires someone respects them. So everyone respects some logician.

The translation of the premisses and negated conclusion shouldn't cause trouble by now (just note that the second premiss means that for *any* pair, if the first admires the second, then the first respects the second):

(1)   ∀x∃y(Fy ∧ Lxy)
(2)   ∀x∀y(Lxy ⊃ Mxy)
(3)   ¬∀x∃y(Fy ∧ Mxy)

Checking off (3), the next steps are more or less automatic.

(4)   ∃x¬∃y(Fy ∧ Mxy)   √
(5)   ¬∃y(Fy ∧ May)   √
(6)   ∀y¬(Fy ∧ May)

We *could* instantiate (6) using 'a', but that would be pointless. So let's next instantiate (1) and (2) instead:

(7)   ∃y(Fy ∧ Lay)
(8)   ∀y(Lay ⊃ May)

Next, instantiating the existential quantification first (following our rule of thumb), and remembering to use a new constant, we get

(9)   (Fb ∧ Lab)

Using this new constant to instantiate the two remaining universal quantifiers at (6) and (8), we get

(10)   ¬(Fb ∧ Mab)
(11)   (Lab ⊃ Mab)

Finally the tree closes using the familiar connective rules:

(12)   Fb
(13)   Lab

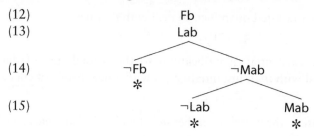

(14)   ¬Fb   ¬Mab
                         *

(15)   ¬Lab   Mab
                                             *   *

Let's next revisit an example from §5.4:

C   No girl loves any sexist pig. Caroline is a girl who loves whoever loves her; Henry loves Caroline; hence Henry is not a sexist pig.

Rendered into **QL** using an obvious translation dictionary, the tree will start:

(1)   ∀x(Gx ⊃ ¬∃y(Fy ∧ Lxy))
(2)   (Gm ∧ ∀x(Lxm ⊃ Lmx))
(3)   Lnm
(4)   ¬¬Fn

And the next few steps are straightforward. Just unpack the conjunction in (2), checking it off, and then instantiate (1) with respect to 'm':

(5)   Gm
(6)   ∀x(Lxm ⊃ Lmx)
(7)   (Gm ⊃ ¬∃y(Fy ∧ Lmy)) √

(8)   ¬Gm   ¬∃y(Fy ∧ Lmy)   √
(9)   *   ∀y¬(Fy ∧ Lmy)

That leaves us with two universal quantifications, at (6) and (9), to instantiate. Instantiating (9) by 'n' will give us 'Fn' in the scope of a negation, which should contradict the earlier '¬¬Fn'. And it does …

(10)   ¬(Fn ∧ Lmn)   √

(11)   ¬Fn   ¬Lmn
               *

The obvious thing to do now is to instantiate (6) with respect to 'n' as well, to conclude the tree

(12)   (Lnm ⊃ Lmn)   √

(13)   ¬Lnm   Lmn
               *   *

We are done: the inference is valid.

Our fourth sample argument is of a type which was recognized by medieval logicians as being valid, but couldn't be shown to be so within their logic of general propositions:

D   All horses are mammals; so any horse's head is a mammal's head.

Modern logicians can do better than their predecessors here. In fact, as soon as we see how to translate the conclusion into **QL**, finding a tree to show that **D** is indeed q-valid is straightforward.

First, then, the translational problem. 'Any horse's head is a mammal's head' is equivalent to 'For any *x*, if *x* is the head of some horse, then *x* is the head of some mammal', i.e. 'for any *x*, if (for some horse *y*, *x* is the head of *y*), then (for some mammal z, x is the head of z)'. So, let's fix on the dictionary

'F' means … *is a horse*
'G' means … *is a mammal*
'M' means … *is the head of* …

and set the domain to be e.g. earthly things. Then C's conclusion can be rendered

$$\forall x\{\exists y(Fy \wedge Mxy) \supset \exists z(Gz \wedge Mxz)\}$$

The tree will therefore commence

(1)	$\forall x(Fx \supset Gx)$
(2)	$\neg\forall x\{\exists y(Fy \wedge Mxy) \supset \exists z(Gz \wedge Mxz)\}$ √
(3)	$\exists x\neg\{\exists y(Fy \wedge Mxy) \supset \exists z(Gz \wedge Mxz)\}$ √
(4)	$\neg\{\exists y(Fy \wedge May) \supset \exists z(Gz \wedge Maz)\}$ √
(5)	$\exists y(Fy \wedge May)$
(6)	$\neg\exists z(Gz \wedge Maz)$

That's pretty much automatic given our two rules of thumb for tree-building. We dealt first with the negated quantifier at (2) to give (3). That leaves us with the universally quantified (1) and the existentially quantified (3), so we apply the 'existentials before universals' rule. Stripping off the quantifier '$\exists x$', and replacing every occurrence of 'x' with a new name gives us (4). Note the bracketing in (4); it is a negated conditional. So applying the relevant rule gives us (5) and (6). Again, we first deal with the negated quantifier, and check off (6) to yield

(7)	$\forall z\neg(Gz \wedge Maz)$

Now we have two universally quantified propositions at (1) and (7), and the existential quantification at (5). So next instantiate (5) with a new name:

(8)	$(Fb \wedge Mab)$ √
(9)	Fb
(10)	Mab

and then the rest is straightforward. Instantiating (1) and (7) with respect to 'b' (why try 'b' rather than 'a'?), and then unpacking the connectives, we finish …

(11)	$(Fb \supset Gb)$ √
(12)	$\neg(Gb \wedge Mab)$ √

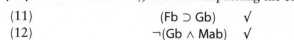

(13)

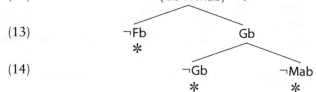

(14)

Finally in this section, we'll return to §5.1, and again consider

E   Everyone loves a lover; Romeo loves Juliet; so everyone loves Juliet.

We saw that this is valid if the first premiss is interpreted to mean 'everyone loves anyone who is a lover' (i.e. 'everyone x is such that, take anyone y, if y loves someone then x loves y'). Using 'L' for *loves*, etc., we begin the tree

(1)   ∀x∀y(∃z Lyz ⊃ Lxy)
(2)   Lmn
(3)   ¬∀xLxn

Following the rule of thumb 'negated quantifiers first', we can check off (3) and add:

(4)   ∃x¬Lxn

Remembering the other tip, 'existentials before universals', we instantiate (4):

(5)   ¬Lan

But now what? Let's try instantiating the two quantifiers in (1) – one at a time! – with 'a' and 'n', which will make the consequent of the conditional 'Lan', preparing the way for a contradiction with (5). So we get

(6)   ∀y(∃z Lyz ⊃ Lay)
(7)   (∃z Lnz ⊃ Lan)   √

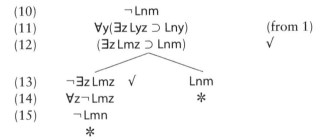

(8)   ¬∃z Lnz   √   Lan
(9)   ∀z¬Lnz   *

How do we carry on from here? Consider the argument we gave to show that this inference is valid in §5.1. We in fact had to use the first premiss twice there, at lines (4) and (7) of the informal proof. We might expect to have to use the premiss twice to close off this tree. So let's press on, and instantiate (1) again

(10)   ¬Lnm
(11)   ∀y(∃z Lyz ⊃ Lny)   (from 1)
(12)   (∃z Lmz ⊃ Lnm)   √

(13)   ¬∃z Lmz   √   Lnm
(14)   ∀z¬Lmz   *
(15)   ¬Lmn
              *

With a bit of effort, again we are done. The inference E is q-valid (and so valid).

It's worth pausing just a moment to compare this last demonstration of validity with the simple informal proof in §5.1. Evidently, the tree proof is a *lot* less 'natural' and direct than the informal one.

But remember: trees are *not* being sold as the best way of formally mimicking ordinary styles of deduction (cf. §20.3): they are not a 'natural deduction' proof system. All that is being claimed is that trees are *one* well-motivated and reasonably manageable way of showing that inferences are q-valid when they are.

For yet more examples of our tree-rules at work, see Chapter 35.

29.3 Extending the (∀) rule

As we noted before (§28.3), our stated rules for tree-building certainly look *sound* (i.e. an argument warranted by a closed tree built following our rules is indeed q-valid). If we start a tree by assuming that $A_1, A_2, ..., A_n, \neg C$ are all true on some valuation q, then the quantifier rules for extending the tree are warranted by the principles (V1) to (V5) about valuations (§27.6). Hence, if the tree *does* end up in contradictions, there can be no such valuation q; and then the target inference $A_1, A_2, ..., A_n \therefore C$ must be q-valid. We'll turn this rough and ready proof-idea into a tidier affair in the next chapter.

But is our basic set of rules *complete* (i.e. do we have enough tree-building rules to find a closed tree corresponding to *every* q-valid inference)?

Consider the inference '∀xFx ∴ ∃xFx'. If we try to construct a tree, we'll get as far as

(1)	∀xFx	
(2)	¬∃xFx	√
(3)	∀x¬Fx	

Then we have to stop, as there is no further rule to apply. So the tree doesn't close. On the other hand, the inference *is* q-valid according to our account. Take any q-valuation q. If q makes '∀xFx' true, then every object in q's domain needs to be in the extension of 'F'. Since domains are non-empty, that means at least one thing is in the extension of 'F'. So q must make '∃xFx' true too.

Now, this last point only holds because we chose to say that domains always have at least one member (§27.2). Suppose we'd taken the other tack and allowed empty domains. Then '∀xFx ∴ ∃xFx' would no longer be q-valid. Given an empty domain it is vacuously true that, if any object is in the domain, it is in the extension of 'F' (that's a general conditional claim with an antecedent which is always false). So '∀xFx' is true in an empty domain. But '∃xFx' will be false.

In sum, our tree-rules are currently out of harmony with our definition of q-validity. Our rules *don't* warrant the inference '∀xFx ∴ ∃xFx'; our definition of q-validity *does*. What to do?

There is a forking of the paths here. You might think the obvious response is to backtrack hastily, revise our definition of validity, adopting instead the idea of q_0-validity (defined in terms of q_0-valuations, which are exactly like q-valuations except that they allow empty domains). And if we do that, it can be proved that our current set of rules *would* then be complete. There are closed trees using our existing rules for all and only the q_0-valid arguments. So that looks like a happy restoration of harmony. Unfortunately, there is a snag.

Consider the inferences '∀xFx ∴ Fn' and ' Fn ∴ ∃xFx'. These are both straightforwardly q_0-valid (take any valuation of the vocabulary in the premises and conclusion: that will involve assigning an object as q-value to 'n'; so *these* valuations won't have empty domains). Then we have '∀xFx ∴ Fn' and ' Fn ∴ ∃xFx' are q_0-valid and '∀xFx ∴ ∃xFx' isn't. Or putting it more starkly, we have A q_0-entails B, B q_0-entails C, but A does *not* q_0-entail C. Which goes clean against our normal assumption that entailments can be chained into longer arguments and our ultimate conclusion will still follow from our initial premiss(es). What to do?

Well, we could adopt a notion of entailment which allows exceptions to that fundamental assumption. Or we could be revisionary in another way and reject '∀xFx ∴ Fn' as a valid entailment in the absence of an explicit premiss to the effect 'n exists'. But most logicians prefer to stick to the two assumptions that we can unrestrictedly chain entailments and endorse '∀xFx ∴ Fn'.

We are going to follow the majority here (though see Chapter 36 for some more comments). Hence we shall *not* revise our definition of q-validity to allow empty domains. That means '∀xFx ∴ ∃xFx' remains q-valid. It also means, however, that we need a further tree-building rule if we are to be able to warrant every q-valid argument by means of a closed tree.

One popular option here is to relax the rule (∀) rather radically. The rule currently says that if W is of the form $\forall v C(...v...v...)$, then we can add $C(...c...c...)$ to any open path on which W appears, where c is a constant already on the path. Suppose instead that we are allowed to instantiate W with *any* constant, old or new. Then we can close the tree which we had trouble with before:

(1)	∀xFx	
(2)	¬∃xFx	√
(3)	∀x¬Fx	
(4)	Fc_0	
(5)	$\neg Fc_0$	
	✳	

Here we've just arbitrarily chosen to use the constant 'c_0'. Since we are assuming that there is at least one thing in the domain, pick such an object, and we can surely use a new name to dub it. Instantiating the universal quantifications at (1) and (3) with a name of this object then exposes a contradiction.

But our suggested generous rule rather ruins the memorable symmetry between our rules (∀) and (∃), summed up in the slogan 'instantiate universals with old constants, existentials with new ones'. There's something unattractive too about relaxing (∀) to allow us to instantiate universals with *any* constant, and then immediately having to issue the warning that you'd be daft ever to instantiate a universal with a random constant except in the very special circumstances that there are no other constants in sight to play with.

So rather than go too radical, here's our preferred revision of the rule (∀). We'll keep the old rule for when there *are* constants already on the relevant path, and just add a clause to deal with the special case when we want to instantiate a universal quantification and there are *no* preceding constants:

> (∀') W is of the form $\forall v C(...v...v...)$: add $C(...c...c...)$ to an open path on which W appears, where c is *either* a constant on that path (and which hasn't been used to instantiate W before), *or* is a new constant (and there are no other constants appearing on the path). W is *not* checked off.
>
> $$\forall v C(...v...v...)$$
> $$|$$
> $$C(...c...c...) \quad [c \text{ old, or unprecedented}]$$

NB: Rule (∀′) still allows us to construct our mini-tree warranting '∀xFx ∴ ∃xFx'.

We will show in the next chapter that this minor extension of our original rule does indeed give us a sound and complete set of rules. *So for the rest of this book, we'll assume the revised rule (∀′) is in play.*

29.4   What can be learnt from open trees?

We have emphasized before that the fact that a quantifier tree doesn't close doesn't *always* show the argument under consideration is invalid. An injudicious sequence of choices of the rules to apply can mean that a tree trails off down an infinite path, when better choices would close off the tree (see §25.5 again).

However, in simple cases, a tree can terminate with no more rules left to apply, but leaving an open branch. If there *is* an open branch on such a completed tree, then we *can* read off from the tree a q-valuation which makes the premises true and conclusion false and which thus demonstrates q-invalidity. Consider, for example,

E   ∀x(Fx ⊃ Hx), ∀x(Gx ⊃ Hx) ∴ ∀x(Fx ⊃ Gx)

Starting in the obvious way, and omitting the unofficial line numbers, we get:

$$
\begin{array}{ll}
\forall x(Fx \supset Hx) & \\
\forall x(Gx \supset Hx) & \\
\neg\forall x(Fx \supset Gx) & \checkmark \\
\exists x\neg(Fx \supset Gx) & \checkmark \\
\neg(Fa \supset Ga) & \checkmark \\
Fa & \\
\neg Ga & \\
(Fa \supset Ha) & \checkmark \\
(Ga \supset Ha) & \checkmark \\
\end{array}
$$

```
                    (Fa ⊃ Ha)        √
                    (Ga ⊃ Ha)        √
                 ┌──────────┴──────────┐
               ¬Fa                     Ha
                *                 ┌─────┴─────┐
                                ¬Ga          Ha
```

The tree has open branches. The only unchecked wffs which aren't atomic wffs or negations of atomic wffs are the universal quantifications at the top of the tree. But we have no new names available to instantiate them with. So we can't use the (∀′) rule again. There are no more rules to apply.

In the case of **PL** trees, if a branch stays open, we can read off a valuation which makes all the wffs on the branch true. *The trick is to pick a valuation which makes the primitives on the branch, i.e. the atoms and negated atoms, all true.* Then truth percolates up the branch. That makes the initial premises and negated conclusion all true, and the valuation therefore shows that the inference is invalid (cf. §§16.2, 19.2). Let's try the very same trick here.

In the present case, it doesn't matter which open branch we choose, the primitives on the branch are 'Fa', '¬Ga', and 'Ha'. We need to choose a reference for the name 'a'. Let's assign it the first number, i.e., 0. And then, to give the primitives the right values, we need the extensions of 'F' and 'H' to *include* 0, and the extension of 'G' *not* to include 0.

A domain must of course always contain any named objects and/or objects which are involved in the extension of predicates. So in the present case, we know the domain must contain at least 0. What else should we put in the domain? *Nothing*. We don't want any extra objects in the domain which we haven't any information about (and which might therefore have unwanted effects on the truth-values of quantified wffs). So, keeping the domain as small as possible, we get:

> Domain = {0}, i.e. the set containing just 0.
> Reference of 'a' = 0.
> Extension of 'F' = extension of 'H' = {0}.
> Extension of 'G' = { }, i.e. the empty set.

This makes the primitives all true. And it is readily checked that this indeed makes *all* the wffs on an open branch of our tree true; and in particular, it makes the premisses of E true and conclusion false (compare the discussion of the same argument in §28.5).

Here's another example:

> F $\forall x(Fx \lor Gx)$, $\forall x(Gx \supset Hx)$ $\therefore$ $(\forall xFx \lor \forall xHx)$

Using our rules of thumb, we get the following (again omitting line numbers):

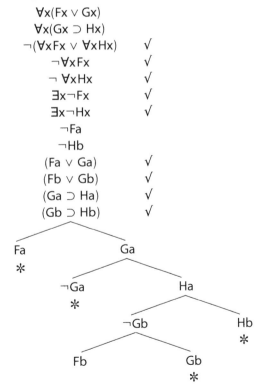

Here, the initial steps are pretty automatic. Then we instantiate the two universal quantifications with every name in sight, and unpack the resulting wffs. There are no more rules to apply, and the tree stays open. So let's try to construct a countervaluation for the argument. Take the open path, where we have

¬Fa, Fb, Ga, ¬Gb, Ha, ¬Hb

There are two names here to which we need to assign q-values. Let's take them in order to refer to the first two numbers. Again we take the domain of discourse to be no larger than it needs to be to include those references for the names. The resulting valuation which makes the primitives all true is then:

> Domain = {0, 1}
> Reference of 'a' = 0
> Reference of 'b' = 1
> Extension of 'F' = {1}
> Extension of 'G' = {0}
> Extension of 'H' = {0}

It is easily checked again that this valuation makes all the wffs on the open branches true. In particular, it makes the premises of F true and conclusion false. Why does it make, e.g., the universal quantification (1) true? Because it makes an instance true for each object in the domain (to emphasize, there are no more objects in the domain than the named ones).

When does this trick work, i.e. this trick of reading valuations from the primitives on an open branch on a QL tree and arriving at a countervaluation for the argument in question? For a start, it works for arguments where the propositions don't feature quantifiers embedded one inside the other. Instantiate all the existential quantifiers, thereby introducing a collection of new names; instantiate all the universal quantifications with respect to all the names thereby introduced (plus any others in the original premises and conclusion). And if the tree doesn't close, construct the 'smallest domain' valuation which contains one object per name and which makes the primitives on some open branch true, and you'll get the desired countermodel.

We saw, however, that when an argument involves embedded quantifiers life gets more complicated. It may be the case that the only countermodels which show that the inference is invalid are infinite ones; see the discussion of §28.5 G. Yet it may also be the case that, when we *do* find an infinite branch that never closes, it doesn't show that the argument is invalid, and we can't read off a genuine countervaluation from the branch; see again the discussion of the example in §25.5. Looking more closely at the infinite open branch in that last case, we see that it in fact was constructed by ignoring its lines (3) and (5). No surprise then that choosing a valuation that makes all the primitives on that branch true won't make (3) and (5) true too! We can only form countermodels from infinite branches when the trees have been constructed more *systematically*, without ignoring any complex wffs on an open branch; but that's not entirely straightforward (more on this in the next chapter).

29.5 Summary

- We summarized our original set of tree-building rules for quantifier trees, and argued for a modification of the rule (∀) to match our restriction to non-empty domains.

- A couple of useful heuristic tips – not strict rules, but guides for applying the rules in a sensible order: *eliminate negated quantifiers first* and then *instantiate existentials before universals*.

- In some simple cases at any rate, we can read countervaluations from open trees (choose a 'smallest domain' valuation that makes the primitives on an open branch true).

Exercises 29

A   Use **QL** trees to evaluate the entailment claims (1) to (10) in Exercises **28A**.

B   Using trees, show the following arguments are valid:

1.   Some philosophers admire Jacques. No one who admires Jacques is a good logician. So some philosophers are not good logicians.

2.   Some philosophy students admire all logicians; no philosophy student admires any rotten lecturer; hence, no logician is a rotten lecturer.

3.   There's a town to which all roads lead. So all roads lead to a town.

4.   Some good philosophers admire Frank; all wise people admire any good philosopher; Frank is wise; hence there is someone who both admires and is admired by Frank.

5.   Any true philosopher admires some logician. Some students admire only existentialists. No existentialists are logicians. Therefore not all students are true philosophers.

6.   Everyone loves a lover; hence if someone is a lover, everyone loves everyone!

7.   If anyone speaks to anyone, then someone introduces them; no one introduces anyone to anyone unless they know them both; everyone speaks to Frank; therefore everyone is introduced to Frank by someone who knows him. [*Use* 'Rxyz' to render 'x introduces y to z'.]

8.   Any elephant weighs more than any horse. Some horse weighs more than any donkey. If a first thing weighs more than a second thing, and the second thing weighs more than a third, then the first weighs more than the third. Hence any elephant weighs more than any donkey.

C   Redo the first three examples of §29.2 as *signed* trees (as in §28.2).

QL trees vindicated

In this chapter, we prove the 'Basic Result' about **QL** trees: there is a closed **QL** tree for an argument if and only if the inference is q-valid. Like Chapter 19, this can certainly be skipped by those with no taste for abstract metalogical argument. Still, the Basic Result is often located the other side of the threshold that divides elementary logic from the serious stuff. So it is nice to show that the work we've already done in fact allows us to give a fairly painless proof.

30.1 Soundness

(Revision homework: re-read §§19.1, 20.4, 27.4, 27.6, 28.2, 28.3!)

We first need to show that **QL** tree proofs are sound, i.e. reliable. If a properly constructed **QL** tree closes, then the argument under scrutiny is indeed q-valid. That's equivalent to saying

(S) If an inference is *not* q-valid, then a corresponding **QL** tree will never close.

A quick reminder: we will say that a path through a tree is *satisfiable* if there is a q-valuation which makes all the wffs on the path true. A satisfiable path plainly cannot contain contradictory wffs, so can't close.

Proof of (S) Suppose that the inference $A_1, A_2, ..., A_n \therefore C$ is not q-valid. Then there is some valuation q which ensures that

$$A_1 \Rightarrow_q T, A_2 \Rightarrow_q T, ..., A_n \Rightarrow_q T, \neg C \Rightarrow_q T$$

and the initial 'trunk' of the corresponding tree is satisfiable. Now consider what happens as we extend our tree, step by step, applying the tree-building rules. *At each stage, a satisfiable path will get extended to a longer satisfiable path.* So a tree with a satisfiable trunk will never close.

To show that a satisfiable path always gets extended to a longer satisfiable path, we just have to inspect the building rules one at a time. It is easy to see that the result holds for the rules telling us how to unpack wffs whose main logical operator is one of the connectives (compare §19.1). Take, for example, the *non-branching* rule that unpacks a wff $(A \land B)$ into its conjuncts. A q-valuation q which satisfies a path including $(A \land B)$ will of course continue to satisfy the path we get by adding A and B, since q will make both these added new wffs true as well. Or take, for another example, the *branching* rule which tells us to extend a path containing $(A \vee B)$ by adding a fork with A on one branch and B on the other. If q satisfies a path including $(A \vee B)$ then at least one of A and B must be true on q, hence q will continue to satisfy at least one of the extended paths. You can easily check that the same is true for extensions by means of other connective rules.

Next, suppose q satisfies a path including a wff of the form $\neg \forall v C$; then applying the relevant rule means adding $\exists v \neg C$. This added wff must also be true on q (that's obvious, but also see (V3) of §27.6). So q continues to satisfy the extended path. Likewise for continuing a path using the $(\neg \exists)$ rule.

That leaves us just two rules to consider. First take the $(\exists)$ rule. Suppose q satisfies a path including a wff of the form $\exists v C(...v...v...)$. The $(\exists)$ rule says we can add $C(...c...c...)$ where c is a constant new to the path. By (V3) of §27.6, q must have some extension q^+ which is just like q except it also assigns a reference to c and it makes $C(...c...c...)$ come out true. And by (V5) all the wffs previously on the path (which didn't involve c and were true on q) remain true on q^+. So q^+ makes all the wffs on the extended path true, both the old ones and the new addition. Hence the newly extended path is satisfied by q^+; so the path is still satisfiable.

Lastly, take the revised $(\forall')$ rule. Suppose q satisfies a path including a wff of the form $\forall v C(...v...v...)$. The $(\forall')$ rule says we can add $C(...c...c...)$ in two sorts of case:

Case (i): c is a constant already on the path. Then by (V1), q must also make $C(...c...c...)$ true, so the extended path remains satisfiable; q continues to do the trick.

Case (ii): c is a new constant. Then by (V1), q must have some extension q^+ which assigns a reference to c and makes $C(...c...c...)$ come out true. By (V5) again, all the wffs previously on the path remain true on q^+. So the extended path is satisfiable; in this case, the extended valuation q^+ does the trick.

So *every* legitimate way of extending a satisfiable path leads to a longer satisfiable path (though the valuation doing the satisfying may alter). Hence a tree with a satisfiable trunk will never close. QED.

Of course, all we are doing here is setting out very carefully the reasoning which we more briskly used to motivate the quantifier tree rules for unsigned trees in §28.3, once we had noted how our original tree rules parallel the natural rules for making *signed* trees for arguing about q-valuations.

30.2 Completeness: strategy

We are aiming next to show that the rules for **QL** trees give us a *complete* proof system, in the sense that any q-valid inference at all can be shown to be valid by some closed tree. We know that we might fail to construct a closed tree for an argument which actually is q-valid because, by misadventure, we manage instead to construct some never-closing branch. Still, the completeness claim is that for any q-valid inference there is *some* closed tree. That's equivalent to saying:

(C)   If *every* QL tree starting with sentences $A_1, A_2, ..., A_n, \neg C$ stays open, then the inference $A_1, A_2, ..., A_n \therefore C$ is not q-valid.

We are going to prove (C) by proving two results:

(Sys)   There's a way of systematically applying the tree-building rules such that the resulting (possibly infinite) QL tree either (i) closes or else (ii) has an open path the wffs on which form a syntactically consistent saturated set.

(Sat)   Every syntactically consistent saturated set of closed wffs is satisfiable.

We'll explain the jargon in a moment. However, you don't need to understand it to see that these two results will give us exactly what we want. If *every* possible tree starting from the sentences $A_1, A_2, ..., A_n, \neg C$ stays open, then by (Sys) there must in particular be a systematically built tree which has an open path, the wffs on which form a consistent saturated set (the wffs are closed sentences, as on any legal tree). By (Sat) there is a q-valuation which makes all the wffs in that set true. This q-valuation in particular makes all of the trunk $A_1, A_2, ..., A_n,$ $\neg C$ true. So the inference $A_1, A_2, ..., A_n \therefore C$ is not q-valid. QED.

So to establish (C), we'll prove (Sat) in the next section, and (Sys) in §30.4.

30.3   Consistent, saturated sets are satisfiable

(Further revision homework: re-read §§19.2, 19.3!)

As before, we'll say that a set of wffs Σ is syntactically consistent if it contains no pairs of wffs of the form $A, \neg A$.

And a set Σ of closed QL wffs is *saturated* if the following conditions all obtain:

(a)   if a wff of the form $\neg\neg A$ is in Σ, then so is A;

(b)   if a wff of the form $(A \wedge B)$ is in Σ, then so are both A and B;

(c)   if a wff of the form $\neg(A \vee B)$ is in Σ, then so are both $\neg A$ and $\neg B$;

(d)   if a wff of the form $\neg(A \supset B)$ is in Σ, then so are both A and $\neg B$;

(e)   if a wff of the form $(A \vee B)$ is in Σ, then so is at least one of A and B;

(f)   if a wff of the form $\neg(A \wedge B)$ is in Σ, then so is at least one of $\neg A$ and $\neg B$;

(g)   if a wff of the form $(A \supset B)$ is in Σ, then so is at least one of $\neg A$ and B;

(h)   if a wff of the form $(A \equiv B)$ is in Σ, then so are either both A and B or both $\neg A$ and $\neg B$;

(i)   if a wff of the form $\neg(A \equiv B)$ is in Σ, then so are either both A and $\neg B$ or both $\neg A$ and B;

($\neg\forall$) if a wff of the form $\neg\forall v C$ is in Σ, then so is $\exists v \neg C$;

($\neg\exists$) if a wff of the form $\neg\exists v C$ is in Σ, then so is $\forall v \neg C$;

($\forall$)   if a wff of the form $\forall v C(...v...v...)$ is in Σ, then so is $C(...c...c...)$ for every constant c that appears anywhere in Σ (and further, at least one such instance $C(...c...c...)$ appears in Σ),

($\exists$)   if a wff of the form $\exists v C(...v...v...)$ is in Σ, then so is $C(...c...c...)$ for some constant c.

Here's a simple example of a saturated set:

> $((\forall x \exists y Rxy \land \forall x \neg Fx) \lor \exists y \forall x Rxy), (\forall x \exists y Rxy \land \forall x \neg Fx), \forall x \exists y Rxy, \forall x \neg Fx,$
> $\exists y Rmy, \exists y Rny, \neg Fm, \neg Fn, Rmn, Rnn.$

Now, when we introduced saturated sets of **PL** wffs, we explained the label by remarking that these sets are 'saturated with truth-makers'. Every non-primitive wff W in such a **PL** set (i.e. every wff that isn't either atomic or the negation of an atomic wff) has truth-makers in the set. In other words, take any non-primitive W, then there are one or two simpler wffs W_1 and W_2 also in the set, whose truth on a valuation is enough to make W true on that valuation.

Something similar holds for a saturated set of **QL** sentences, Σ. Take any non-primitive W in Σ; then – if we are careful about how we choose the domain of discourse – there are one or more wffs W_i in Σ whose joint truth on a q-valuation is enough to make W true on that valuation.

That's obvious for any complex wff that isn't a universal quantification. In each such case, the relevant condition defining saturation plainly guarantees that if a wff W of a certain form is in the set, so are one or two wffs whose truth on a valuation would be enough to make W true too.

So what about a wff W of the form $\forall v C(...v...v...)$? Condition ($\forall$) tells us that if W is in Σ, so is at least one substitution instance, and maybe very many more. However, we noted before (§22.4) that even if a lot of substitution instances $C(...c...c...)$ are true, $\forall v C(...v...v...)$ need not be. Some nameless thing in the domain may not satisfy the condition expressed by C. *But suppose we choose a q-valuation for the wffs in Σ whose domain has a distinct object for each of the constants c that appears in Σ, but has nothing else in it.* With this special choice of domain in play, if all the instances $C(...v...v...)$ are true – one for each constant in Σ – then so is $\forall v C(...v...v...)$.

In our mini-example above, suppose we make the primitives '$\neg Fm$', '$\neg Fn$', 'Rmn', 'Rnn' all true, and put nothing in the domain but the objects picked out by 'm' and 'n'; then all the wffs in the set are true.

In sum: by choosing the right domain we can ensure that every non-primitive wff in the saturated set Σ has truth-makers in the set. The more complex wffs will have simpler truth-makers, which have yet simpler truth-makers, ... until we bottom out at the level of primitives. And now the argument will go pretty much as for the **PL** case. We can define a q-valuation q, designed to make all the *primitives* in Σ true. Having chosen the domain correctly, these true primitives will then make true the next level of wffs in Σ, and these will be truth-makers for the next most complex wffs, and so on upwards. Hence q will make true *all* the wffs in Σ, showing that the set is indeed satisfiable. Which proves (Sat), i.e. that every syntactically consistent saturated set of closed wffs Σ is satisfiable.

To really nail down this line of proof, then, we just need to confirm

(C_1) for any syntactically consistent saturated set of closed wffs Σ, there is a q-valuation which makes the *primitives* true;

(C_2) we can choose the domain of a q-valuation which makes the primitives in Σ true so that it will make *all* the wffs in Σ true.

For the sake of elegance, though, we will first prove a restricted version of these claims, which assumes that the constants in Σ are all of the form c_k – i.e. constants with numerical subscripts – and also that there are no zero-place predicates in Σ. Lifting these restrictions afterwards is very easy: it just makes for a bit of mess.

Proof of (C_1)   Consider the following q-valuation ('the *chosen* valuation'):

> The domain: the set of numbers k such that c_k is in Σ.
>
> Constants: each c_k has as q-value the corresponding number k.
>
> Predicates: If F is a monadic predicate, then its extension is the set of numbers k such that the wff Fc_k is in Σ; if R is a dyadic predicate, then its extension is the set of pairs of numbers $\langle j, k \rangle$ such that the wff Rc_jc_k is in Σ; if S is a three-place predicate, then its extension is the set of triples of numbers $\langle i, j, k \rangle$ such that the wff $Sc_ic_jc_k$ is in Σ; etc.

This valuation evidently makes each (un-negated) atomic wff in Σ true. For example, if Fc_k is in Σ, then by definition the q-value of c_k (i.e. the number k) is in the extension of F. So Fc_k is true. Similarly for other atomic wffs. Suppose next that the negation of an atomic wff appears in Σ, e.g. $\neg Fc_j$. Since Σ contains no contradictions, the corresponding bare atomic wff Fc_j can't appear in Σ. On the chosen valuation q, then, the number j is *not* in the extension of F; so Fc_j is false and hence $\neg Fc_j$ is true. Similarly for other negated atomic wffs in Σ. Hence q makes *all* primitives in Σ true. QED.

Proof of (C_2)   Suppose that the chosen valuation q does *not* make all the wffs in Σ true. Choose one of the *least complex* of these false wffs (i.e. a false wff with the least number of connectives and/or quantifiers). Since the primitives are all true by (C_1), this selected false wff is non-primitive. And since Σ is saturated, this wff must satisfy the appropriate condition for belonging to a saturated set. So now we just consider cases.

Suppose the selected false wff is of the form $(A \wedge B)$: then both A and B must be in Σ, and being less complex must by assumption be true on q. But that's impossible; we can't have $(A \wedge B)$ false on q yet A and B each true. Again, suppose the selected false wff is of the form $(A \vee B)$: then either A or B must appear in Σ, and being less complex must by hypothesis be true on q. That's impossible too; we can't have $(A \vee B)$ false yet one of its disjuncts true on q. The other cases where conditions (a) to (i) apply are similar.

Suppose the selected false wff is of the form $\neg \forall v C$: then by condition $(\neg \forall)$ the equally complex corresponding wff $\exists v \neg C$ must also be false and in Σ, so we turn our attention to that instead. Similarly suppose the selected false wff is of the form $\neg \exists v C$: then the corresponding wff $\forall v \neg C$ must also be false and in Σ, and we turn our attention to that.

Suppose next that the selected false wff is of the form $\exists v C(...v...v...)$. Then by condition $(\exists)$ an instance $C(...c...c...)$ for some constant c is in Σ; and being less complex must be true on q. But then $\exists v C(...v...v...)$ must be true on q after all.

Suppose finally that the selected false wff is of the form $\forall v C(...v...v...)$. Then by condition $(\forall)$ the instance $C(...c...c...)$ is in Σ for each constant c appearing anywhere in Σ; and being shorter, those instances must all be true on q. But we've chosen the domain of q to contain no more than the references of all the constants in c. So that means that C holds true of every object in the domain of q. So $\forall v C(...v...v...)$ must be true on q after all.

In sum, there is no way that a non-primitive can be false on q. So q makes every wff in Σ true. QED.

To finish off, we just need to relax those two suppositions which we made for the sake of an elegant argument. First, suppose that Σ contains some of the six constants 'a', 'b', etc., which lack numerical subscripts. Then form the chosen valuation by assigning them the numbers 0, 1, etc., as references, and then assign c_k the number $k + 6$. Make compensating adjustments elsewhere; so, for example, if Fc_k is in Σ, put $k + 6$ into the extension of F. Then this will make all primitives true as before. Not so elegant, but it works!

Second, suppose that Σ contains some zero-place predicates (in effect, propositional letters). These need truth-values as q-values. We just follow the lead of our proof of the analogous result for **PL** and assign a zero-place predicate the q-value *true* just if it is in Σ. With this addition the chosen valuation will still make all primitives true. (And, as a final wrinkle, if Σ *only* contains zero-place predicates and so no constants and no quantifiers, you can choose anything you like to be the domain, since it plays no further part in evaluating the wffs.)

So we've established (C_1) and (C_2). And, to repeat, they together entail the desired conclusion (Sat): every syntactically consistent saturated set Σ of closed **QL** wffs is satisfiable. Note carefully that nothing in the argument for (Sat) depends on the *size* of Σ. The argument still goes through just the same if Σ has infinitely many members.

30.4 Systematic trees

We now want to show (Sys): there's a way of systematically applying the tree-building rules such that the resulting tree either closes or has an open path, the wffs on which form a syntactically consistent saturated set.

Now, our systematic procedure may never terminate, and hence may describe an infinite tree. So we are going to need the following simple result about infinite trees, an instance of what's called 'König's lemma' (a 'lemma' is a mini-theorem, a stepping stone to a more weighty result):

(KL) Any infinite tree built by our rules will have an infinite branch.

That's because at any fork there are only two branches, so we can't make an infinite tree of finite height by (so to speak) infinite sideways sprawl. If that remark doesn't convince you, here's a proof:

Proof of (KL)   Call a node n which is above an infinite number of nodes (on paths through n) *exciting*; other nodes are *dull*. Suppose a node d is followed by a dull node – or two dull nodes if the branch forks there. Then d

will have only a finite number of descendants, and be dull too. Hence if a node is exciting, then at least one of the nodes immediately below it must be exciting too. The top-most node n_1 of an infinite tree is exciting by assumption. So there will be a node n_2 immediately below it which is exciting; and a node n_3 immediately below n_2 which is exciting; and so on for ever, giving us an infinite branch $n_1 \to n_2 \to n_3 \to \dots.$ QED.

Next, then, we will describe one possible systematic method of tree building (there are various options here, but this is as simple as any):

A systematic procedure Start at the top of the tree, and work methodically downwards examining the wffs on the tree (at any level of the branching tree, look at the nodes at that level from left to right). You can of course ignore wffs that are on closed branches. If the wff under examination has a connective as the main logical operator, apply the appropriate tree-building rule of type (a) to (i) – see the box in §29.1. If the wff is of the form $\neg \forall v C$ or $\neg \exists v C$, apply the relevant rule for negated quantifiers. If the wff is of the form $\forall v C(\dots v \dots v \dots)$, then add $C(\dots c \dots c \dots)$ to each relevant path *for every constant c that appears on that path* (or instantiate with a new constant if there are no other constants on the path). Finally, if the wff is of the form $\exists v C(\dots v \dots v \dots)$, then add $C(\dots c \dots c \dots)$ to each relevant path where c is new to those paths, *and then instantiate again with this new name c every previously visited universally quantified wff on those paths, and add those new instantiations to the foot of every relevant path too.*

The basic idea, of course, is that we keep using the rule ($\forall$) to instantiate any universal quantifiers with *every* name on its path, systematically revisiting them if we add a new name to the path because we've just instantiated an existential quantifier. And we keep on going for ever, unless the tree closes off, or we run out of rules to apply. Call the result a *completed systematic tree*. ('Completed' doesn't mean finished in a finite number of steps: we are, so to speak, taking a God's eye view of the procedure! If you don't like this dramatic idiom, drop it; just think of completed trees as ordered sets of wffs with the right structure.)

This completed systematic tree is either finite or infinite. In the finite case, the tree is obviously either closed or open (either every branch ends in absurdity, or at least one doesn't). What about the infinite case? Can something weird happen then? No: if a tree is infinite then, by König's lemma, there will be an infinite path through the tree – and an infinite path has to be open (for if it eventually contains A and $\neg A$, that path would be finitely terminated after all). So a completed tree is either finite and closed, or it has an open path. And whether the open path is infinite or not, the wffs on it will form a saturated set, by construction. For compare the tree-building rules in §29.1 with our definition of saturation. Clause by clause they match, so that the application of our tree-building rules (with as many return visits to universals as there are names on the branch) is *exactly* what we need to produce saturation.

Which proves (Sys). In sum, our systematic way of tree-building produces a closed tree, or one with an open path, whose wffs form a syntactically consistent, saturated set (finite or infinite).

30.5 Completeness completed

Putting that all together: we've proved (Sat) in §29.3, (Sys) in §29.4; and as we already remarked at the outset, those results trivially entail that any q-valid inference at all can be shown to be valid by some closed tree.

So our tree proof system for **QL** is complete in the sense of §20.4. Which is a really beautiful metalogical theorem.

Of course, we have made proving completeness easy for ourselves by choosing a tree-based proof system. For the key concept of *saturated* sets – which underlies the neatest completeness proofs for all the various alternative proof systems for **QL** – is so easy to motivate if we've encountered trees. But the relative ease of our proof doesn't make it trivial. It is a deep and important result, with some intriguing consequences (topics for exploration in more advanced logic books).

Returning to our turnstile symbolism (§20.4), if there is a **QL** tree-proof starting from $A_1, A_2, \ldots, A_n$ and $\neg C$, we can write $A_1, A_2, \ldots, A_n \vdash_{QL} C$ (and drop the subscript if it is clear which proof system is in question). Likewise, if the inference from $A_1, A_2, \ldots, A_n$ to C is q-valid, we can write $A_1, A_2, \ldots, A_n \vDash_q C$ (and again drop the subscript if it is clear which semantic notion of validity is in question). What we have shown in this chapter, in briskest headline form, is the syntactic turnstile and the semantic turnstile relate the same wffs: that is,

$$A_1, A_2, \ldots, A_n \vdash_{QL} C \text{ if and only if } A_1, A_2, \ldots, A_n \vDash_q C.$$

30.6 Summary

- Our rules for extending trees are provably *sound*; i.e. if the original trunk $A_1, A_2, \ldots, A_n, \neg C$ is satisfiable, so is at least one path on the tree extended using the rules. Since a satisfiable path can't close, that means that if a tree starts with a satisfiable trunk, it will stay open. Equivalently, if a tree closes, then its trunk won't be satisfiable, i.e. the inference from $A_1, A_2, \ldots, A_n$ to C is q-valid.

- A saturated set of (closed **QL**) wffs has truth-makers for every non-primitive wff in the set, assuming that there is one object in the domain per constant in the set.

- (Sat) Every syntactically consistent saturated set of closed wffs is satisfiable.

- (Sys) We can systematically apply the tree-building rules so that the resulting tree is either closed or has an open path whose (closed) wffs form a syntactically consistent saturated set.

- (Sat) plus (Sys) entails that there is a closed tree corresponding to every q-valid inference, so our tree-based proof-system is *complete*.

Interlude

Developing predicate logic

In the last ten chapters:

- We informally introduced the language QL which provides an elegant and well-motivated symbolism for unambiguously encoding quantified propositions. QL does ride roughshod over the distinctions marked in ordinary language between e.g. saying what *kind* a thing belongs to, saying it has a certain *property*, and saying what it is *doing* (§23.5). It also involves ignoring the differences between e.g. 'all', 'any' and 'every' (§21.2). But the payoff is a certain uniformity of treatment of closely related arguments. For example, consider the arguments (1) 'All men are mortal, Socrates is a man, hence Socrates is mortal', (2) 'Anyone who is white is pale, Socrates is white, hence Socrates is pale', (3) 'Everyone who walks moves, Socrates is walking, hence Socrates is moving'. Despite dealing respectively with kinds, properties and activities, and despite the different vernacular quantifiers, there is *something*, surely, in common here in the inferential moves. And they can all be rendered into QL in the same way: '∀x(Fx ⊃ Gx), Fn ∴ Gn'.

- We noted that standard QL treats restricted quantifiers with a certain artificiality. That's because a QL valuation fixes on a certain domain of discourse as being in play throughout an argument, so that we need to use conjunctions and conditionals when we want to generalize over just parts of that fixed domain. The artificiality buys us ease of logical manipulation.

- We introduced QL trees as a way of showing valid QL inferences are indeed valid (we start by assuming the premisses and negation of the target conclusion, and derive contradictions). In contrast to the logical apparatus available to our ancient and medieval predecessors, QL and its proof methods allows us to deal smoothly with arguments involving multiply quantified propositions.

- We then developed the formal syntax and semantics of QL: in particular, we explained the idea of a q-valuation (which specifies a domain of discourse, and assigns objects in the domain to constants and sets of *n*-tuples of objects from the domain to predicates). We explained how a q-valuation of the vocabulary in a closed wff without free variables fixes the truth-value of the wff.

- Next came the idea of q-validity. An argument is q-valid if every q-valuation which makes the premisses true makes the conclusion true. The q-valid arguments are those that are valid in virtue of their quantificational logical form (i.e. in virtue of the way that quantifiers and connectives are distributed in the premisses and conclusion). This gives us another sharp (but narrow) notion of 'validity in virtue of form': compare §13.6.

- We noted (but didn't prove) that there is no mechanical test for q-validity. However, we can establish q-validity by assuming an inference has a countermodel and deriving a contradiction, tree-style.

- We then looked at some more examples of trees in action, and also saw how we can in *some* cases use trees that don't close to construct countermodels to show that an inference is invalid.

- Finally, we confirmed that our system of tree-building rules enables us to establish the q-validity of all (but only) the genuinely q-valid arguments.

The foundations of one version of classical predicate logic (i.e. first-order quantification theory) are therefore in place. So after all this, where next?

First, we are going to back-track a little. When we were motivating the semantics of **QL**, we claimed that what *normally* matters about an ordinary-language name – as far as the truth or falsity of what we say using the name is concerned – is the *reference* of the name. This claim is worth exploring at greater length, and our next chapter on 'extensionality' does so.

We then extend our treatment of predicate logic in an absolutely standard way by adding a special predicate for *identity* (the notion involved in ordinary claims like 'George Orwell is the one and the same as Eric Blair', 'the Morning Star is one and the same as the Evening Star'). This notion of identity has the kind of generality and topic-neutrality to make it arguably a *logical* concept. Adding an identity predicate to the resources of **QL** very significantly increases the expressive power of the language. For example, in plain **QL**, we can express 'at least one thing is F'; in our extended language **QL$^=$**, we can express 'exactly one thing is F' (or equivalently, 'there is an F, and any F is one and the same thing as it'), 'at least two things are F', 'everyone except n is F', etc.

Having described this augmented language, we will then make a corresponding addition to our logical proof devices, adding rules for making **QL$^=$** trees.

Another standard topic on our agenda is the treatment of 'definite descriptions', i.e. referring expressions like 'the current President' or 'the greatest living logician'. The standard way of dealing with definite descriptions within **QL$^=$** is via 'Russell's Theory of Descriptions'.

We'll conclude by briefly considering a further extension of **QL$^=$** to include expressions for functions.

Extensionality

QL is a *fully extensional* language; an ordinary language like English isn't. This chapter discusses what this means and whether it matters.

31.1 Interpretations vs. valuations

From time to time, we have informally talked about *sets* (sets of propositions, sets of objects forming a domain, etc.). Sets are collections of things considered independently of their order and of the way they are specified. In other words, *A* is one and the very same set as *B* if anything which is a member of *A* is a member of *B* and vice versa. Sets are identical if they have the same members.

Consider then the set of humans and the set of terrestrial featherless bipeds. Anyone who is a human is a terrestrial featherless biped and vice versa. So we have just one set here. In other words, the predicates, '... is human' and '... is a terrestrial featherless biped' have the same set as their shared extension. Likewise, take the set of humans over ten feet tall and the set of humans over 200 years old. No one falls into either set. Vacuously, if anyone is a member of one they are a member of the other. So again these are the same set, in this case the *empty* set. Hence the two predicates '... is a human over ten feet tall' and '... is a human over 200 years old' also share the same extension.

But of course, the sentences 'Jack is a human over ten feet tall' and 'Jack is a human over 200 years old' don't *mean* the same; the predicates here may share the same extension, but they differ in meaning. Likewise the two predicates, '... is human' and '... is a terrestrial featherless biped' have different meanings even if they have the same extension. *If predicates mean the same, they must share extensions; but not vice versa.*

Now, with some exceptions, the extension of a predicate is what matters for fixing the truth of sentences that contain it. So, for the purposes of logic – which is all about truth-preserving arguments – our semantic story can concentrate on these extensions. But when we are trying to interpret what someone is saying, we need to know more than the extensions of their predicates. On the ordinary understanding of the term, an interpretation of a language needs to specify the meaning of a predicate (in other terms, it needs to give the *sense* of the predicate, or to indicate what *concept* is expressed by the predicate).

It is pretty much agreed on all sides that we need a meaning/extension distinction for predicates. More controversially, a similar distinction seems to apply to names as well. Take for example 'Superman' and 'Clark Kent', two different names of the same person. It is quite possible for someone not to know that these *are* two names for the same person. Thus Lois Lane has both names in her repertoire, but she thinks of Superman as the handsome superhero in tights, and of Clark Kent as the bespectacled reporter. The names, for Lois, are associated with different ways of thinking about the same person, different 'modes of presentation' (as Frege famously puts it); and she doesn't make the connection. That's how she can assert 'Superman can fly' yet deny 'Clark Kent can fly'. That's also how 'Superman *is* Clark Kent' can be informative for her.

In interpreting Lois's behaviour, and trying to see it as rational, we need to distinguish when she is thinking about the person she is interacting with as *Superman* and when she is thinking about him as *Clark Kent*. Why isn't she alarmed when she sees the person in front of her leap out of a window? Because she takes him as being Superman (not Clark Kent). To report Lois as thinking that Clark Kent can fly when she is reporting the man's marvellous deeds and referring to him as 'Superman' would be to misrepresent her take on the world.

In short, for the purposes of *interpretation*, we need to know more than the reference of the names. We need to know something of the modes of presentation Lois associates with the names, or (to use Frege's term) need to know the *sense* the name has for her. Just as two predicates can have different senses but the same extension, two names can have different senses but the same reference, as do 'Superman' and 'Clark Kent' for Lois. However, as far as the *truth* of her assertions is concerned, it doesn't seem to matter whether Lois says 'Superman is *F*' or 'Clark Kent is *F*'; the claims will stand or fall together depending on whether that single individual falls into the extension of the predicate. The truth (at least for most fillings for '*F*') depends only on *who* is being referred to, not *how* he is being thought of.

By the way, don't for a moment get distracted by the fact that we have gone along with a vivid fictional example here. We could equally well, though perhaps much less memorably, use examples like 'George Eliot' vs. 'Mary Ann Evans', or Frege's own example of 'The Morning Star' vs. 'The Evening Star' (which he treats as two names for the planet Venus).

What goes for English also goes in principle for uses of artificial languages like **QL**. Pretend that Lois uses a QL-style code in her diary, using 'm' to record encounters when rescued by a superhero and 'n' to record the doings of her reporter colleague. Then to capture the thoughts she is expressing in her coded notes, we would need to translate 'm' as 'Superman' and 'n' as 'Clark Kent', rather than vice versa. But this close translation, matching sense with sense, would not normally matter for determining the *truth* of what she writes: if our focus is on truth, then what matters is whom she is referring to. Hence our claim that a q-valuation apt for logical purposes, which assigns truth-relevant properties to expressions of **QL**, need only assign references (not senses) to constants.

31.2   Extensional and intensional contexts

Suppose the q-valuation q assigns the two names m and n the same reference. Then it is easily seen that a **QL** sentence of the form $C(...m...m...)$, involving the name m, will be true on q if and only if the corresponding sentence $C(...n...n...)$, where n appears instead of m, is also true on q. It is only the shared reference of m and n that matters for truth. Hence, co-referential names can be intersubstituted in **QL** wffs *salva veritate* (i.e., preserving the truth).

Again, suppose q assigns the two predicates F and G the same extension. Then similarly, a **QL** sentence $C(...F...F...)$ involving the predicate F will be true on q exactly when the corresponding sentence $C(...G...G...)$ is true on q. In other words, co-extensional predicates can be intersubstituted *salva veritate*.

We'll call a context within which pairs of co-referential names and pairs of co-extensional predicates can be intersubstituted preserving the truth an *extensional* context. A language where all sentences are extensional contexts is called an *extensional language*. **QL** is in that sense an extensional language.

Now, very many sentences of a natural language like English are extensional. What matters for their truth is the reference of names and the extensions of predicates (that's what motivated the idea of a q-valuation for their **QL** analogues). But English is not a *fully* extensional language. Some ordinary-language constructions yield non-extensional or *intensional* contexts. In other words, we can form pairs of sentences $C(...m...m...)$, $C(...n...n...)$ which take different truth-values, despite the fact that m and n are names denoting the same thing. Similarly, we can form pairs $C(...F...F...)$, $C(...G...G...)$ which take different truth-values, despite the fact that F and G have the same extension.

We will look below at two main types of example (arising in psychological and modal contexts). First, however, we'll limber up by saying something about the quotational contexts that may – or may not! – provide a useful comparison case.

31.3   Quotation

'Bill' and 'Clinton' are names picking out the same man. In many, many contexts we can interchange the names, *salva veritate*. Take, for example,

> (1)   Bill kissed Monica.
> (2)   Clinton kissed Monica.

These must either both be true or both be false. But compare

> (3)   'Bill' has four letters.
> (4)   'Clinton' has four letters.

Here swapping the names 'Bill' and 'Clinton' leads from truth to falsehood.

But how is that possible? How can swapping the names make a difference? The answer is absolutely obvious and trivial. In (3), the expression

> Bill

is no longer functioning to pick out the man, but is part of a longer expression, i.e. part of the expression

> 'Bill'

– including the quotation marks – that serves to pick out someone's *name* (see §10.2 again for a reminder about our conventions governing quotation marks). Likewise, in (4), the expression

> Clinton

is part of a longer expression, i.e. part of the expression

> 'Clinton'

that serves to pick out (not the same *man*, but) a different *name*. No wonder swapping the expressions appearing as parts of the complete quotational expressions in (3) and (4) won't always preserve truth.

Evidently, we get similar effects with quoted predicates. So, quotational occurrences of expressions, names and predicates, aren't extensional.

Or at least, that's one way of putting it. But maybe it is a bit misleading. Maybe we shouldn't really say that the word 'Bill' has an *occurrence* in (3) at all, any more than the word 'rat' has an occurrence in 'Socrates walked'. The sequence of letters 'r'+'a'+'t' appear within that sentence, but not as a separate word. Likewise, it might be said, the sequence of letters 'B'+'i'+'l'+'l' appears in (3) but not as a separate word, but only as part of the longer denoting expression "'Bill'".

We won't pause to try to decide the best way of thinking about quotational contexts. What is clear is that we can't substitute *salva veritate* within them.

31.4 Intentional contexts are intensional

Our first more serious kind of example of a non-extensional context is suggested by the discussion of §31.1 above. The names 'Superman' and 'Clark Kent' are co-referential: and in simple, straightforward, contexts they can be intersubstituted preserving the truth. For instance, compare

> (5) Superman can fly.
> (6) Clark Kent can fly.

These stand or fall together, because Superman *is* Clark Kent. Yet, as we told the story, the following sentences *do* differ in truth-value:

> (7) Lois believes that Superman can fly.
> (8) Lois believes that Clark Kent can fly.

Swapping 'Superman' and 'Clark Kent' now leads from truth to falsehood. But how is that possible? How can swapping the names here make a difference?

This time the answer is non-obvious and non-trivial. One suggestion might be that (7) and (8) can be analysed along the lines of

> (7*) Lois accepts-as-true 'Superman can fly'.
> (8*) Lois accepts-as-true 'Clark Kent can fly'.

So, on this analysis, 'Superman' has a disguised quotational occurrence in (7). And the reason we can't substitute co-referential names *salva veritate* in (7) is akin to the reason we can't substitute co-referential names in (3).

This would be a very neat account of the failure of substitutivity if (7) could indeed be analysed as (7*). But, as it stands, the analysis is surely wrong. After all, Lois could have beliefs, including the belief that Superman can fly, even if she is a monoglot French speaker and so doesn't accept as true *any* English sentence. But maybe we can try to patch things up, and provide a more sophisticated 'disguised quotation' analysis of (7) which still explains the failure of substitutivity in the same basic way.

An alternative line goes as follows (the idea here goes back at least to Frege). When we say 'Lois believes that Superman can fly' we are *not* relating her to a person; after all, she could have this belief even if Superman is a figment of her imagination. Rather, we are relating her to a way-of-thinking, the mode of presentation *Superman* which is a constituent of her thought. And if we instead say 'Lois believes that Clark Kent can fly', we'd be saying something different about the way she is thinking. In this sort of context, then, the truth of the whole sentence depends on the *sense* of the embedded name. So substituting a name with a different sense usually won't preserve truth (even if the names happen to have the same reference).

This second account shares with the quotational account the core idea that an occurrence of 'Superman' after 'believes that ...' no longer simply serves to pick out the man. The quotational account has it that the reference shifts away from the man to the *word* 'Superman'; the second account (at least in the classic Fregean version) has it that reference shifts to the *sense* of the word 'Superman'. Either way, that explains the failures of substitution.

We get similar effects with predicates. Consider the predicates '... is a unicorn' and '... is a dragon'. These have the same extension, namely the empty set. But plainly we can't swap them in quotational contexts, and always hope to preserve truth. Neither can we do so after 'believes that ...'. For example, compare

> (9)   Helen believes that Pegasus is a unicorn.
> (10) Helen believes that Pegasus is a dragon.

The first could be true and the second false.

How can that be? Well, again two stories could be run here. Obviously a quasi-quotational analysis of 'believes that ...' contexts would explain the failure of substitutivity. If (9) and (10) can be analysed on the following lines –

> (9*)   Helen accepts-as-true 'Pegasus is a unicorn'.
> (10*) Helen accepts-as-true 'Pegasus is a dragon'.

– then changing the quoted predicate changes the sentence that Helen accepts, and she might accept one sentence without accepting the other. Alternatively, the Fregean line would be (very roughly) that when we say 'Helen believes that Pegasus is a unicorn' we are indicating that her thought involves the concept *unicorn*. If we swap the predicate for one with a different sense, which expresses a different concept, then we change what we are saying about how Helen is thinking. No wonder then that swapping co-extensional predicates with different senses won't necessarily preserve truth in this case.

What goes for 'believes that …' contexts goes for other contexts governed by psychological verbs which take 'that'-clauses. For example, compare

> (11) Lois hopes that Superman loves her.
> (12) Lois hopes that Clark Kent loves her.

Or compare

> (13) Lois expects that Superman will save the world again.
> (14) Lois expects that Clark Kent will save the world again.

Again, substitutivity of co-referential names typically won't preserve truth. The contexts are non-extensional – i.e. *intensional*.

In the philosophy of mind, states attributed by psychological 'that'-clauses are standardly called *intentional* states. We can likewise call contexts created by such clauses *intentional contexts*. So, in a slogan, intentional contexts (with a '*t*') are intensional (with an '*s*').

Is **QL**'s lack of intentional contexts a problem? You might think that it is an expressive shortcoming of our logical symbolism. But arguably it is an *advantage*. In an extensional language, the reference of a constant stays fixed. Not so in a language like English. For example, a name like 'Superman' might sometimes pick out a superhero, and sometimes (as in 'Lois believed that Superman loved her') pick out a name or the sense of a name. Perhaps the intensionality here relates to a kind of ambiguity. So we might argue that it is good to be forced to recast what the English says using by intentional idioms into an unambiguous formalism like **QL**. Doing so would make it clear that sometimes we are speaking of heroes, sometimes of their names, sometimes of the sense of their names, so it reveals frankly and unambiguously what is going on.

31.5 Modal contexts are intensional

Here's a second major kind of case where extensionality fails in ordinary language. Consider the following two sentences

> (15) Necessarily, anything which is a human is a human.
> (16) Necessarily, anything which is a terrestrial featherless biped is a human.

The first is trivially true; the second is false (there could have been non-human featherless bipeds on earth). But the second results from the first by substituting one predicate for a co-extensional one. Likewise, consider

> (17) It is logically possible that something exists which is a unicorn.
> (18) It is logically possible that something exists which is a square circle.

Again, it seems that the first is true, the second is false. But again, the second results from the first by substituting a predicate for a co-extensional one.

Notions of necessity, possibility and impossibility – notions of what could and couldn't be the case – are standardly called *modal* notions. Contexts in the scope of modalities are intensional.

Why so? On one natural line of thought, 'Necessarily *P*' (for example) depends for its truth on whether *P* holds across other possible worlds (see §12.5). If *P* involves some predicate *F*, then whether *P* is true at some other world depends on the extension *F* has *at that world*. Suppose *G* is some predicate that happens to have the same extension as *F* in the actual world. Typically, *F* and *G* can still have differing extensions *at some other worlds*. So swapping *F* for *G* won't in the general case preserve the truth-value of *P* at other worlds. So the swap can change the value of 'Necessarily *P*'.

For example, suppose *P* is 'Anything which is a human is a human'. That's true however the world is, or (as it is often put) is true at all possible worlds. So 'Necessarily *P*' is true. And now consider what happens when we replace the first occurrence of the predicate *F* = 'human' with the predicate *G* = 'terrestrial featherless biped'. These predicates are co-extensional (in the actual world), but their extensions can peel apart in other worlds: just imagine a possible world where naked ostriches have evolved. So swapping the predicates can change truth-values at other worlds. While the original *P* is true at all worlds, after swapping predicates we get 'Anything which is a terrestrial featherless biped is a human' which is false at some worlds. So that's how the swap can change the value of 'Necessarily *P*'.

Is QL's lack of intensional modal contexts a shortcoming? Perhaps so. But then everyday modal contexts are not easy to understand. Perhaps forcing ourselves to recast what English tries to say using modal idioms into an extensional formalism like QL (which would involve e.g. doing some honest quantification over possible worlds) will again reveal frankly and unambiguously the commitments of modal discourse. But that's not a discussion we can even begin to pursue here.

31.6 Summary

- QL is an extensional language: co-referential names and co-extensional predicates can always be intersubstituted *salva veritate*.

- English contains intentional contexts (as in 'Lois believes Superman can fly') which are intensional, i.e. non-extensional.

- Modal contexts (as in 'It is possible that unicorns could exist') are also non-extensional.

- It is moot whether QL's lack of such non-extensional contexts is an expressive lack or a methodological advantage.

Identity

The idea of identity – involved, for example, in the claims that Superman is identical to (is one and the very same as) Clark Kent, or that the Evening Star is identical to the Morning Star – has the kind of generality and topic-neutrality that makes it of interest to logic. In this chapter we briefly describe the fundamental principles governing the identity relation.

32.1 Numerical vs. qualitative identity

Compare two types of cases:

(A) Jill and her twin sister Jane are (as we say) identical; you cannot tell them apart. You and I have identical sweaters. Simultaneously in London and Paris, Jack and Jacques are wrapping up identical bottles of Chanel No. 19 to send to their respective beloveds.

(B) Since Jill and Jane are twins, Jill's mother is identical to Jane's mother (she is one and the same person). There are two sweaters, but you and I bought them from the identical branch of Marks and Spencer. And as luck would have it, Jack's beloved is one and the very same as Jacques's; the identical woman is the object of both men's desires.

In type (A) cases, we have what is often called *qualitative identity*. Jill and Jane are distinct beings sharing the same intrinsic properties (well, in the real world, maybe not strictly *all* the same properties: but let's pretend that they are perfect duplicates). Likewise for the two distinct but identically similar sweaters and the two bottles of perfume. In the (B) cases, however, we don't have mere duplicates but rather instances of *numerical identity*. It is one and the very same entity (the very same mother, the very same shop, the very same beloved woman) that is in question each time.

Our topic in this chapter is the notion of identity in the latter sense of strict numerical identity.

It is momentarily rather tempting to say that our concern is with the idea of 'one object being the very same thing as another'. But on a moment's reflection, *that* cannot possibly be the right way to put it. If there really is 'one object' and 'another object' then there are *two* of them, and so they of course cannot be numerically identical after all!

32.2   Equivalence relations

We initially defined a relation as whatever is expressed by a predicate with two or more places (§3.1). So, at least in that sense, what is expressed by '... is one and the very same as ...' is indeed a two-place *relation*.

Moreover, strict numerical identity is an *equivalence* relation: as we shall also see, it is in a good sense the *'smallest'* equivalence relation. And the basic principle governing numerical identity is *Leibniz's Law* or, as it is often called, the 'Indiscernibility of Identicals'. The rest of this chapter explains these assertions. This section makes a start by defining the idea of an equivalence relation.

We first need to define three features a (two-place) relation R can have.

- We have already defined what it is for a relation R to be *transitive* (§28.6). To repeat, a relation R is transitive just if, whenever a has R to b, and b has R to c, then a has R to c. (Examples: 'is heavier than', 'is an ancestor of', and 'is part of' express transitive relations.)

- A relation R is *symmetric* just if, whenever a has R to b, then b has R to a. (Examples: 'is married to', 'is the opposite sex to', 'is adjacent to' express symmetric relations.)

- A relation R is *reflexive* just if, for any a, a has R to a. (Examples: 'is as tall as', 'has a parent in common with', 'drives the same make of car as' express reflexive relations.)

In each case, we'll assume that the objects a, b, c are ones which it makes sense to think of standing in the relation R. We don't, for example, want to deny that *being as tall as* is a reflexive relation, just because it makes no sense to say e.g. that the number seven isn't as tall as itself. We might say: *being as tall as* is reflexive in its domain of application.

Relations can have these features in various combinations. For example 'is heavier than' is transitive, but not symmetric or reflexive. (Single-premiss) entailment is also transitive and not symmetric, but is reflexive. Again, 'has a parent in common with' is reflexive, symmetric, but non-transitive, while 'is not taller than' is reflexive, transitive, but non-symmetric.

We are interested now, though, in relations which have all three features:

> An equivalence relation is a relation which is transitive, symmetric and reflexive.

Examples: 'is the same age as', 'has the same surname as', 'is a full sibling of', 'is parallel to', 'is exactly similar to', 'is truth-functionally equivalent to' express equivalence relations.

Suppose we have got a domain of objects and an equivalence relation R defined over those objects. Because of reflexivity, every object in the domain is R

to something, if only itself. Note too that if *a* is *R* to *b* and *a* is *R* to *c* then, by symmetry, *b* is *R* to *a*; and so by transitivity *b* is *R* to *c*. In other words, if *a* is *R* related to two things they are *R* related to each other. If you think about it, that means that *R* carves up the domain into non-overlapping groups of things, where the members of a group all stand in relation *R* to each other: these are *R*'s so-called *equivalence classes*.

For an example, take the domain of USA citizens, and take the equivalence relation *having the same legal surname as*. This carves up the domain into equivalence classes of people who share the same surname. Everyone is in one such class, and no one is in more than one class. (Some of these equivalence classes are large: there are a lot of Smiths. Others may contain just one member – a philosophy student once legally adopted the surname 'qFiasco', and it could well be that there is still only one qFiasco!)

For another example, take the domain of positive integers, and consider the relation *differs by a multiple of 3 from*. This is an equivalence relation (its symmetry and transitivity are obvious; and for reflexivity, note that a number differs by a zero multiple of three from itself). And the relation carves up the numbers into the equivalence classes $\{0, 3, 6, 9, ...\}$, $\{1, 4, 7, 10, ...\}$ and $\{2, 5, 8, 11, ...\}$.

32.3 The 'smallest' equivalence relation

Let's now add some informal symbolism to English. When *a* is indeed one and the very same thing as *b*, then we will write '$a = b$' (read '*a* is identical to *b*'). Note that it is perfectly appropriate to borrow the familiar mathematical symbol here. For when in ordinary arithmetic we write, say, '$2 + 3 = 10 - 5$' we are claiming that the number which is the result of adding 2 and 3 is *the very same* as the number which results from subtracting 5 from 10. The left-hand and the right-hand of the equation aren't just picking out qualitatively similar things; the expressions on either side pick out the very same number.

What '$=$' expresses is evidently reflexive, symmetric and transitive. Play along with the familiar fiction again. Then, first, Clark Kent is, of course, one and the very same as Clark Kent. More generally,

- For any *a*, $a = a$.

Second, if Clark Kent is one and the very same as Superman, then Superman must be one and the very same as Clark Kent. Generalizing:

- For any *a* and *b*, if $a = b$, then $b = a$.

And third, if Clark Kent is one and the very same as Superman, and Superman is one and the very same as the Superhero from Krypton, then Clark Kent is one and the very same as the Superhero from Krypton. Generalizing again:

- For any *a*, *b* and *c*, if $a = b$ and $b = c$, then $a = c$.

As with any equivalence relation, the identity relation carves up a domain of objects into equivalence classes. But of course, in this case, the equivalence classes will all contain exactly *one* item each (the only being related to Clark

Kent by the identity relation is that very person: to be sure, we may call him by a variety of names, but they all pick out one and the same individual). The identity relation, then, is the equivalence relation which partitions a domain into the smallest possible equivalence classes, with each member of the domain in its own separate class.

Note, then, that the extension of the identity relation defined over a domain is just the set of pairs $\{\langle a, a \rangle, \langle b, b \rangle, \langle c, c \rangle, ...\}$ (running through each object in that domain). But *any* equivalence relation defined over that domain also has in its extension *those* pairs, because every equivalence relation is reflexive. What is distinctive about the identity relation is that these are the *only* pairs in its extension. In other words, given a domain, the extension of the identity relation is contained in the extension of every equivalence relation defined over that domain.

So in that sense, identity is the 'smallest' equivalence relation.

32.4   Leibniz's Law

The crucial principle governing strict numerical identity is this:

> (LL)   If *a* and *b* are identical (i.e. are one and the same thing), then whatever property *a* has, *b* has.

There can be no exceptions to this. If the object *a* is one and the very same thing as the object *b*, then *a*'s having the property *P* just is the identical state of affairs as *b*'s having the property *P*.

Hence, if Superman is Clark Kent, then whatever properties Superman has, Clark has. If Superman can fly, then so can Clark (though to look at him in the guise of a reporter, you might not realize it!). If Superman loves Lois, so does Clark. Likewise, if Jill's mother is one and the very same person as Jane's mother, then Jill's mother and Jane's mother have the same properties. If Jill's mother is a logician, so is Jane's; if Jill's mother has three children, so does Jane's. And so on.

Our label '(LL)' here is short for 'Leibniz's Law'. The principle is also often called the *Indiscernibility of Identicals* – if *a* and *b* really are the same thing, then of course you can't tell 'them' apart. It is trivially equivalent to another uncontroversial principle, which we can call the 'Diversity of the Dissimilar':

(DD) If *a* has some property that *b* lacks, then *a* is not identical to *b*.

If Superman can fly and Clark can't, then that settles the matter: Clark isn't the superhero. If Jill's mother has red hair and Jane's mother hasn't, then they are different people.

Don't confuse the correct principle of the Indiscernibility of Identicals with the reverse principle of the Identity of Indiscernibles. Leibniz himself held that if *a* and *b* share all the same properties then they must be one and the very same thing. If we are generous enough about what counts as a 'property' this may be defensible. However, the Identity of Indiscernibles looks false if we mean ordinary, *intrinsic*, properties (properties that don't consist in relations to other things). For example, the two peas in the pod can surely, at least in principle, be qualitative duplicates in each and every intrinsic respect, can be qualitatively identical, yet still be two distinct things, differing only e.g. in their spatial location and their spatial *relations* to other things.

32.5 Leibniz's Law and co-referential designators

Note that Leibniz's Law is a statement about *things*: if a and b are the same thing, then a and b have the same properties. But it implies a claim about language, and the interchangeability of referring expressions.

Let's say that a *designator* is (roughly speaking) an expression whose standard function is to refer to or pick out a particular individual person or thing. Names are paradigm designators: but other instances are expressions like 'that table', 'Jill's father', 'the man in the corner drinking a martini'. Two designators 'a' and 'b' are *co-referential* if they actually refer to the same thing: in that case, the identity claim '$a = b$' will be true. For example, the pair of designators 'Jack' and 'Jill's father' are co-referential if the two designators pick out the same man, and so Jack *is* Jill's father. With the terminology thus understood, (LL) implies the following linguistic principle:

> (LL') If 'a' and 'b' are functioning as co-referential designators in the claims $C(...a...a...)$ and $C(...b...b...)$, and also the context C expresses the same property in each case, then these claims say the same about the same thing, and so must have the same truth-value.

How does (LL') square with the fact that we cannot exchange the normally co-referential designators 'Superman' and 'Clark Kent' in the context 'Lois believes that Superman can fly' and preserve the truth-value? Because, as we noted in the last chapter (§31.4), in *this* sort of context the designators are *not* serving to pick out their normal common reference.

Here's another apparent exception. Suppose Jack is known to his friends as 'Fats'. So 'Fats' and 'Jack' are co-referential designators. And no doubt 'Fats is so-called because of his size' is true. But we surely can't infer that 'Jack is so-called because of his size'. Is that an objection to (LL')?

No. The predicate 'is-so-called-because-of-his-size', when attached to the name 'Fats', says that Fats (i.e. Jack) is called '*Fats*' because of his size; while 'is-so-called-because-of-his-size', when attached to the name 'Jack', says that Fats (i.e. Jack) is called '*Jack*' because of his size. So although we have 'C(Fats)' true and 'C(Jack)' false with the same linguistic filling for the context C each time, the repeated expression 'is-so-called-because-of-his-size' in fact expresses *different* properties in the two cases. That's why (LL') can't be applied.

To repeat, (LL) is an exceptionless principle. And apparent exceptions to its linguistic correlate (LL') are indeed merely apparent, involving cases where the designators aren't in fact co-referential or where the property expressed by the context C shifts.

32.6 Summary

- We must distinguish qualitative from strict numerical identity.

- Qualitative and strict numerical identity are both transitive, symmetric and reflexive relations, i.e. are both equivalence relations. But strict numerical identity is the 'smallest' equivalence relation – it relates an object to nothing other than itself.

- The key principle governing identity is (LL) Leibniz's Law: if *a* and *b* are identical then whatever property *a* has, *b* has.

- There is a linguistic version of this principle, (LL'): if '*a*' and '*b*' are functioning as co-referential designators in the claims *C(...a...a...)* and *C(...b...b...)*, and also the context *C* expresses the same property in each case, then these claims must have the same truth-value. Apparent exceptions – e.g. the behaviour of designators in intentional contexts – can be explained away.

Exercises 32

In addition to the properties of relations already defined, we say that a relation *R* is *Euclidean* just if, whenever *a* has *R* to *b*, and *a* has *R* to *c*, then *b* has *R* to *c*. *R* is *asymmetric* if, whenever *a* has *R* to *b*, then *b* does *not* have *R* to *a*. *R* is *irreflexive* if no object has *R* to itself. These properties, along with the properties of being *reflexive*, *symmetric* and *transitive*, can all be captured by QL wffs. For example, *R*'s being symmetric is captured by

$$\forall x \forall y (Rxy \supset Ryx)$$

Give wffs that similarly capture *R*'s having each of the other five properties. Then give both informal arguments and QL trees to show

     1.   If *R* is asymmetric, it is irreflexive.
     2   If *R* is transitive and irreflexive, it is asymmetric.
     3.   If *R* is an equivalence relation, it is Euclidean.
     4.   If *R* is Euclidean and reflexive, it is an equivalence relation.

What about

     5.   If *R* is transitive and symmetric, it is reflexive?

The language QL$^=$

We now add to **QL** a symbol for the identity relation; and then we explore the expressive power of the resulting augmented language **QL$^=$**.

33.1 Adding identity to QL

It would be consistent with the basic design of **QL** to introduce a predicate letter, say 'I', which belongs to the category of two-place predicates, and takes two names to form an atomic wff such as 'Imn', to be read '*m* is identical to *n*'.

But it is absolutely standard (and a lot more perspicuous) to add the now familiar identity sign '=' to the basic alphabet of our language, and to allow this symbol to be written in the usual way *between* a pair of terms (constants or variables), yielding expressions like 'm = n', 'x = y' and so on. There's no need to introduce brackets with such expressions, any more than there is a need to bracket other expressions involving two-place predicates. We'll call the resulting language **QL$^=$**.

If we are going to allow these new kinds of expression, we must of course expand our rules for forming atomic wffs. So the syntactic rules for forming atomic wffs in **QL** (§26.3) can be updated to become:

> (A1$^=$) For all $j \geq 0$, a j-place predicate followed by j terms is an atomic wff of **QL$^=$**.
> (A2$^=$) A term followed by '=' followed by a term is an atomic wff of **QL$^=$**.
> (A3$^=$) Nothing else is an atomic wff.

The rules for building complex **QL$^=$** wffs remain exactly as for **QL**.

What about the semantics of **QL$^=$**? The basic idea is trivial: a wff like 'm = n' should come out true just so long as the references of the constants 'm' and 'n' are the same. But, for a systematic semantics in the style of Chapter 27, we need to allow wffs involving free variables, and then give rules for evaluating them on (extended) q-valuations which assign q-values to variables. What we need is:

> (Q0$^+$) An atomic wff of the form $t_1 = t_2$ is true on the valuation q just if the terms t_1 and t_2 are assigned the same objects as q-values by q.

Add that to (Q0) in §27.4. Then everything else remains semantically as in **QL**. In particular, we can apply the notions of q-validity, q-logical truth, etc. to cover wffs and inferences expressed in **QL⁼**. It's as easy as that!

The general principles about reflexivity, symmetry and transitivity of identity can now be expressed thus:

(Ref) $\forall x\, x = x$
(Sym) $\forall x \forall y (x = y \supset y = x)$
(Tran) $\forall x \forall y \forall z ((x = y \wedge y = z) \supset x = z).$

And it is easy to show that these are, as we'd expect, q-logical wffs of **QL⁼** (see Chapter 35).

What about Leibniz's Law? Informally, this says: if x and y are identical, then for any property X, if x has property X, y has it too. So what we'd *like* to be able to say in our formal language is something like

(LL) $\forall x \forall y \forall X (x = y \supset (Xx \supset Xy))$

where we are not only quantifying over *objects* ('take any things, x, y, ...') but also quantifying over *properties* ('for any property X'). However, we can't do that in **QL⁼**. For in this language we only have the first 'order' of quantifier, i.e. *we only have quantifiers that run over objects in the domain*. We don't have quantifiers running over properties. For this reason, by the way, languages like **QL** and **QL⁼** are standardly called *first-order* languages, and their logic is therefore called *first-order logic* (contrast second-order logic, where we *can* quantify over properties too).

The fact that **QL** and **QL⁼** are first-order languages is arguably a major limitation: for example, some intuitive mathematical reasoning does seem to be essentially second-order, and so can't be captured in first-order terms. But we can't explore this tricky issue here.

Lacking second-order quantifiers that run over properties, the best we can do in **QL⁼** to approximate Leibniz's Law is to say something like this:

> Each way of filling out the following schema is true,
>
> (LS) $\forall v \forall w (v = w \supset (C(...v...v...) \supset C(...w...w...)))$
>
> where v, w are variables, $C(...v...v...)$ is a wff with just v free, and $C(...w...w...)$ is the result of substituing w for each occurence of v.

The idea, roughly, is that whatever **QL⁼** wff we put for C, then if $v = w$, $C(...v...v...)$ implies $C(...w...w...)$. Which is indeed true, because **QL⁼** is an extensional language (cf. §31.2).

Instances of the Leibnizian schema (LS) include, say,

$\forall x \forall y (x = y \supset (Fx \supset Fy))$
$\forall y \forall z (y = z \supset ((Lay \wedge Lby) \supset (Laz \wedge Lbz)))$
$\forall x \forall y (x = y \supset (\forall z(Lxz \supset z = m) \supset \forall z(Lyz \supset z = m)))$

But note that in saying that all instances of (LS) are true, we are in effect only committing ourselves to the claim that identical objects share *every feature expressible in QL⁼*, which is a weaker claim than the full-strength Leibniz's Law which says that identicals share *every* feature.

33.2 Translating into QL⁼

Adding the identity sign gives us a surprisingly rich extension to our expressive resources. Let's run through some examples. Interpret QL⁼ so that

'a' means *Angharad*
'b' means *Bryn*
'm' means *Mrs Jones*
'F' means ... *speaks Welsh*
'G' means ... *is a girl*
'L' means ... *loves* ...

and where the domain of quantification consists of human beings. We'll translate the following:

1.   Angharad is none other than Mrs Jones.
2.   If Angharad speaks Welsh, and Mrs Jones doesn't, then they are different people.
3.   Everyone except Angharad loves Bryn.
4.   Someone other than Bryn loves Angharad.
5.   Only Mrs Jones loves Bryn.
6.   Only Angharad and Bryn love Mrs Jones.
7.   Angharad only loves people who love her.
8.   Angharad loves at most one person.
9.   At most one girl speaks Welsh.
10.   Whoever loves Mrs Jones loves no-one else.
11.   If Bryn loves some Welsh-speaker other than Angharad, then he loves Mrs Jones.
12.   Every girl other than Angharad loves someone other than Bryn.

(Try translating at least the first few of these before reading on.)

1.   'Angharad is none other than Mrs Jones' is naturally read as simply an assertion of identity (so note, we don't always have to translate 'none' by a quantifier). The translation is

     a = m

2.   Similarly, 'they are different people' is a denial of identity. The translation is

     ((Fa ∧ ¬Fm) ⊃ ¬a = m)

   Some allow a wff like '¬a = m' to be written 'a ≠ m'; but we won't be using that shorthand in this book. NB, the negation sign negates 'a = m', not 'a'!

3.   'Everyone except Angharad loves Bryn' is equivalent to 'Everyone is such that, if they are *not* Angharad, they love Bryn'. Whence,

     ∀x(¬x = a ⊃ Lxb)

   Someone who says 'Everyone except Angharad loves Bryn' perhaps strongly implies that Angharad doesn't love Bryn: but, arguably she doesn't actually *assert* that. For it would be quite consistent to add '... and maybe Angharad does too: I just don't know.'

4.   Similarly 'Someone other than Bryn loves Angharad' says that there is someone who is not identical to Bryn and loves Angharad: i.e.

$\exists x(\neg x = b \wedge Lxa)$

5.   What about 'Only Mrs Jones loves Bryn'? Well, this says (on the natural reading) that Mrs Jones loves Bryn and no-one else does, i.e. anyone who isn't identical to Mrs Jones doesn't love Bryn. So we have

$(Lmb \wedge \forall x(\neg x = m \supset \neg Lxb))$

or equivalently,

$(Lmb \wedge \forall x(Lxb \supset x = m))$

6.   'Only Angharad and Bryn love Mrs Jones' is similar: it says that they both love Mrs Jones, and anyone who isn't one of them doesn't love her. Hence,

$((Lam \wedge Lbm) \wedge \forall x(\neg(x = a \vee x = b) \supset \neg Lxm))$

or equivalently,

$((Lam \wedge Lbm) \wedge \forall x(Lxm \supset (x = a \vee x = b)))$

7.   Once we note that 'only' is often translated using the identity predicate, there is a danger of getting distracted and forgetting that many uses of 'only' in quantified sentences *don't* have anything to do with identity, and can be translated using unaugmented QL. 'Angharad only loves people who love her' is a case in point. It says just that whoever Angharad loves is someone who loves her. Translating yields:

$\forall x(Lax \supset Lxa)$

8.   'Angharad loves at most one person' is equivalent to: if we pick someone Angharad loves, and then again pick someone Angharad loves, we must have picked the same person twice. So we can translate this as

$\forall x \forall y((Lax \wedge Lay) \supset x = y)$

9.   Similarly, we can render 'At most one girl speaks Welsh' by

$\forall x \forall y(\{(Fx \wedge Gx) \wedge (Fy \wedge Gy)\} \supset x = y)$

10.   If we use an English/QL⁼ mixture of the kind that we allowed ourselves in §24.1, then 'Whoever loves Mrs Jones loves no-one else' is semi-translated by

$\forall x(Lxm \supset x$ loves no-one other than Mrs Jones$)$

But 'x loves no-one other than Mrs Jones' can be translated as

$\forall y(\neg y = m \supset \neg Lxy)$

or equivalently,

$$\forall y(Lxy \supset y = m)$$

And plugging that in the half-completed translation, we finally get:

$$\forall x(Lxm \supset \forall y(Lxy \supset y = m))$$

As always, we could alternatively translate the 'no-one' quantification using the existential quantifier:

$$\forall x(Lxm \supset \neg \exists y(Lxy \land \neg y = m))$$

11. The antecedent 'Bryn loves some Welsh-speaker other than Angharad' is translated by

$$\exists x((Lbx \land Fx) \land \neg x = a)$$

So the whole conditional translates as follows:

$$(\exists x((Lbx \land Fx) \land \neg x = a) \supset Lbm)$$

12. Finally, we'll again take 'Every girl other than Angharad loves someone other than Bryn' in stages. As a half-way house we have this:

$$\forall x((Gx \land \neg x = a) \supset x \text{ loves someone other than Bryn})$$

Then 'x loves someone other than Bryn' translates as

$$\exists y(Lxy \land \neg y = b)$$

So putting those together we get

$$\forall x((Gx \land \neg x = a) \supset \exists y(Lxy \land \neg y = b))$$

33.3 Numerical quantifiers

We have just seen how to translate 'there is at most one F':

(1) $\forall x \forall y((Fx \land Fy) \supset x = y)$

We know from before how to translate 'there is at least one F':

(2) $\exists x Fx$

Putting these together, then, we can translate 'there is exactly one F' (i.e. there is at least one and at most one F) by the conjunction:

(3) $(\exists x Fx \land \forall x \forall y((Fx \land Fy) \supset x = y))$

Here's an alternative, simpler, translation that will do as well:

(4) $\exists x(Fx \land \forall y(Fy \supset y = x))$

(Convince yourself that those two indeed come to the same!) And here's an even simpler version:

(5) $\exists x \forall y(Fy \equiv y = x)$

We can formally show that (4) and (5) are q-equivalent, i.e. q-entail each other (see §35.1). But this result too should look intuitively right. For suppose x is the one and only F. Then take anything y you like, if it is identical to x it is F (which gives you one direction of the biconditional); and if y is F it must be none other

than x again (which gives you the other direction of the biconditional). Incidentally, you will find that (4) – or equivalently (5) – is quite often abbreviated as

(6)   ∃!xFx

But we won't adopt that notation here.

Now consider the following wffs,

(7)   ∃x∃y((Fx ∧ Fy) ∧ ¬x = y)
(8)   ∀x∀y∀z(((Fx ∧ Fy) ∧ Fz) ⊃ ((x = y ∨ y = z) ∨ z = x))

The first says that there is an F and also another, distinct F. In other words, (7)
says that there are at least two *F*s.

Note – by the way of a quick aside – that we can, if we want, similarly translate a proposition like 'Some philosophers are logicians' into **QL=** and retain the
indication that there is more than one logical philosopher, by writing e.g.

(7*)   ∃x∃y(((Gx ∧ Hx) ∧ (Gy ∧ Hy)) ∧ ¬x = y)

(8), to continue, says that if we try to pick three things which are all F, then at
least two of the *F*s will be identical – i.e., there are at most two *F*s. The conjunction of (7) with (8) will therefore say that there are at least two *F*s and at most
two *F*s, or in other words that there are *exactly* two *F*s. Or we could equally
well, and more neatly, translate that by

(9)   ∃x∃y({(Fx ∧ Fy) ∧ ¬x = y} ∧ ∀z{Fz ⊃ (z = x ∨ z = y)})

which says that there is an F and another distinct F and any F you choose has to
be one of those.

Similarly

(10)   ∃x∃y∃z({((Fx ∧ Fy) ∧ Fz) ∧ ((¬x = y ∧ ¬y = z) ∧ ¬z = x)}
        ∧ ∀w{Fw ⊃ ((w = x ∨ w = y) ∨ w = z)})

says that you can find three *F*s, all different, and any F you choose is one of them
– in other words there are exactly three *F*s.

We can obviously keep going in a similar vein, expressing numerical quantifications ('there are at least n *F*s', 'there are at most n *F*s', 'there are exactly n *F*s')
using just the familiar quantifiers, the connectives, and identity.

But the link between logic and elementary arithmetic is closer than a mere
point about **QL=**'s capacity to express numerical quantifiers. For suppose we
abbreviate the forms (4), (9) and (10) respectively by '∃₁vFv', '∃₂vFv', '∃₃vFv'.
Then consider the long wff we can then abbreviate as

(11)   ({(∃₂vFv ∧ ∃₁vGv) ∧ ¬∃x(Fx ∧ Gx)} ⊃ ∃₃v(Fv ∨ Gv))

What does this say? That if there are two F and one G (and nothing is F and G),
then we have three things that are *F-or-G*. For example, if there are two things in
my left hand, and one thing in my right, then there are three things in my hands
altogether. Or, as we learnt to say in nursery school, two and one makes three.

So here we find that a bit of simple applied arithmetic is also expressible in
QL=. But more than that, not only is this truth expressible in logical terms, but it
is in fact a q-logical truth: it can be shown, with a bit of effort, to be true on all

q-valuations. Which raises an intriguing, deep and difficult question: just how much arithmetic *is*, so to speak, just logic in disguise?

33.4 Summary

- We add to QL the identity sign – governed by the syntactic rule that a term followed by '=' followed by a term is an atomic wff – to get QL$^=$.

- The identity sign is governed by the semantic rule that an atomic wff of the form '$t_1 = t_2$' is true on the valuation q iff the terms t_1 and t_2 have the same reference according to q.

- QL$^=$ has the resources to express more complex quantifiers like 'Everyone other than Angharad ...', 'Exactly three people ...', 'No one else ...' etc.

Exercises 33

A   More Welsh affairs! Using the same translation manual as §33.2, translate the following into QL$^=$:

  1.   Angharad and Mrs Jones are one and the same.
  2.   At least one Welsh speaker loves Bryn.
  3.   Either Angharad or Bryn is Mrs Jones.
  4.   Someone other than Bryn loves Angharad.
  5.   Mrs Jones loves everyone but Angharad.
  6.   Some girls only love Bryn.
  7.   Some girls only love people who love them.
  8.   Only if she loves Bryn is Angharad loved by him.
  9.   Exactly one girl who loves Bryn speaks Welsh.
  10. If Angharad isn't Mrs Jones, then at least two people love Bryn.

B   Take the domain of quantification to be the (positive whole) numbers, and let 'm' denote the number zero, 'n' denote the number one, let 'Mxy' mean that *x immediately follows y in the number series* (equivalently, *x is an immediate successor of y*), and 'Rxyz' mean that *x equals y plus z*. Then translate the following from QL$^=$ into natural English:

  1. $\forall x \forall y \forall z((Mxz \land Myz) \supset x = y)$
  2. $\forall y \exists x Mxy$
  3. $\neg \exists z Mmz$
  4. $\forall x \forall y \forall z((Mxy \land y = z) \supset Mxz)$
  5. $\forall x \forall y(Rxym \supset x = y)$
  6. $\forall x \forall y(Rxyn \supset Mxy)$
  7. $\forall x \forall y \forall z(Rxyz \equiv Rxzy)$
  8. $\exists x(Rxxx \land \forall y(Ryyy \supset y = x))$
  9. $\forall x \forall y \exists z(Rzxy \land \forall v_0(Rv_0xy \supset v_0 = z))$
  10. $\forall z \exists x((Mxz \land \forall y(Myz \supset y = x)) \land Rznx)$
  11. $\exists x \exists y(\{[((Mxm \lor Mxn) \land (Mym \lor Myn))] \land \neg x = y\}$
       $\land \forall z\{(Mzm \lor Mzn) \supset (z = x \lor z = y)\})$

Descriptions and existence

In this chapter, we extend our discussion of the expressive resources of **QL**= by considering how 'definite descriptions' – i.e. expressions of the kind *the F* – can be translated using Russell's Theory of Descriptions. We will also see how various statements about existence can be translated.

34.1 Definite descriptions

Consider the expressions 'the man in the corner drinking a martini', 'the oldest woman in the world', 'the Queen of England', 'the smallest prime number', 'the present King of France', 'the largest prime number'. Such expressions – more generally, expressions of the type *the (one and only) F* – are standardly called *definite descriptions*. A definite description is a designator, or more accurately it *aims* to designate a particular thing. As the last two cases remind us, definite descriptions may in fact fail to pick out any entity: there may be nothing which is *F*, and so nothing which is *the F*.

Imagine we are at a party. I tell you that a number of professors from different disciplines are here, and say 'The philosopher is over there, talking to George'. Then it certainly seems that you can infer that there is (at least) one philosopher present.

Generalizing, if a claim of the form 'The *F* is *G*' is true, then so is the corresponding claim of the form 'There is at least one *F*' (ignoring for the moment some rogue cases, where *G* is filled by e.g. 'non-existent' or 'fictitious').

Suppose we want to reflect this sort of logical entailment inside **QL** or **QL**=. How can we do that?

Take '*F*' to mean ... *is a philosopher* and '*G*' to mean ... *is talking to George*; and let the domain be, say, people present at the party. And imagine first that we render the designator 'the philosopher' by a simple constant 'n'. Then the valid inference 'The philosopher present is talking to George; so there is at least one philosopher present' would get translated by something like 'Gn ∴ ∃xFx', which is of course *not* q-valid. The translation of 'the philosopher' by the simple constant 'n' misses logically relevant structure: it ignores the key fact that the definite description embeds the predicate 'philosopher'.

Moral: if we want to capture the logical behaviour of definite descriptions, we need some way of translating them into our formalized language that similarly embeds the relevant predicate.

How shall we proceed? Well, remember that we can express the proposition that there is one and only one philosopher at the party as follows (compare §33.3, (4)):

    (1) $\exists x(Fx \land \forall y(Fy \supset y = x))$

Suppose we want to add that this one and only philosopher is talking to George. Then we can just add another conjunct inside the scope of the quantifier:

    (2) $\exists x((Fx \land \forall y(Fy \supset y = x)) \land Gx)$

But now note, the claim (2) – there is one and only one philosopher (here at the party), and he is talking to George – seems to come to more or less the same as this: *the philosopher (here at the party) is talking to George*. For if there is one and only one philosopher present, there is indeed someone to be *the* philosopher; and (2) says that that person is talking to George. Note too that (2) pretty trivially q-entails

    (3) $\exists x Fx$

So the rendition (2) *does* enable us to reflect in $QL^=$ the valid inference from 'The philosopher (here at the party) is talking to George' to 'There is a philosopher (here at the party)'.

Instead of (1), we could equally well use

    (1′) $\exists x \forall y(Fy \equiv y = x)$

to render the proposition that there is one and only one philosopher (compare §33.3, (5)). And hence, instead of (2), we could equally well use the snappier

    (2′) $\exists x \forall y((Fy \equiv y = x) \land Gx)$ *or* $\exists x(\forall y(Fy \equiv y = x) \land Gx)$

to render the proposition that the philosopher is talking to George.

And a third equivalent way of saying the same is the more long-winded

    (2″) $(\{\exists x Fx \land \forall x \forall y((Fx \land Fy) \supset x = y)\} \land \forall x(Fx \supset Gx))$

i.e. 'There is at least one philosopher *and* there is at most one philosopher *and* whoever is a philosopher is talking to George'. In the next chapter we will develop the resources to show that (2), (2′) and (2″) are indeed q-equivalent, i.e. they q-entail each other: but for the moment let's take this intuitively correct claim on trust.

Generalizing – and fixing on translations like (2) or (2′) rather than (2″) for reasons of brevity – these remarks suggest the following thesis:

> We can render a sentence of the kind 'The F is G' into $QL^=$ by corresponding wffs of the form
>
>   (R) $\exists v((Fv \land \forall w(Fw \supset w = v)) \land Gv)$
>   (R′) $\exists v \forall w((Fw \equiv w = v) \land Gv)$

(In stating the thesis this way, the metalinguistic '*F*' and '*G*' do double duty in a now familiar way, first holding the place for English expressions, and then holding the place for the respective translations of these expressions into QL$^=$: but the intended idea should be clear enough.)

The claim that *The F is G* can be rendered into QL$^=$ by (R) or (R') is due to Bertrand Russell. It is *part* of his famed *Theory of Descriptions*. But this Theory says quite a lot more. For Russell also claims:

- Translating *The F is G* by (R) is not merely the best available option, given the limited resources of QL$^=$, but is *fully accurate*, in the sense of getting it exactly right about the conditions under which *The F is G* is true and the conditions under which it is false.

- Moreover, the QL$^=$ translation reflects the true underlying *logical form* of sentences involving definite descriptions. Descriptions really are, in some sense, a kind of quantifier in disguise.

Let's briefly take these additional Russellian claims in turn.

Plausibly, as we said, if (R) is true, then *The F is G* is true. But do the two match on falsehood?

Suppose that there is no *F*. Then anything of the form '$\exists v(Fv \wedge \ldots)$' will be false. Hence the QL$^=$ rendition of *The F is G* comes out plain false when there is no *F*. Would we want to say the same, however, about *The F is G*? Russell himself uses the example 'The present King of France is bald'. Is *this* straightforwardly false? Maybe this sort of claim is better described as not even making the starting line for consideration for being true or false, since the definite description fails to pick out a referent. In other words, we might want to say that the claim is *neither true nor false*.

Compare 'Mr Brixintingle is bald', where the apparent proper name 'Mr Brixintingle' is just a fake, invented to look vaguely name-like, but in fact attached to no-one. *That* sentence fails to make any determinate claim, and so plausibly lacks a truth-value. In other words, non-referring names arguably create *truth-value gaps*. The anti-Russellian suggestion is that non-referring descriptions do so too. And if that is right, 'The present King of France is bald' – neither true nor false – is *not* equivalent to 'There is one and only one King of France, and he is bald', or to a rendition like (2) – which *is* straightforwardly false.

Now, it is agreed on all sides that a claim of the form *The F is G* can fail to be true in more than one way. It can fail (a) because there isn't an *F*; it can fail (b) because there are too many *F*s (so we shouldn't talk of *the F*); it can fail (c) when there is a unique *F* but it isn't *G*. The Russellian will claim that his rendition (R) captures this very nicely: *The F is G* can be *false* in three different ways, depending on which of the three conjuncts in the analysis (R) isn't satisfied. The anti-Russellian disagrees, and will instead say that only in case (c) is the claim actually false; in cases (a) and (b) *The F is G* is neither true nor false.

But what does this disagreement really come to? How do we decide such issues?

It is *very* unclear what to say, and the principles that should guide debate here are intensely disputed. We certainly can't pursue these matters any further in this book. We just have to note that there is a minefield hereabouts!

There is another minefield around claims to do with 'logical form'. Suppose that (R) *does* get it right about when *The F is G* is true and when it is false. It remains quite unclear what cash-value we can give to the thesis that (R) somehow reveals the true semantic structure of vernacular claims involving definite descriptions. And it is again quite obscure how we are to evaluate such claims. In Chapter 36, we'll say just a little more that is germane to these vexed issues, but without any hope of resolving them.

So for the moment let's set aside both of Russell's more ambitious claims, and stick to basics. This much, at any rate, is uncontroversial: *given the limited resources of* QL$^=$, *wffs of the Russellian form (R), (R'), or their equivalents, are the best we can do in rendering sentences involving definite descriptions.* We'll therefore use the schemas (R) or (R') henceforth, whenever we need to translate descriptions.

34.2 Descriptions and scope

One arguable resemblance between definite descriptions and quantifiers (not in itself enough to warrant treating the first as a disguised form of the second, but a suggestive resemblance none the less) is that descriptions exhibit what plausibly can be construed as *scope* phenomena (compare §21.3). For example, just as ordinary-language quantifiers mixed with negation can produce scope ambiguity, so ordinary-language definite descriptions mixed with negation can produce ambiguity.

Suppose we are discussing famous baldies. You challenge me to name ten. I reel off a list beginning

> Yul Brynner is bald,
> Bruce Willis is bald,
> the actor who plays 'Captain Picard' in *Star Trek* is bald,
> the King of France is bald,

and you interrupt, saying

(A)   No! That last one's not right. The King of France isn't bald – he doesn't even exist!

Your protest is surely correct, and the way you have phrased it seems natural enough. But now compare a second occasion. This time we are discussing people who *aren't* bald. So this time I reel off a list of famous non-baldies:

> The Queen of England isn't bald,
> George W. Bush isn't bald,
> Bill Clinton isn't bald,

(B)   the King of France isn't bald,

and again you interrupt, this time saying

No! That's not right. You can't say that – he doesn't even exist!

Again, you are surely right. But how can it be that my assertion (B) is flawed, when you, using exactly the same words in (A), were saying something correct?

We will have to discern an ambiguity. 'The King of France isn't bald' can be used to express two different claims, one true, the other false. And Russell's Theory of Descriptions gives us a very neat diagnosis of the apparent ambiguity here; it treats it as a straightforward scope ambiguity. On one reading, the negation has wide scope, governing the whole sentence, and the resulting message can be rendered

(1) $\neg \exists x((Fx \wedge \forall y(Fy \supset y = x)) \wedge Gx)$

This captures the truth that you asserted in (A). On the other reading, the negation has narrow scope, governing just the predicate 'bald', and the resulting message can be rendered

(2) $\exists x((Fx \wedge \forall y(Fy \supset y = x)) \wedge \neg Gx)$

And this is the falsehood which I apparently asserted in (B).

Here's another case of scope ambiguity, a modal case worth mentioning even though it takes us beyond the resources of pure $\mathbf{QL^{=}}$. Consider:

(3)   The President might have been a woman.

This again has two readings. Are we saying there's a possible scenario in which a different political history leads to a President in that scenario who is a woman? Or are we saying of the actual President, that he (that very person) might have been a woman – contemplating some kind of sex change, perhaps? If we imagine adding a modal operator '◇' to be read as 'It might have been the case that', then the Russellian will represent this ambiguity as between giving '◇' wide or narrow scope, along the following lines:

(4) $\Diamond \exists x((Fx \wedge \forall y(Fy \supset y = x)) \wedge Gx)$

or

(5) $\exists x((Fx \wedge \forall y(Fy \supset y = x)) \wedge \Diamond Gx)$

34.3   More translations

A few more quick examples. We'll use the translation manual

   'a' means *Angharad*
   'b' means *Bryn*
   'm' means *Mrs Jones*
   'F' means ... *speaks Welsh*
   'G' means ... *is a girl*
   'L' means ... *loves* ...
   'M' means ... *is taller than* ...

and where the domain of quantification again consists of human beings, and translate the following into $\mathbf{QL^{=}}$:

1. The Welsh speaker who loves Mrs Jones is taller than her.
2. Bryn loves the girl who loves him.
3. Bryn only loves the girl who loves him.
4. The tallest girl speaks Welsh.
5. The girl who loves Bryn loves the Welsh speaker who loves Angharad.

1. The first involves the definite description 'The Welsh speaker who loves Mrs Jones', and 'x is a Welsh speaker who loves Mrs Jones' is translated by '$\{Fx \wedge Lxm\}$'. So following the Russellian schema (R), the complete translation will run

$$\exists x(((\{Fx \wedge Lxm\} \wedge \forall y(\{Fy \wedge Lym\} \supset y = x)) \wedge Mxm)$$

If we'd used schema (R'), we'd get the brisker

$$\exists x \forall y((\{Fy \wedge Lym\} \equiv \supset y = x) \wedge Mxm)$$

2. This is similar: the definite description this time is 'The girl who loves Bryn', and 'x is a girl who loves Bryn' is translated by '$\{Gx \wedge Lxb\}$'. Using (R) again we arrive at

$$\exists x(((\{Gx \wedge Lxb\} \wedge \forall y(\{Gy \wedge Lyb\} \supset y = x)) \wedge Lbx)$$

We'll leave the (R')-style version to be done as an exercise.

3. To translate this, we need to add that for anyone other than this girl, x, Bryn does not love x; or equivalently, anyone he *does* love is none other than x again. Whence

$$\exists x[(((\{Gx \wedge Lxb\} \wedge \forall y(\{Gy \wedge Lyb\} \supset y = x)) \wedge Lbx) \wedge \forall z(Lbz \supset z = x)]$$

4. The tallest girl is *the* girl such that she is taller than all other girls. Now, 'y is a girl who is taller than all other girls' can be rendered '$\{Gy \wedge \forall z((Gz \wedge \neg z = y) \supset Myz)\}$'. Plugging *that* into the simpler (R') schema, we get

$$\exists x \forall y((\{Gy \wedge \forall z((Gz \wedge \neg z = y) \supset Myz)\} \equiv y = x) \wedge Fx)$$

This time, we'll leave the longer (R)-style version as an exercise.

5. Our last example involves *two* descriptions. Abbreviate 'loves the Welsh speaker who loves Angharad' by 'C'. And to reduce clutter, let's again use schema (R') straight off. Then we have, as half-way translations,

The girl who loves Bryn is C
$$\mapsto \exists x \forall y((\{Gy \wedge Lyb\} \equiv y = x) \wedge Cx)$$

And (using different variables, so we don't get into a tangle),

$$Cx \mapsto \exists z \forall w((\{Fw \wedge Lwa\} \equiv w = z) \wedge Lxz)$$

So, if we plug the one into the other, we get

$$\exists x \forall y((\{Gy \wedge Lyb\} \equiv y = x) \wedge \exists z \forall w((\{Fw \wedge Lwa\} \equiv w = z) \wedge Lxz))$$

The last translation does the trick! However, the messiness of this $QL^=$ rendition compared to the relative simplicity of the original English might very arguably begin to cast some doubt on the Russellian thesis that the Theory of Descriptions captures the true underlying 'logical form' of the vernacular claim.

34.4　Existence statements

We'll consider three types of existential statement:

- examples like 'Tame tigers exists' and 'Unicorns don't exist', i.e. statements of the kind '*F*s do/don't exist', with *F* filled by some more or less complex specification of a kind of thing;

- examples like 'The Queen of England exists' and 'The King of France doesn't exist', i.e. statements of the kind 'The *F* does/doesn't exist', with 'The *F*' a definite description;

- examples like 'Gwyneth Paltrow exists' (she's a living breathing actress) and 'Lois Lane doesn't exist' (not really, she's just a fictional character).

How do we render such claims into $QL^=$?

The first case is easy. 'Tame tigers exist' is equivalent to 'There are some tame tigers', so can be rendered as

(1)　　$\exists x(Fx \wedge Gx)$

with the obvious interpretation of the predicates. Or if you want explicitly to capture the thought that there is more than one, as signalled by the English plural, then you can write:

(2)　　$\exists x \exists y(\{(Fx \wedge Gx) \wedge (Fy \wedge Gy)\} \wedge \neg x = y)$

Likewise, 'Unicorns don't exist', which is equivalent to 'It is not the case that there are some unicorns', can be rendered by

(3)　　$\neg \exists x Hx$

What about 'The Queen of England exists' and 'The present King of France doesn't exist'?

The simplest suggestion is this. To say that the Queen of England exists is just to say that there is one and only one (present) Queen of England. Rendering that into $QL^=$ (using 'F' for ... *is a Queen of England*), we'll get

(4)　　$\exists x(Fx \wedge \forall y(Fy \supset y = x))$

And if we negate the whole thing –

(5)　　$\neg \exists x(Fx \wedge \forall y(Fy \supset y = x))$

– we get something which, changing the translation manual, will render the claim that the present King of France does not exist.

Unsurprisingly, then, existence claims involving predicative expressions and definite descriptions get rendered into $QL^=$ using the existential quantifier. But what about the third type of existence claim, the type involving names?

Take that interpretation of $QL^=$ which reads 'm' as a name of Gwyneth Paltrow. Can't we simply express the thought that Gwyneth exists, i.e. that there is indeed someone who is Gwyneth, as follows? –

(6)   ∃x x = m

At first sight, that looks satisfactory; but on reflection, there's a problem. For note, *(6) is a q-logical truth!*

Why so? Well, consider any q-valuation q of the vocabulary in '∃x x = m'. By definition, q has to assign an object m to be the reference of 'm'. And so q will always have an extension q^+ which also assigns m to the variable 'x'; hence q will always have an extension q^+ which makes 'x = m' true. So, by definition, '∃x x = m' is true on any q-valuation of its vocabulary. But if '∃x x = m' is a q-logical truth, then (we might surmise) it can't really be an ideal rendition of 'Gwyneth Paltrow exists' (which of course *isn't* a logical truth).

To press the point further, consider the claim 'Lois Lane doesn't exist'. That's another truth. But we can't render it into $QL^=$ by writing something like

(7)   ¬∃x x = n

since (7) is *false* on all q-valuations of its vocabulary.

Of course, these results about the status of (6) and (7) directly reflect the stipulation that q-valuations assign a reference to any constant that is in play. By contrast, ordinary language has 'empty names', names which fail to pick out an object, yet which have a role in the language and don't produce mere nonsense. In the Fregean terminology of §31.1, it seems that ordinary names can have sense even if they lack reference. That is why it is contentful to be told that Lois Lane or Santa Claus don't exist, but that there *is* such a person as Gwyneth.

There are two reactions we can have to all this.

- One line is to proceed to extend QL and $QL^=$ to allow meaningful but non-denoting names. This yields a so-called *free logic*, i.e. a logic free of certain existence assumptions.

- The other line is to cling to the simplicity of $QL/QL^=$ (on the grounds that there is, after all, a sense in which non-denoting names are linguistically defective when used in serious, fact-stating discourse, and we are in the business of constructing a logically ideal language).

Now, adopting a free logic does involve significant revisions and complications; so, rightly or wrongly, the second reaction is the standard one. But if we do stick to the standard semantic story about $QL/QL^=$, then we must be aware of the implications of the built-in assumption that every constant we actually use has an object as its q-value, i.e. has a reference.

To summarize: in our languages QL and $QL^=$, we require any constant we use to have a reference in the domain of discourse. The logical truths of those languages are then truths that are guaranteed *once that assumption of reference is in place*. In particular, a wff like '∃x x = n' always comes out true on a q-valuation. In this respect, there is arguably significant slippage between the semantic nature of some $QL^=$ wffs and the apparently corresponding ordinary language sentences. And because of that, there will be some q-valid (and hence valid) $QL^=$ arguments whose most tempting English renditions are not valid. (Compare the argument of §28.4.)

A quick final remark. Apprentice philosophers often come across the slogan 'Existence isn't a predicate'. If this is read as saying that a language can't have a predicative expression (i.e. an expression with one slot waiting to be filled up by a designator) which applies to all the things assumed to exist in its domain of quantification, then the slogan misleads. In QL$^=$, '$\exists x \, x = \ldots$' would do the job perfectly: it is true of everything in the domain. The same expression works just as well in some free logics too. But if the philosophical slogan is intended to bring out the close relation between vernacular existence talk and the use of something fundamentally different from an ordinary predicate, namely the operation of existential quantification, then the slogan is apt.

34.5   Summary

- Sentences of the form *The F is G* are best rendered into QL$^=$ by a corresponding wff of the form '$\exists v((Fv \land \forall w(Fw \supset w = v)) \land Gv)$', or equivalently by '$\exists v \forall w((Fw \equiv w = v) \land Gv)$'.

- This uncontentious claim, part of Russell's Theory of Descriptions, should be distinguished from Russell's much more contentious claims that this rendition gets the truth-conditions of *The F is G* exactly right, and moreover in some sense reveals the true logical form of vernacular statements involving descriptions.

- However, Russell's analysis plausibly diagnoses certain apparent ambiguities involving descriptions as being scope ambiguities.

- Existential claims are naturally rendered using the existential quantifier (though there are problems about cases involving proper names, given the assumption built into the semantics of QL and QL$^=$ that names always have referents).

Exercises 34

Yet more Welsh affairs! Using the same translation manual as §34.3, translate the following into QL$^=$:

1. The Welsh speaker loves Mrs Jones.
2. The girl who loves Angharad does not loves Bryn.
3. Angharad loves the girl who loves Bryn.
4. The Welsh speaker who loves Mrs Jones is either Angharad or Bryn.
5. Someone other than the girl who loves Bryn is taller than Angharad.
6. The one who loves Angharad is the one she loves.
7. Only if she loves him does Bryn love the girl who speaks Welsh.
8. The girl other than the girl who loves Bryn is Angharad.
9. The shortest Welsh speaker loves Bryn.
10. The shortest Welsh speaker loves the tallest Welsh speaker.

35

Trees for identity

We now expand our treatment of quantifier trees to cover arguments involving identity, expressed in $\mathbf{QL}^=$. We first introduce and motivate two rules dealing with wffs involving identity. We explore some introductory examples and also show that these rules can be used to demonstrate that the basic laws of identity are q-logical truths. Then we consider some examples involving definite descriptions and numerical quantifiers.

35.1 Leibniz's Law again

Take the argument: 'Bertie is clever. Bertie is none other than Russell. So Russell is clever.' That's valid. Translating, we have

> **A**   Fm, m = n ∴ Fn

Starting a signed tree in the standard way, we suppose that there is a valuation q such that

(1) $Fm \Rightarrow_q T$
(2) $m = n \Rightarrow_q T$
(3) $\neg Fn \Rightarrow_q T$

How are we to proceed now? Consider again Leibniz's Law, in the form we met in §32.5:

> (LL′) If 'a' and 'b' are functioning as co-referential designators in the claims $C(...a...a...)$ and $C(...b...b...)$, and also the context C expresses the same property in each case, then these claims say the same about the same thing, and so must have the same truth-value.

We noted that, in English, different occurrences of the same predicate need not attribute the same property (remember the behaviour of '... is so-called because of his size'). But in our tidy artificial language $\mathbf{QL}^=$ we just don't have that troublesome kind of predicate. And our official semantics indeed implies

> (A)   If 'm' and 'n' are co-referential designators according to q, then the claims 'Fm' and 'Fn' must have the same truth-value on q.

Further, it is immediate that

(B)   The atomic wff 'm = n' is true on the valuation q just if 'm' and 'n' are assigned the same q-values by q.

(That follows from (Q0$^+$) in §33.1.) Putting (A) and (B) together, they imply

(C)   If the claims 'Fm' and 'm = n' are true on the valuation q, then 'Fn' must also be true on q.

Hence, given (1) and (2), we can appeal to (C) and continue our tree-argument

(4) $Fn \Rightarrow_q T$

which immediately contradicts (3), and our tree closes.

We can evidently generalize the (A)–(B)–(C) reasoning; and this warrants the following corresponding rule for building unsigned trees:

(L)   If W_1 is of the form $C(...m...m...)$, involving one or more occurrences of the constant m, and W_2 is of either the form $m = n$ or the form $n = m$, then we can add $C(...n...n...)$, formed by substituting some or all of the occurrences of m in W_1 by occurrences of n, to the foot of any open branch containing both W_1 and W_2. Schematically,

$$C(m)$$
$$m = n \text{ or } n = m$$
$$|$$
$$C(n)$$

Two comments on this:

- (L) differs from all our previous rules for tree-building in taking *two* wffs as input. The inputs can appear, of course, in either order earlier on the branch. Since the conclusion added to the branch does not exhaust the combined contents of the input wffs, we *don't* check them off.

- Note that previous rules involving substitution – the universal quantifier instantiation rule (∀), the existential instantiation rule (∃) – tell us to uniformly replace *every* occurrence of one symbol (a variable) with another symbol (a constant). By contrast, our rule (L) can be more permissive. It allows us to replace one or more occurrences of one constant with another. So take, for example, the wff

  ∀x(Raxa ⊃ Fa)

  We can focus on the middle occurrence of 'a', and take all ∀x(Rax_⊃ Fa)' to be 'context'. Then, given 'a = b' and applying rule (L), we can infer

  ∀x(Raxb ⊃ Fa)

  Alternatively, we can focus on the first and last 'a's, take the context to be '∀x(R_xa ⊃ F_)', and use (L) to infer

  ∀x(Rbxa ⊃ Fb).

  Or, of course, we can infer the result of simultaneously replacing all three occurrences of 'a'. Each inference is plain valid.

So now let's take some simple examples of our rule (L) in operation.
    First, consider the argument

    B   Either Alice or Mrs Brown is the winner. Alice isn't the winner. So
             Alice isn't Mrs Brown.

It is trivial to render this into **QL$^=$**, and the resulting tree is equally straightforward:

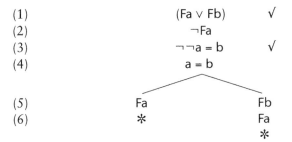

(1)	(Fa ∨ Fb)	√
(2)	¬Fa	
(3)	¬¬a = b	√
(4)	a = b	

(5)	Fa	Fb
(6)	*	Fa
		*

Here, the only step worth remarking on is (6), where we have used (L) to derive
'Fa' from (5) 'Fb' and (4) 'a = b'. The tree closes, hence the intuitively valid argument (B) is, as we'd expect, q-valid.
    For another example, consider the following:

    C   Only Russell is a great philosopher. Bertrand is a great philoso-
             pher who smokes. Hence Russell smokes.

Put 'F' for *great philosopher*, 'G' for *smokes*, etc., and we can render this:

$$(Fm \land \forall x(Fx \supset x = m)), (Fn \land Gn) \therefore Gm$$

A tree can then be constructed as follows:

(1)	(Fm ∧ ∀x(Fx ⊃ x = m))	√
(2)	(Fn ∧ Gn)	√
(3)	¬Gm	
(4)	Fm	
(5)	∀x(Fx ⊃ x = m)	
(6)	Fn	
(7)	Gn	
(8)	(Fn ⊃ n = m)	√

(9)	¬Fn	n = m	
(10)	*	Gm	(by (L), from 7, 9)
		*	

(Check that you follow each step.) Again we have a valid argument.
    What about the simple inference

    D   George is Mr Orwell; Mr Orwell is Eric Blair; hence George is
             Eric Blair?

With the obvious translation scheme, the tree for this inference is

(1)	m = n	
(2)	n = o	
(3)	¬m = o	
(4)	m = o	(by (L), from 1, 2)

*

To see that the move to (4) is warranted, think of 'm = n' as of the form C(n), with the context C being 'm = _'; and then from C(n) and 'n = o' we infer the corresponding C(o), i.e. 'm = o'.

Generalizing that last example, recall the transitivity law

(Tran)   ∀x∀y∀z((x = y ∧ y = z) ⊃ x = z)

This is, as we'd expect, a q-logical truth. But how can we use a tree to show this? We start the tree with the negation of (Tran) and aim for absurdity.

(1)	¬∀x∀y∀z((x = y ∧ y = z) ⊃ x = z)	√
(2)	∃x¬∀y∀z((x = y ∧ y = z) ⊃ x = z)	√
(3)	¬∀y∀z((m = y ∧ y = z) ⊃ m = z)	√
(4)	∃y¬∀z((m = y ∧ y = z) ⊃ m = z)	√
(5)	¬∀z((m = n ∧ n = z) ⊃ m = z)	√
(6)	∃z¬((m = n ∧ n = z) ⊃ m = z)	√
(7)	¬((m = n ∧ n = o) ⊃ m = o)	√
(8)	(m = n ∧ n = o)	√
(9)	¬m = o	
(10)	m = n	
(11)	n = o	
(12)	m = o	(by (L), from 10, 11)

*

The moves up to line (11) are all automatic. We drive in (1)'s initial negation sign to get (2), where we can do nothing else but instantiate the existential quantifier to get (3). That pattern repeats twice again until we get to (8), when we apply the familiar connective rules to get us to (9), (10) and (11). The final step is exactly as in the previous tree.

Next, consider the following inference:

E   Only Angharad and Bryn love Mrs Jones; someone who loves Mrs Jones speaks Welsh; so either Angharad or Bryn speaks Welsh.

For simplicity, let's use the simple 'Gx' to render 'x loves Mrs Jones' (for the internal structure of '... loves Mrs Jones' isn't actually used in the argument, so we needn't explicitly represent it). Then the argument goes into **QL=** as

((Ga ∧ Gb) ∧ ∀x(Gx ⊃ (x = a ∨ x = b))), ∃x(Gx ∧ Fx) ∴ (Fa ∨ Fb)

The corresponding closed tree starts more or less automatically

(1)	((Ga ∧ Gb) ∧ ∀x(Gx ⊃ (x = a ∨ x = b)))	√
(2)	∃x(Gx ∧ Fx)	
(3)	¬(Fa ∨ Fb)	√
(4)	(Ga ∧ Gb)	

(5) $\forall x(Gx \supset (x = a \lor x = b))$
(6) $\neg Fa$
(7) $\neg Fb$

Next, following our rule of thumb 'instantiate-existentials-before-universals', we check off (2) and instantiate with a new name to continue as follows:

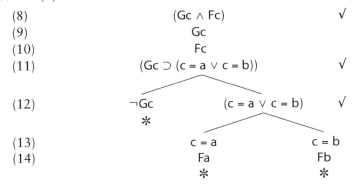

(8) $(Gc \land Fc)$ ✓
(9) Gc
(10) Fc
(11) $(Gc \supset (c = a \lor c = b))$ ✓

(12) $\neg Gc$ $(c = a \lor c = b)$ ✓
          ✳

(13) $c = a$ $c = b$
(14) Fa Fb
         ✳   ✳

Here, on the left branch, 'Fa' is derived from (10) and (13); similarly for 'Fb' on the right branch.

Finally in this section, let's partially fulfil the promise we made in §33.3 to show that the following wffs are equivalent:

F $\exists x(Fx \land \forall y(Fy \supset y = x))$
G $\exists x \forall y(Fy \equiv y = x)$

We will give a tree to show that **F** q-entails **G**:

(1) $\exists x(Fx \land \forall y(Fy \supset y = x))$ ✓
(2) $\neg \exists x \forall y(Fy \equiv y = x)$ ✓
(3) $\forall x \neg \forall y(Fy \equiv y = x)$
(4) $(Fa \land \forall y(Fy \supset y = a))$ ✓
(5) Fa
(6) $\forall y(Fy \supset y = a)$

So far, so automatic (we dealt with the negated quantifier at (2), and then instantiated the only existential quantifier). We now instantiate (3) with the name 'a':

(7) $\neg \forall y(Fy \equiv y = a)$ ✓
(8) $\exists y \neg(Fy \equiv y = a)$ ✓
(9) $\neg(Fb \equiv b = a)$ ✓
(10) $(Fb \supset b = a)$ ✓   (from 6)

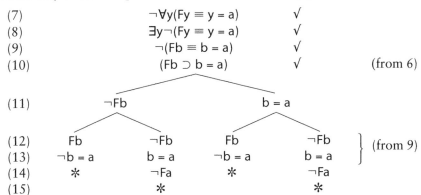

(11) $\neg Fb$ $b = a$

(12) Fb $\neg Fb$ Fb $\neg Fb$ } (from 9)
(13) $\neg b = a$ $b = a$ $\neg b = a$ $b = a$
(14)   ✳ $\neg Fa$ ✳ $\neg Fa$
(15)   ✳   ✳

We use (L) at line (14) to derive '¬Fa' from the previous two wffs on the path: the other paths close off without appealing to (L). We will leave it as an exercise to show that, conversely, G q-entails F.

35.2   Self-identity

Consider next the law of self-identity

(Ref) $\forall x\, x = x$

This is again a q-logical truth. How can we construct a tree to show this? As usual, by assuming it is false on an evaluation and deriving an absurdity. Commencing the tree, we get:

(1)	$\neg \forall x\, x = x$	√
(2)	$\exists x \neg x = x$	√
(3)	$\neg a = a$	

Now, we haven't yet exposed a formal contradiction in the sense of a pair of wffs of the form W and ¬W. Still, getting to something of the form $\neg n = n$ is surely just as bad! For plainly this is itself an outright logical absurdity. Nothing can possibly fail to be identical to itself. *So let's adopt a new rule for closing branches, which treats denials of self-identities on a par with contradictions.*

Henceforth, then, our rules for **QL**$^=$ trees will be the **QL** rules, together with the new rule (L) for extending trees, *and now also our new rule for closing branches* – so we need to replace the structural rule (2) given in §29.1 with this:

> (2$^=$) Inspect any not-yet-closed path down from the top of the tree to see whether it involves either an explicit contradiction (i.e. for some W, it contains both W and also the corresponding formula ¬W), or a wff of the form $\neg n = n$. If it does, then we *close* that path with a star as an absurdity marker, and ignore it henceforth.

With our last new rule in place, we can also prove e.g. that identity obeys the symmetry law

(Sym) $\forall x \forall y (x = y \supset y = x)$.

For consider the following tree:

(1)	$\neg \forall x \forall y (x = y \supset y = x)$	√
(2)	$\exists x \neg \forall y (x = y \supset y = x)$	√
(3)	$\neg \forall y (a = y \supset y = a)$	√
(4)	$\exists y \neg (a = y \supset y = a)$	√
(5)	$\neg (a = b \supset b = a)$	√
(6)	$a = b$	
(7)	$\neg b = a$	
(8)	$\neg b = b$	(by (L) from 6, 7)
	*	

Here, think of '¬b = a' at (7) as being formed from a frame C (i.e. '¬b = _') filled by 'a', and then from C(a) and 'a = b' we get C(b) by Leibniz's Law. We can close the tree as we've arrived at an absurd denial of self-identity.

Two comments on this last proof. First, note that the wffs at lines (6) and (7), i.e. 'a = b' and '¬b = a' are *not* a formal contradiction: for they are not strictly a pair of the form W and ¬W. Second, we *could* have introduced a symmetry rule for identity, allowing us to move from a wff of the form $m = n$ to one of the corresponding wff $n = m$. With that rule in play, we could have completed the last tree by applying the symmetry rule to (6) to derive

(8')   b = a
                                                *

and the tree would have closed off without appeal to our new closure rule. But with our new closure rule in play, an explicit symmetry rule would be redundant.

For another simple example, take the argument

H   Only Bertrand is a great philosopher. Russell is a great philosopher. Hence Bertrand is Russell.

With an obvious rendition of the argument into **QL$^=$**, the tree starts

(1)   (Fm ∧ ∀x(Fx ⊃ x = m))   √
(2)   Fn
(3)   ¬m = n   √
(4)   Fm
(5)   ∀x(Fx ⊃ x = m)
(6)   (Fn ⊃ n = m)   √

(7)   ¬Fn   n = m
(8)   *   ¬m = m   (by (L) from 3, 7)
                                         *

Here we have completed the right-hand branch by dancing the same kind of two-step as before – i.e. we have treated '¬m = n' as C(n), and used 'n = m' to infer the corresponding C(m), and then used our new structural principle for closing trees.

35.3 Descriptions again

In this section, we'll consider some arguments involving the use of Russell's theory of descriptions. Consider firstly

I   The author of the Iliad wrote the Odyssey. Homer wrote the Iliad. Hence, Homer wrote the Odyssey.

Taking 'the author of the Iliad' to mean 'the person who wrote the Iliad', this is a valid argument.

Let's use 'F' to render 'wrote the Iliad', 'G' for 'wrote the Odyssey'. Then the argument can be rendered

∃x((Fx ∧ ∀y(Fy ⊃ y = x)) ∧ Gx), Fm ∴ Gm

And the corresponding tree is

(1)	$\exists x((Fx \wedge \forall y(Fy \supset y = x)) \wedge Gx)$ $\checkmark$
(2)	Fm
(3)	$\neg Gm$
(4)	$((Fa \wedge \forall y(Fy \supset y = a)) \wedge Ga)$ $\checkmark$
(5)	$(Fa \wedge \forall y(Fy \supset y = a))$ $\checkmark$
(6)	Ga
(7)	Fa
(8)	$\forall y(Fy \supset y = a)$
(9)	$(Fm \supset m = a)$ $\checkmark$

(10)	$\neg Fm$	$m = a$
(11)	$*$	$\neg Ga$ (by (L) from 3, 10)
		$*$

Note, by the way, that had we rendered the designator 'the author of the Iliad' by a simple unstructured name 'n' – so the first premiss is translated 'Fn' – we wouldn't have been able to prove validity.

Another example:

J   Homer, and no-one else, wrote the Iliad. Homer wrote the Odyssey. Hence, the author of the Iliad wrote the Odyssey.

Translating, using the same key, we can start the tree

(1)	$(Fm \wedge \forall x(Fx \supset x = m))$
(2)	Gm
(3)	$\neg \exists x((Fx \wedge \forall y(Fy \supset y = x)) \wedge Gx)$

Unpacking (1) and eliminating the negated quantifier in (3), we immediately get

(4)	Fm
(5)	$\forall x(Fx \supset x = m)$
(6)	$\forall x \neg ((Fx \wedge \forall y(Fy \supset y = x)) \wedge Gx)$

Instantiating the universal quantification in (6) with 'm' – what else? – we naturally continue

(7)	$\neg ((Fm \wedge \forall y(Fy \supset y = m)) \wedge Gm))$ $\checkmark$

(8)	$\neg (Fm \wedge \forall y(Fy \supset y = m))$ $\checkmark$	$\neg Gm$
		$*$

(9)	$\neg Fm$	$\neg \forall y(Fy \supset y = m)$
(10)	$*$	

Now, we've almost got an explicit contradiction between the right-hand wff at (9) and (5): but not quite. '$\forall x(Fx \supset x = m)$' and '$\neg \forall y(Fy \supset y = m)$' involve different variables and aren't strictly speaking a pair of the form W and $\neg W$. If we'd had the foresight to render the initial premiss as '$(Fm \wedge \forall y(Fy \supset y = m))$' – which

would have been quite legitimate – then we *would* have had an explicit contradiction by line (9). As things are, we have to continue as follows:

| (10) | $\exists y \neg (Fy \supset y = m)$ | (from 9). |
| (11) | $\neg (Fa \supset a = m)$ | |

Instantiating (5) we have

| (12) | $(Fa \supset a = m)$ |
| | ✳ |

And we are done. Note that, in this last case, we *haven't* had to use the special
rules for identity to show this argument is valid. This is sometimes the way: use
of the identity relation may be naturally involved in the translation of the premisses and conclusion of an argument; but it may so happen that showing the
resulting inference is q-valid only uses the fact that it is a relation and not the
further special logical properties of this relation.

Here, though, is another case where we do have to appeal to the special laws
of identity. Consider

> **K**   The author of the Iliad wrote the Odyssey. Hence at most one
> person wrote the Iliad.

According to Russell's Theory of Descriptions, the premiss is only true if there is
a unique author, and is false if there is more than one author. So, on Russell's
view, the argument is a valid one. Well, be that as it may about the vernacular
argument, its closest rendition into QL$^=$ is certainly q-valid. For consider:

(1)	$\exists x((Fx \wedge \forall y(Fy \supset y = x)) \wedge Gx)$	
(2)	$\neg \forall x \forall y((Fx \wedge Fy) \supset x = y)$	✓
(3)	$\exists x \neg \forall y((Fx \wedge Fy) \supset x = y)$	✓
(4)	$\neg \forall y((Fa \wedge Fy) \supset a = y)$	✓
(5)	$\exists x \neg ((Fa \wedge Fy) \supset a = y)$	✓
(6)	$\neg ((Fa \wedge Fb) \supset a = b)$	✓
(7)	$(Fa \wedge Fb)$	
(8)	$\neg a = b$	
(9)	Fa	
(10)	Fb	

So far, we've unpacked (2) in automatic steps. Let's now instantiate (1), using yet
another new name, check it off and unpack the resulting triple conjunction:

(11)	$((Fc \wedge \forall y(Fy \supset y = c)) \wedge Gc)$	✓
(12)	$(Fc \wedge \forall y(Fy \supset y = c))$	✓
(13)	Gc	
(14)	Fc	
(15)	$\forall y(Fy \supset y = c)$	

Where now? Well, we plainly need to extract information from this last universal quantification. It would be pretty pointless to instantiate it with respect to 'c'
(why?). But we have two other names to play with, yielding

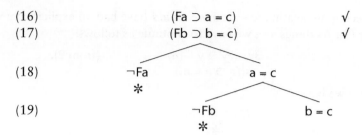

(16)   (Fa ⊃ a = c)   √
(17)   (Fb ⊃ b = c)   √

(18)   ¬Fa   a = c
              *

(19)   ¬Fb   b = c
                         *

The right-hand branch is still open: what is there left to make use of? Well, note that at (8) we have '¬a = b', and now on the same branch we have both 'a = c', and 'b = c', and those wffs are clearly an inconsistent triad. Take 'a = c' as C(c), and use the 'b = c' to derive C(b) by (L), and the branch closes thus …

(20)   a = b
                                             *

We can similarly, and rather more simply, show that the following two arguments are also valid:

L   The author of the Iliad wrote the Odyssey. Hence at least one person wrote the Iliad.

M   The author of the Iliad wrote the Odyssey. Hence whoever wrote the Iliad wrote the Odyssey.

Those cases are left as exercises. But we will demonstrate the validity of

N   A least one person wrote the Iliad. At most one person wrote the Iliad. Whoever wrote the Iliad wrote the Odyssey. Hence, the author of the Iliad wrote the Odyssey.

The tree starts straightforwardly enough, following our familiar rules of thumb:

(1)   ∃xFx   √
(2)   ∀x∀y((Fx ∧ Fy) ⊃ x = y)
(3)   ∀x(Fx ⊃ Gx)
(4)   ¬∃x((Fx ∧ ∀y(Fy ⊃ y = x)) ∧ Gx)   √
(5)   ∀x¬((Fx ∧ ∀y(Fy ⊃ y = x)) ∧ Gx)
(6)   Fa
(7)   ¬((Fa ∧ ∀y(Fy ⊃ y = a)) ∧ Ga)   √

(8)   ¬(Fa ∧ ∀y(Fy ⊃ y = a))   ¬Ga

So far, so automatic. The right hand branch quickly closes, for we have

(9)   (Fa ⊃ Ga)   (from 3)

(10)   ¬Fa   Ga
                               *   *

Meanwhile, the left-hand branch can be continued from (8) like this:

(9)	¬Fa	¬∀y(Fy ⊃ y = a)	✓	(from 8)
(10)	*	∃y¬(Fy ⊃ y = a)	✓	
(11)		¬(Fb ⊃ b = a)	✓	
(12)		Fb		
(13)		¬b = a		

Inspection reveals that we haven't yet made any appeal to (2). So let's instantiate it with respect to the two names in play, thus:

(14)	∀y((Fb ∧ Fy) ⊃ b = y)	(from 2)
(15)	((Fb ∧ Fa) ⊃ b = a)	✓

(16)   ¬(Fb ∧ Fa)   b = a
                                               *

(17)   ¬Fb   ¬Fa
                *   *

Once more, we arrive at a closed tree.

Note, by the way, that we could put together the proofs for examples **K** to **M** to show that

(R)   ∃x((Fx ∧ ∀y(Fy ⊃ y = x)) ∧ Gx)

entails each of the conjuncts in the more long-winded triple conjunction

({∃xFx ∧ ∀x∀y((Fx ∧ Fy) ⊃ x = y)} ∧ ∀x(Fx ⊃ Gy))

And the proof for **N** shows that this triple conjunction entails (R) – thus justifying the claim (§34.1) that these two forms of Russell's Theory of Descriptions for rendering 'The *F* is *G*' into **QL**= are equivalent.

35.4 'One and one make two'

Finally in this chapter, consider the following inference:

O   ∃x(Fx ∧ ∀y(Fy ⊃ y = x)), ∃x(Gx ∧ ∀y(Gy ⊃ y = x)), ¬∃x(Fx ∧ Gx)
          ∴ ∃x∃y(([{Fx ∨ Gx} ∧ {Fy ∨ Gy}] ∧ ¬x = y) ∧
                                              ∀z[{Fz ∨ Gz} ⊃ {z = x ∨ z = y}])

which renders 'There is exactly one *F*, there is exactly one *G*, nothing is both *F* and *G*, so there are exactly two things which are *F* or *G*' (i.e. one thing and another thing make two altogether).

This inference is provably q-valid. To warrant **O** via a tree requires a little effort, but it is fun (of a kind) to show that it can be done. We start:

(1)	∃x(Fx ∧ ∀y(Fy ⊃ y = x))
(2)	∃x(Gx ∧ ∀y(Gy ⊃ y = x))
(3)	¬∃x(Fx ∧ Gx)
(4)	¬∃x∃y(([{Fx ∨ Gx} ∧ {Fy ∨ Gy}] ∧ ¬x = y) ∧
	∀z[{Fz ∨ Gz} ⊃ {z = x ∨ z = y}])

First, then, eliminate the two negated quantifiers, and then instantiate the two existential quantifiers (what else?):

(5) $\forall x \neg (Fx \wedge Gx)$
(6) $\forall x \neg \exists y(([\{Fx \vee Gx\} \wedge \{Fy \vee Gy\}] \wedge \neg x = y) \wedge$
                                        $\forall z[\{Fz \vee Gz\} \supset \{z = x \vee z = y\}])$
(7) $(Fa \wedge \forall y(Fy \supset y = a))$
(8) $(Gb \wedge \forall y(Gy \supset y = b))$

Now we might as well unpack the conjunctions in (7) and (8):

(9) Fa
(10) $\forall y(Fy \supset y = a)$
(11) Gb
(12) $\forall y(Gy \supset y = b)$

Next, instantiate (6) with a name, and deal with the resulting negated wff.

(13) $\neg \exists y(([\{Fa \vee Ga\} \wedge \{Fy \vee Gy\}] \wedge \neg a = y) \wedge \forall z[\{Fz \vee Gz\} \supset \{z = a \vee z = y\}])$
(14) $\forall y \neg(([\{Fa \vee Ga\} \wedge \{Fy \vee Gy\}] \wedge \neg a = y) \wedge \forall z[\{Fz \vee Gz\} \supset \{z = a \vee z = y\}])$
(15) $\neg(([\{Fa \vee Ga\} \wedge \{Fb \vee Gb\}] \wedge \neg a = b) \wedge \forall z[\{Fz \vee Gz\} \supset \{z = a \vee z = b\}])$

We now have to disentangle (15), which is a negated conjunction.

(16) $\neg([\{Fa \vee Ga\} \wedge \{Fb \vee Gb\}] \wedge \neg a = b)$ $\neg \forall z[\{Fz \vee Gz\} \supset \{z = a \vee z = b\}]$

(17) $\neg[\{Fa \vee Ga\} \wedge \{Fb \vee Gb\}]$ $\neg \neg a = b$

(18) $\neg\{Fa \vee Ga\}$ $\neg\{Fb \vee Gb\}$
(19) $\neg Fa$ $\neg Fb$
(20) $\neg Ga$ $\neg Gb$
                *   *

Continuing the middle branch at (17):

(18′) $a = b$
(19′) Ga from 11, 18′
(20′) $\neg(Fa \wedge Ga)$ from 5

(21′) $\neg Fa$ $\neg Ga$
                            *   *

And on the rightmost branch, we obviously have to continue like this:

(17″) $\exists z \neg[\{Fz \vee Gz\} \supset \{z = a \vee z = b\}]$
(18″) $\neg[\{Fm \vee Gm\} \supset \{m = a \vee m = b\}]$
(19″) $\{Fm \vee Gm\}$
(20″) $\neg\{m = a \vee m = b\}$
(21″) $\neg m = a$
(22) $\neg m = b$

(23) Fm Gm

So far, in fact, this tree has almost written itself. We've applied the obvious building moves at each step. But now, checking back, we note that we haven't yet made use of the two universal quantifications at (10) and (12). Instantiating (10), we can quickly close off the left of the two remaining branches:

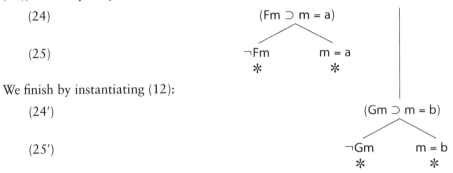

(24)   (Fm ⊃ m = a)

(25)   ¬Fm   m = a
                   *   *

We finish by instantiating (12):

(24')   (Gm ⊃ m = b)

(25')   ¬Gm   m = b
                   *   *

Phew! So the arithmetic inference O indeed turns out to be a piece of pure logic.

And that is probably *quite* enough by way of examples of (closed) QL= trees! As for open trees, the remarks about QL open trees in §29.3 carry over with only minor changes: to read off countermodels from finished trees with open branches, we just must make sure that whenever we have a wff of the form $m = n$ on the open branch, the names involved are given the same object as q-value. But we won't delay over this.

35.5   Soundness and completeness again

Finally, we should add for the record that there is the predictable Basic Result about QL= trees too. Our collection of rules is sound and complete, so that a closed QL= tree correctly indicates a q-valid argument, and there is a closed QL= tree corresponding to every q-valid argument involving identity.

To show this, we essentially need to re-run the arguments of Chapter 30. But we do need a few new tricks to handle identity, which make things just a bit more complicated. We'll cut ourselves some slack and won't give the details here.

35.6   Summary

- We have introduced two additional rules for quantifier tree-building, to govern trees with identity, namely a version of Leibniz's Law for extending trees, and a new rule for closing a branch when it contains an absurdity of the form ¬$n = n$.

- These two additional rules allow us to warrant valid QL= inferences by the tree method, including inferences where Russell's Theory of Descriptions is used to render propositions, and inferences using numerical quantifiers.

- Our rules for QL= trees give us not only a sound but also a complete proof-system for establishing q-validity.

Exercises 35

A   Use **QL**$^=$ trees to show the following inferences are valid:

1.   Jack is Fingers. Fingers is never caught. Whoever is never caught escapes justice. So Jack escapes justice.

2.   There is a wise philosopher. There is a philosopher who isn't wise. So there are at least two philosophers.

3.   Whoever stole the goods, knew the safe combination. *Someone* stole the goods, and only Jack knew the safe combination. Hence Jack stole the goods.

4.   For every number, there's a larger one. No number is larger than itself. So for every number, there's a distinct number which is larger than it.

5.   No one who isn't Bryn loves Angharad. At least one person loves Angharad. So Bryn loves Angharad.

6.   Exactly one person admires Frank. All and only those who admire Frank love him. Hence exactly one person loves Frank.

7.   The present King of France is bald. Bald men are sexy. Hence whoever is a present King of France is sexy.

8.   Angharad loves only Bryn and Caradoc, who are different people. So Angharad loves exactly two people.

9.   Juliet kisses exactly two philosophers. Hence there is more than one philosopher.

B   Show that the following wffs are q-logical truths

1. $\forall x \forall y (x = y \supset (Fx \supset Fy))$

2. $\forall y \forall z (y = z \supset (\forall x(Lxy \land Fy) \supset \forall x(Lxz \land Fz)))$

Thinking about the structure of those proofs, conclude that each way of filling out the following schema from §33.1 does indeed yield a q-logical truth:

(LS) $\forall v \forall w (v = w \supset (C(...v...v...) \supset C(...w...w...)))$

C   Returning to the labelled wffs and inferences earlier in this chapter, show that

1.   G q-entails F.

2.   L is q-valid.

3.   M is q-valid.

Finally, confirm that the two Russellian ways of translating *The F is G*, via the schemas (R) and (R') of §34.1, are indeed equivalent, by showing

4.   '$\exists x \forall y((Fy \equiv y = x) \land Gx)$' q-entails '$\exists x((Fx \land \forall y(Fy \supset y = x)) \land Gx)$'.

5.   '$\exists x((Fx \land \forall y(Fy \supset y = x)) \land Gx)$' q-entails '$\exists x \forall y((Fy \equiv y = x) \land Gx)$'.

Functions

One of the first quantificational formal languages that you are likely to meet if you go on to slightly more advanced work on logic is the language of arithmetic. And this – as you would expect – has symbols for e.g. the addition and multiplication functions. So it would be odd to leave our introductory exploration of quantificational logic without briefly saying something about how to extend the language QL$^=$ to include such expressions for functions. Our discussion will also give us another angle on the theory of descriptions.

36.1   Functions re-introduced

We've already touched on the basic idea of a function in §11.2. A function maps an object or objects to some unique thing. For example, 'the age in years (at January 1, 2009) of x' expresses a function defined over living people, correlating each person with a single number. And 'the product of x and y' expresses a function defined over numbers, correlating two numbers with another number.

Let's quickly review an assortment of elementary points about functions, and fix nomenclature and informal notation.

- We call the input(s) to a function – the object(s) to which the function is applied – the *argument(s)* of the function. The unique output of the function for given argument(s) is the *value* of the function. For example, the number 35 is the value of the function *the age in years (at January 1, 2009)* for the argument Gwyneth. The number 35 is also the value of the *product* function for the arguments 5 and 7.

- The value of a function can be a quite different sort of entity from the argument: 35 is a number, Gwyneth is a woman. But of course the value can equally well be the same sort of thing as the argument(s): the *product* function maps numbers to numbers.

- A function can have more than one argument. For example, the arithmetical addition and multiplication functions each take two arguments. Or consider the family of truth-functions. The material conditional, for example, maps two truth-values (those of its antecedent and consequent) to a truth-value; the trumped-up $ truth-function of §11.7 takes three truth-values as arguments; etc.

- Mathematicians notationally represent functions in various ways. For instance, the *square root* function is expressed by the special prefixed symbol in '√9'; the *product* function is usually expressed by the 'infix' multiplication symbol, i.e. the symbol placed in the middle of '5 × 7'; and the *square* function is indicated by the 'postfix' superscript in '9^2'. But the generic way of representing functions is by symbolism like '$f(\ )$' or '$g(\ ,\ )$' etc., with one or more slots waiting to be filled in by designators for the arguments. Suppose that '$f(\ )$' expresses the function that maps a number to its square. Then '$f(2)$' is a complex designator denoting the number 4, '$f(7)$' denotes 49, and so on. Likewise, if '$g(\ ,\ )$' expresses the product function, then '$g(5, 7)$' denotes the number 35.

- When an argument and value are objects of the same kind, we can apply that function again to the results of previous applications. Thus '$f(f(2))$' denotes the square of $f(2)$, i.e 16. Similarly, '$g(f(2), f(3))$' denotes 36.

So far, so elementary! Now for two further points that will need more careful explanation. First,

- We must distinguish the *sense* or *interpretation* of a function expression from its *extension*; the latter is what fixes the truth-value of sentences involving the function expression.

Compare our distinction between the sense and extension of predicates (§30.1). The sense is what is grasped by someone who understands the predicate; but it is the extension which fixes the truth-values of sentences involving the predicates (if we forget about intensional contexts). Similarly here: the sense of a function sign is what you have to grasp in order to count as understanding the expression. We can typically think of this as a *rule* or *procedure* for correlating objects with other objects. For example, to grasp the sense of 'square of ( )', you need to understand the rule that we arrive at the value of the function for argument n by multiplying n by itself. The rule determines a set of pairings of arguments and values, thus $\{\langle 0, 0\rangle, \langle 1, 1\rangle, \langle 2, 4\rangle, \langle 3, 9\rangle, \langle 4, 16\rangle, \ldots\}$. This set of pairs is the extension of the function.

Just as two predicate expressions can have different senses but the same extension, so too for function expressions. Here's a very simple arithmetical example. Consider '$(x + 1)^2$' and '$x^2 + 2x + 1$'. These expressions have different senses, for they indicate different calculation procedures for associating values with arguments. However, the resulting argument-to-variable pairings are the same.

We needn't argue about whether the word 'function' is ordinarily used to talk of a mapping procedure or to talk of the resulting pairings of arguments and values. What matters is that we note the relevance of the sense/extension distinction here. The *extension* of a function is what fixes truth-values. For example, '$f(m) = n$' is true if the pair $\langle m, n\rangle$ is in the extension of 'f'; 'For all x, $f(x) > x$' is true if for any pair $\langle m, n\rangle$ in the extension of 'f', $n > m$. And so on.

Finally, note that even when a rule for processing an argument is well-defined across some domain, it may not always deliver a value. Take for example the *eldest-son-of* function. The rule for getting a value can be applied to any adult human: we look to see whether he or she has sons, and if so pick out the eldest. But this procedure may well deliver a null return. Compare the *square-of* function: take any integer as argument and this function always delivers a value. We therefore need to note the following distinction:

- Some functions are *total* functions, i.e. have a value for all arguments from the relevant domain. Others are strictly *partial* functions, which have no value for certain arguments.

36.2 Adding functions to QL⁼

It is syntactically pretty straightforward to add function expressions to QL⁼. And it is semantically unproblematic too, *so long as we restrict ourselves to total functions defined over the whole domain, with values in that domain* (the real significance of this restriction will become clear in §36.4). An outline sketch of the resulting language QL^f will suffice.

So, first we'll want to add to QL⁼'s vocabulary some function expressions taking various numbers of arguments. We can start off with 'naked' lower case letters like 'f', 'g', 'h', ... (And then, if we want to ensure we don't run out of possible function expressions, we can again use superscripts and subscripts, so 'f^j_k' is the kth j-argument function symbol.)

We will next need to complicate our definition of a *term* (compare §26.3).

(1) An individual constant or individual variable is a term.
(2) If f is a function symbol taking one argument and t is a term, then $f(t)$ is a term. More generally, an n-argument function symbol followed by a left-hand bracket, following by n terms (separated by commas), followed by a right-hand bracket, is also a term.
(3) Nothing else is a term.

This allows expressions like the following to count as terms:

$$f(m), f(y), g(m, n), g(x, f(m)), g(f(y), f(m)), g(g(m, n), f(m)), \ldots$$

If a term doesn't include any variables, it is a *closed* term (compare the idea of a closed wff, §26.4).

The syntactic rules for building up wffs can stay as before, except that 'atomic wff' is now perhaps something of a misnomer for the basic building block wff consisting of an n-place predicate followed by n terms, given that the terms can themselves be quite complex.

By the way, it is worth commenting that the use of bracketing and commas here in our notation for functions is strictly speaking redundant. However, it is absolutely standard (and you can easily see why: bracket-free expressions like 'ggmnfm' are horribly unreadable).

Now for the semantics. Start with the simplest case, a one-argument function.

A q-valuation will assign this function a set of ordered pairs of objects (i.e. the pairs of argument and value), where these objects are both from the domain (they need to be among the things that count as 'everything'). This set of pairs must have the following two features:

  i.   For each object *m* in the domain of discourse, there is some pair $\langle m, o \rangle$ in the extension, i.e. a pair whose first member is *m*.

  ii. It contains no distinct pairs $\langle m, o \rangle$ and $\langle m, o' \rangle$, where *o* differs from *o'*.

(i) must hold given we are currently restricting our attention to *total* functions. (ii) must hold because a function relates an object as argument to a *unique* object as value.

Similarly for a two-argument function. A q-valuation will assign this sort of function a set of ordered triples of objects from the domain. This set must have the following two features:

  i'. For each pair of objects *m, n* in the domain of discourse – which serve as arguments to the function – there is some triple $\langle m, n, o \rangle$ in the extension.

  ii'. It contains no distinct pairs $\langle m, n, o \rangle$ and $\langle m, n, o' \rangle$, where *o* differs from *o'*.

The generalization to cover functions with *n* arguments is straightforward.

The next stage in setting up the semantics is to give rules for evaluating terms including function expressions. The basic idea is very simple. Suppose a q-valuation already assigns 'm' the object *m* as reference. And suppose that on the same q-valuation, the pair $\langle m, o \rangle$ is in the extension of the one-place 'f' (intuitively, the function maps *m* to *o*). Then the q-value of the term 'f(m)' is of course *o*.

That idea generalizes in the obvious way to case where 'm' is replaced with any term *t*. Suppose a q-valuation assigns *t* the object *n* as reference, and $\langle n, o \rangle$ is in the extension of 'f'. Then again 'f(t)' is *o*. And the idea generalizes too to the case of many-place functions. For example, the q-value of 'g(m, n)' is the unique *o* such that the triple formed by taking the references of 'm' and 'n' followed by *o* is in the extension of 'g'.

And because we are only considering *total* functions defined everywhere in the domain and which have values in the domain, it will always make sense to feed in the value of one function as an argument to another function. So those iterated functional expressions like 'g(f(n), f(m))' will always have a q-value.

Which about wraps things up. We have introduced a new kind of term – one involving a function-expression – and indicated how to work out its q-value, i.e. its reference, on a q-valuation. The old rules for working out the values of basic wffs (predicates followed by the right number of terms) now kick in, followed by the rules for evaluating more complex wffs just as before.

Finally, to get a proof-system for arguments couched in QL^f we just use the same tree rules as before, emended to allow us to intantiate universal quantifiers with closed functional terms. The idea is that if e.g. '∀xGx' is true, then so must be 'Gf(m)', 'Gf(f(m))', 'Gg(m, n)', etc. since all those functional terms pick out objects in the domain.

36.3 Functions and functional relations

Suppose that the two-place relation R is such that every object in the domain is R to one and only one thing. In that case, the extension of R is a set of pairs satisfying the conditions (i) and (ii) in the last section. Call this kind of two-place relation *functional*. Then a one-place function f and a two-place functional relation R can have identical extensions, i.e. a certain set of ordered pairs satisfying our two conditions. In such a case, we'll say that f *corresponds to* R. (Similarly, a two-argument function can correspond to a three-place functional relation, etc.: we won't labour to spell this out.)

A very simple example. Take the domain to be the integers. Then consider the *next number* (or *successor*) function and the functional relation *is immediately followed by* (for short, *precedes*). The function and the relation have exactly the same extension, i.e. the set of pairs $\{\langle 0, 1\rangle, \langle 1, 2\rangle, \langle 2, 3\rangle, \langle 3, 4\rangle, \ldots\}$. The *successor* function therefore corresponds to the relation *precedes*.

Now, *the successor of 3* is, of course, *the number such that 3 precedes it*. And that observation generalizes. Suppose that the one-argument function f corresponds to the two-place functional relation R. Then $f(n)$ is *the object x such that n is R to x*.

That's an obvious point, but it has an important moral. It shows that instead of using a function term like '$f(n)$' to refer to some object, we could use a definite description applied to a corresponding relation expression 'R' to get '*the x such that n is R to x*' and still be talking of just the same thing.

This last point carries over to our formal languages. *Everything you can do in the extended language* QL^f *with symbols for total functions you can in principle do in the original unextended* $\mathrm{QL}^=$ *without functions.* Just use functional relations plus Russell's Theory of Descriptions!

To illustrate this, let's continue with the same basic example. So suppose 'f' in QL^f expresses the *successor function*, let 'n' denote zero, and let 'G' mean *odd*. Then, for example, '$f(n)$' denotes the successor of zero, i.e. the number one. And we can say that this object is odd thus:

$$\mathrm{Gf(n)}$$

But suppose $\mathrm{QL}^=$ has the two-place predicate 'L' meaning *precedes*. Then we can say the same in $\mathrm{QL}^=$ by translating 'The number such that zero precedes it is odd'. Applying the familiar Russellian schema, we'll get

$$\exists x((Lnx \land \forall y(Lny \supset y = x)) \land Gx)$$

Likewise, have the following mappings from QL^f wffs to $\mathrm{QL}^=$ equivalents:

$$\exists z Gf(z) \mapsto \exists z \exists x((Lzx \land \forall y(Lzy \supset y = x)) \land Gx)$$
$$f(n) = m \mapsto \exists x((Lnx \land \forall y(Lny \supset y = x)) \land x = m)$$
$$\forall z(Gz \supset f(z) = m) \mapsto \forall z(Gz \supset \exists x((Lzx \land \forall y(Lzy \supset y = x)) \land x = m))$$

and so on and so forth.

Now, it is already clear that there *is* a real gain in clarity in having the function symbolism available. The gain is very substantial when there's more than one functional symbol to be translated away at the same time. Take for example the simple claim that addition in arithmetic is symmetric, i.e. for all x and y, $x + y = y + x$. This goes into standard QL^f simply enough as something like

$$\forall x \forall y (h(x, y) = h(y, x))$$

if we use 'h' for the two-argument sum function. Or we may simply allow ourselves to borrow the familiar infix arithmetical function symbols in an extension of $QL^=$ so we can write

$$\forall x \forall y (x + y = y + x)$$

But suppose we try to manage instead in unaugmented $QL^=$, using the three-place sum *relation* 'S', where 'Smno' means $m + n = o$. Then to express the symmetry of addition we need, informally,

$$\forall x \forall y (\text{the } v_1 \text{ such that } Sxyv_1 = \text{the } v_2 \text{ such that } Syxv_2)$$

Getting rid of the two descriptions now introduces another *four* quantifiers:

$$\forall x \forall y \exists v_1 \{(Sxyv_1 \wedge \forall z(Sxyz \supset z = v_1)) \wedge$$
$$\exists v_2 [(Syxv_2 \wedge \forall z(Syxz \supset z = v_2)) \wedge v_2 = v_1]\}$$

which is a *lot* less perspicuous!

36.4 Partial functions and free logic

In the light of what we've just said, the question whether to expand $QL^=$ to QL^f really comes down to the question of how important it is for us to have a logical formalism which can deal briskly and perspicuously with total functions (recall: QL^f *only* deals with total functions). It is a trade-off between economy of resources and ease of manipulation.

But now it is worth stressing again that many functions are strictly *partial*. They abound in mathematics: to take the very simplest case, the division function doesn't deliver a value everywhere (there is no output when you divide a number by zero).

You might wonder whether partial functions are much of an issue. To be sure, some definitions only characterize partial functions, like the definition of division. However, can't we just (as it were) fill up the gaps in a partial function by arbitrary stipulation? For example, we could define the total *division** function, where m divided* by n is the same as m divided by n when $n > 0$, and is equal to (say) 0 when n is 0. And we could stipulate that the *eldest son** function has the same value as the eldest son function when that has a value but outputs (say) the number 0 as a default in other cases. So, when we want to be logically tidy, can't we pretend all our functions are 'completed' in this sort of way?

Unfortunately, wishing away partial functions doesn't work. It is, for example, a fairly elementary result in the theory of computation that you can have a partial function f which is computable – i.e. there is a computer program that works out the value of f for every argument for which it is defined – but f's completion-by-stipulation yields a total function f^* which is *not* computable by any possible program. Moral: a general theory of functions has to take partial functions seriously. In short, if we want to treat functions in a logical system, the natural aim will be to cover all functions, total *and* merely partial.

So suppose now, to return to an earlier example, that 's( )' expresses the partial *eldest-son-of* function. Suppose too that 'm' denotes George Bush Snr., 'n' denotes Hillary Clinton, and 'F' means *is male*. We will take the domain to be people. Then 's(m)' is unproblematic: it is a term denoting George W. Bush, and 'Fs(m)' straightforwardly evaluates as true.

What about 's(n)'? Hillary has no son, so this denotes no one, and 's(n)' has no q-value. So what should we say about the value of 'Fs(n)'? A wff consisting of a predicate followed by a term is only true if the q-value of the term is in the extension of the predicate. So such a wff is certainly not *true* if the term lacks a q-value. Does that mean (1) that 'Fs(n)' is *false*? Or does it mean (2) that it *lacks a truth-value*?

Either option is technically workable, and we get two versions of *free logic* – i.e. a logic which is free from the assumption that all closed terms (terms without free variables) denote. But option (2) is arguably more natural. For if we have no general beef against the idea of partial functions, then it is pretty unclear why we shouldn't allow partial evaluation functions too, and hence allow some wffs to come out neither true nor false. However, allowing such truth-value gaps will force major revisions in the very fundamentals of our formal logic. For example, even instances of $(A \lor \neg A)$ needn't always be true, at least if we keep the traditional truth-tables: replace A with a wff which is neither true nor false and the disjunction will lack a truth-value too.

Option (1) is in some ways less radical. On this line, all wffs remain either true or false. But we will still have to revise our logic. For a start, in our example, we could have '∀xFx' true but 'Fs(n)' is false. So we now can't always instantiate universal quantifications with functional terms.

So in fact either version of free logic we have mentioned will be revisionary of the logic of QL^f. Moreover, once we allow non-denoting functional terms, it is difficult to see why we shouldn't allow non-denoting *names* as well, a further departure into a thoroughly free logic.

The situation in sum is this: QL^f is an extension of $QL^=$, which gives us a neat, brisk representation of *total* functions defined over a domain, more manageable than dealing with functions via definite descriptions applied to functional relations. However, the restriction to total functions is rather unnatural (though it enables us to preserve 'classical', two-valued logic). Going any further, and taking partial functions seriously, requires significant innovations.

But we can't explore these matters any further here, except to conclude by revisiting the issue of ...

36.5 Definite descriptions again

Russell, we noted, held that his Theory of Descriptions in some sense revealed the true logical form of our everyday sentences involving definite descriptions. It is far from clear what the cash value of such a claim is. Still, we are now in a position to make some pertinent comments.

Recall the very close connection between functions and descriptions noted in §36.3. We saw that the successor function, for example, can be defined by applying a definite description operator (a 'the' operator) to the predecessor relation. Or consider Russell's own paradigm example 'the (present) King of France' again. This invites treatment as a functional term, denoting the value of the *the king of* function for the argument France. And 'the king of ...' can, in turn, be thought of as equivalent to 'the *x* such that *x* is king of ...' – i.e. as defined in terms of a 'the' operator applied to a two-place functional relation '... is king of ...'.

Suppose that we are dealing with a language which has expressions for both functions and functional relations. *Then it is very natural to add a 'the' operator for making functions out of functional relations.* In §36.3, we were talking about how we might do without functions by using functional relations instead; now we are considering how to move between relations and functions when you *do* have both in the language.

Let's use the symbol 'ι' for our *the* operator – that's a Greek letter *iota*, which is traditional for a description operator, following Russell (though we certainly don't want to get sidetracked here into a discussion of the nuances of various uses of the iota symbol in the literature). The basic idea, therefore, will be that the ι-operator applied to an $n + 1$ place functional predicate R yields the corresponding function of n arguments. And the natural way of implementing the idea is to make 'ι' a variable-binding operator akin to a quantifier. In other words: we use 'ι' with the variable 'x' or 'y' or ..., and this ties it to a slot in R with the same variable in it.

For example, take the two place functional relation 'L'. Then

$$\iota y Lxy$$

means *the thing y such that x has relation L to y.* 'ιyLxy' is therefore a one-place functional expression that works like 'f(x)' (it lacks the brackets, but then the brackets we usually use with function expressions are only there to promote readability). If we replace the free variable with a constant as in

$$\iota y Lmy$$

we get a term that works like 'f(m)'.

Four observations. (1) If we are still working in the context of QL^f, then ι-operators have to be applied with care to make sure that the resulting functions indeed exist as total functions. But if we suppose now that we are working in the context of a free logic which allows partial functions, then – if you will excuse the pun – we can be much more free in the use of our new operator.

(2) We said that applying the ι-operator to an $n+1$ place relation yields an n-place function. Let's allow the case where $n = 0$. So applying the same ι-operator to a *one* place predicate yields a term with zero slots left to be filled up – 'ιxFx' just serves to designate *the x which is F*. So, the ι-operator is also a way of forming definite description terms out of monadic predicates.

(3) Since 'ιxFx' is like a closed term, 'GιxFx' is a sentence. And in *some* appropriate logical systems for a description operator – including some versions of free logic – we will be able to show that

(F)   GιxFx

(i.e. *the thing which is F is G*) is true on a valuation if and only if

(R) $\exists x(Fx \land (\forall y(Fy \supset y = x) \land Gx))$

is also true.

(4) That shows that, even if Russell is right that the rendition (R) has the same truth-conditions as 'The *F* is *G*', it doesn't mean that he is right that descriptions are fundamentally some kind of quantifier. It could be that the description-former 'the' in natural language often works like the 'ι'-operator, i.e. not as a quantifier but as a function-forming operator on relations. In other words, perhaps (R) has the same truth-conditions as 'The *F* is *G*' as a *consequence* of descriptions working on the functional model, rather than because (R) itself reflects the correct story about the 'logical form' of sentences with descriptions.

However, we can't pursue this suggestion any further here. The issue of the proper treatment of definite descriptions remains contentious a century after Russell first proposed his treatment. Which is a salutary closing reminder that issues in logic are not always as cut-and-dried as text books – like this one! – can make them seem.

36.6 Summary

- We could also extend $QL^=$ to include expressions for functions. The simplest extension, QL^f, only allows expressions for *total* functions (i.e. functions that have a value for all arguments in the domain).

- The logic of QL^f is not very different to that of $QL^=$, and it allows a much more economic treatment of functions compared with using functional relations and the theory of descriptions in $QL^=$.

- However, QL^f is arguably an unstable compromise: if we are going to treat functions, shouldn't we also treat *partial* functions (i.e. functions that do not deliver a value for all arguments in the domain)?

- To cope with partial functions, and hence the use of terms which lack reference, will mean adopting some kind of *free* logic, which will have significantly different logical laws from those that govern $QL^=$.

- In a language which has functions, we could introduce a ι-operator (read 'the') that takes us from an $n + 1$-place predicate to a corresponding n-place function. Arguably, the vernacular description-forming operator 'the' works similarly.

Further reading

These notes on further reading are divided into two parts. The first section lists references that are directly tied to points arising in the body of the book; the second suggests some text books for parallel and more advanced reading.

Matters arising

On the nature of propositions (§2.5), see

> A. C. Grayling, *An Introduction to Philosophical Logic* (Blackwell, 3rd edn. 1997), Ch. 2.

Quine-quotes (§10.3) are introduced in

> W. V. O. Quine, *Mathematical Logic* (Harper and Row, 2nd edn. 1951), §6;

they are also well explained e.g. by

> Graeme Forbes, *Modern Logic* (Oxford U.P., 1994), Ch. 2, §6.

Real enthusiasts might be amused by

> George Boolos, 'Quotational ambiguity', in his *Logic, Logic, and Logic* (Harvard UP, 1988).

On the ideas of necessity, analyticity and the a priori (§12.5), see

> A. C. Grayling, *An Introduction to Philosophical Logic*, Ch. 3.

On the troublesome question of the meaning of 'if' (Chapters 14, 15), see

> Frank Jackson, ed., *Conditionals* (Oxford UP, 1991).
> Frank Jackson, *Conditionals* (Blackwell, 1987).

The first of these books is a reader, collecting together a number of key papers on the interpretation of conditionals (read V. Dudman's concluding contribution if you aren't yet persuaded of the complexities of the behaviour of conditionals). The reader reprints a paper by the Jackson himself, which proposes his 'robust material conditional' story about the meaning of indicative 'ifs' (§15.4). This is developed in his own extremely readable short book. Our example 'If kangaroos had no tails, they would topple over' (§15.1) is borrowed from the opening section of a modern classic,

David Lewis, *Counterfactuals* (Blackwell, 1973).

We said in §20.3 that tree proofs can sometimes be radically shortened by adopting the (evidently sound) rule that at any point you can add an instance of the law of excluded middle, i.e. a wff of the form $(C \vee \neg C)$. For discussion, see

George Boolos, 'Don't eliminate cut', in his *Logic, Logic, and Logic*.

For further discussion of the motivation for the quantifier-variable notation for generality (§21.4), see another modern classic,

Michael Dummett, *Frege: Philosophy of Language* (Duckworth, 2nd edn. 1981), Ch. 2.

We mentioned in passing (§23.4) that with a bit of cunning we can do without variables (while retaining the key idea that quantifiers are operators that are applied to predicates, and are quite different from names). For more on this, see

W. V. O. Quine, 'Variables explained away', in his *Selected Logic Papers* (Harvard UP, 1995).

On Frege's claim that names have sense as well as reference (§31.1), see the much-reprinted

Gottlob Frege, 'On sense and reference' – or, as it is called in his *Collected Papers* (Blackwell, 1984), 'On sense and meaning'.

For a lively challenge to the conventional wisdom about how we can substitute co-referential names in 'simple' contexts, see

Jennifer M. Saul, 'Substitution and simple sentences', *Analysis* 57 (1997), pp. 102–108.

For Russell's Theory of Descriptions (§34.1), see the essays in

Gary Ostertag, ed., *Definite Descriptions* (MIT Press, 1998)

especially Chs. 3 (Russell's own statement), 8 and 11. And for a little on free logic (§34.4), see

Graham Priest, *An Introduction to Non-Classical Logic* (Cambridge UP, 2nd edn. 2008), Ch. 13.

Other texts

A *Parallel reading* For a classic text which covers similar ground to this present book, see

Richard Jeffrey, *Formal Logic: Its Scope and Limits* (McGraw-Hill, 3rd edn. 1990).

Another much used student text (which also concentrates on trees but goes rather more slowly than Jeffrey) is

Wilfrid Hodges, *Logic* (Penguin, 2nd edn., 2001).

B *Some introductory texts that range more widely* Three more good texts (at about the same level of difficulty as this book) which cover natural deduction proofs as well as trees are

Merrie Bergmann, James Moor and Jack Nelson, *The Logic Book* (McGraw-Hill, 3rd edn. 1998).

Graeme Forbes, *Modern Logic* (Oxford UP, 1994)

Paul Teller, *A Modern Formal Logic Primer* (Prentice Hall, 1989).

Teller's text is in two volumes but don't be put off by that: it is in fact quite short, and has a particularly approachable, relaxed style. The bad news is that it is out of print; the good news is that an electronic version is freely available at the book's website, http://tellerprimer.ucdavis.edu.

But the best introduction to natural deduction is probably the elegant and sophisticated

Neil Tennant, *Natural Logic* (Edinburgh UP, 2nd edn. 1990).

While Teller, for example, presents a linear natural deduction system with indented subproofs, Tennant goes straight for downwards-joining tree deductions (see our §20.1).

C *Going further in first-order logic* Two good places to start are

David Bostock, *Intermediate Logic* (Oxford UP, 1997)

Ian Chiswell and Wilfrid Hodges, *Mathematical Logic* (Oxford UP, 2007).

Despite their titles, neither book goes a great deal further than this one. But Bostock is very good on the motivations for varying styles of logical system, on the relations between them, and on explaining some key proof ideas. Chiswell and Hodges is also very clear and gives an excellent basis for more advanced work.

A classic text that has to be mentioned here is the wonderful

Raymond Smullyan, *First-Order Logic* (Springer-Verlag, 1968: Dover, 1995).

But this isn't easy, as it packs a great deal into just over 150 pages; but at least read the first third to see where all our work on trees in this book ultimately comes from. Another text that is elegantly done and, like Smullyan, pushes well beyond the bounds of this book is

Dirk van Dalen, *Logic and Structure* (Springer-Verlag, 4th edn. 2004).

These two books should particularly appeal to the mathematically minded.

D *Extending classical logic* One or twice, we've mentioned *modal logic*, i.e. the logic of necessity and possibility. Three recent texts worth looking at (the third is based very much on trees) are

J. C. Beall and Bas C. van Fraassen, *Possibilities and Paradox* (Oxford UP, 2003).

Rod Girle, *Modal Logics and Philosophy* (Acumen, 2000).

Graham Priest, *An Introduction to Non-Classical Logic*, Chs 2, 3.

Another standard text is

Brian Chellas, *Modal Logic: An Introduction* (Cambridge UP, 1980).

The formal and philosophical aspects of *second-order* logic (where we quantify over properties as well as things) are discussed in the terrific

Stewart Shapiro, *Foundations without Foundationalism* (Clarendon Press, 1991).

And for much, much more, see the (mostly excellent) survey articles in the four volumes of the somewhat mistitled

D. Gabbay and F. Guenthner, eds., *A Handbook of Philosophical Logic* (D. Reidel. 1983–89).

E *Philosophical logic* Returning now from more advanced technicalities to philosophical commentary on the sort of formal logic covered in this book, the place to start has to be

Mark Sainsbury, *Logical Forms* (Blackwell, 2nd edn. 2000).

Each chapter ends with an extraordinarily helpful guide through the complications of the relevant literature, and we need not repeat those guides here.

Index